BIOCHAR APPLICATION

BIOCHAR APPLICATION

ESSENTIAL SOIL MICROBIAL ECOLOGY

Edited by

T. KOMANG RALEBITSO-SENIOR, CAROLINE H. ORR
Teesside University, Middlesbrough, United Kingdom

ELSEVIER

AMSTERDAM • BOSTON • HEIDELBERG • LONDON
NEW YORK • OXFORD • PARIS • SAN DIEGO
SAN FRANCISCO • SINGAPORE • SYDNEY • TOKYO

Elsevier
Radarweg 29, PO Box 211, 1000 AE Amsterdam, Netherlands
The Boulevard, Langford Lane, Kidlington, Oxford OX5 1GB, UK
50 Hampshire Street, 5th Floor, Cambridge, MA 02139, USA

British Library Cataloguing-in-Publication Data
A catalogue record for this book is available from the British Library

Library of Congress Cataloging-in-Publication Data
A catalog record for this book is available from the Library of Congress

ISBN: 978-0-12-803433-0

For information on all Elsevier publications
visit our website at https://www.elsevier.com/

Working together
to grow libraries in
developing countries

www.elsevier.com • www.bookaid.org

Publisher: Candice Janco
Editorial Project Manager: Marisa LaFleur
Production Project Manager: Vijayaraj Purushothaman
Designer: Matthew Limbert

Typeset by TNQ Books and Journals

Dedication

To our families

Dedication

Contents

8. Microbial Ecology of the Rhizosphere and Its Response to Biochar Augmentation   199

C.H. ORR, T. KOMANG RALEBITSO-SENIOR AND S. PRIOR

9. Potential Application of Biochar for Bioremediation of Contaminated Systems   221

H. LYU, Y. GONG, R. GURAV AND J. TANG

10. Interactions of Biochar and Biological Degradation of Aromatic Hydrocarbons in Contaminated Soil   247

G. SOJA

11. A Critical Analysis of Meso- and Macrofauna Effects Following Biochar Supplementation   268

X. DOMENE

12. Summation of the Microbial Ecology of Biochar Application   293
C.H. ORR AND T. KOMANG RALEBITSO-SENIOR

Contributors

P. Barakoti Teesside University, Middlesbrough, United Kingdom

A.O. Bayode Teesside University, Middlesbrough, United Kingdom

S. Behrens University of Minnesota, Minneapolis, MN, United States and BioTechonology Institute, St. Paul, MN, United States

F.S. Cannavan University of São Paulo, Piracicaba, São Paulo, Brazil

T.R. Cavagnaro The Waite Research Institute, The University of Adelaide, SA, Australia

R. Chintala Nutrient Management & Stewardship, Innovation Center for U.S. Dairy, Rosemont, IL, United States

L.F. de Souza University of São Paulo, Piracicaba, São Paulo, Brazil

X. Domene Autonomous University of Barcelona, Bellaterra, Barcelona, Spain; Centre for Ecological Research and Forestry Applications (CREAF), Bellaterra, Barcelona, Spain

C.J. Ennis Teesside University, Middlesbrough, United Kingdom

A.-M. Fortuna North Dakota State University, Fargo, ND, United States

M.G. Germano University of São Paulo, Piracicaba, São Paulo, Brazil; Brazilian Agricultural Research Corporation, Embrapa Soybean, Londrina, Paraná, Brazil

Y. Gong Nankai University, Tianjin, China

R. Gurav Nankai University, Tianjin, China

N. Hagemann University of Tüebingen, Tüebingen, Germany

J. Harter University of Tüebingen, Tüebingen, Germany

H. Lyu Nankai University, Tianjin, China

F.M. Nakamura University of São Paulo, Piracicaba, São Paulo, Brazil

E.-L. Ng Future Soils Laboratory, Melbourne, VIC, Australia

C.H. Orr Teesside University, Middlesbrough, United Kingdom

J. Pickering Teesside University, Middlesbrough, United Kingdom

S. Prior Teesside University, Middlesbrough, United Kingdom

T. Komang Ralebitso-Senior Teesside University, Middlesbrough, United Kingdom

T.E. Schumacher South Dakota State University, Brookings, SD, United States

G. Soja AIT Austrian Institute of Technology GmbH, Tulln, Austria

C. Steiner University of Kassel, Witzenhausen, Germany

S. Subramanian South Dakota State University, Brookings, SD, United States

J. Tang Nankai University, Tianjin, China

S.M. Tsai University of São Paulo, Piracicaba, São Paulo, Brazil

Foreword

Biochar is charcoal made from biomass and used as a soil amendment. In late 2014, Elsevier asked me to review a book proposal on biochar by Komang Ralebitso-Senior and Caroline Orr. I had been aware of the potential of biochar to improve agricultural productivity, clean up contaminated land, and promote carbon sequestration to reduce the impact of climate change. As always my advice to investigators was to provide evidence in support of the claims. It seemed to me that a lot had already been written, but few books covered in depth the hard scientific evidence backing the case for biochar use. My dialogue with the authors on their proposal suggested that they had the capacity to produce just the type of book needed for scientists to evaluate the evidence base for the biological activity. If it passed that test, and field trials were positive, then policymakers would be in a better place to recommend various biochar formulations as a means to protect the planet and promote food security. Komang and Caroline have been up to the challenge of producing such a book, and the scientific and agricultural communities now have a very useful volume to consider the evidence.

I have spent a career in research largely trying to establish evidence that might be useful to develop policy for public and environmental good. This is in contrast to curiosity–driven activity, which might also be termed "blue sky" research. The reality is that innovation, which is the process of translating an idea or invention into goods or services for the public good or for which people will pay commercially, is critical in both paths. As Coordinator of the OECD Co-operative Research Programme for Biological Resource Management for Agricultural Sustainability from 1989 to 2006, I was continually inviting topics for bi-national research collaboration and international workshops. Always the challenge was to select innovative ideas, often applying the latest technologies from widely varying topics. For example, these ranged from the use of earth observation in precision agriculture, to the use of the latest DNA/RNA probes to identify relevant biological activity in soil. Many were in the context of climate change, which is having so much impact on the planet, its biota, and its people. Biochar is a potentially innovative topic and product, and can be investigated using innovative technologies such as DNA/RNA probes.

Innovation can be clouded by false claims. While working in the United States and in other countries such as India, I became aware of the "snake oil salesman" who would try to peddle microbial inoculants to improve agricultural production. Most could be dismissed because of lack of evidence but occasionally evidence was produced for the value of a preparation. These included rhizobia to stimulate nitrogen fixation in legumes, and bacteria and fungi that can control pests and diseases, some of the inoculants having been genetically modified. Most importantly, however, molecular markers became increasingly the system of choice to identify establishment and effectiveness. Caroline and Komang's book effectively collates some of the latest scientific investigations, including molecular markers, on biochar action in the environment and will provoke much discussion; I hope it is read widely.

Jim Lynch, *OBE*
Distinguished Professor of Life Sciences, University of Surrey

Acknowledgments

First, we would like to acknowledge those who contributed directly to this manuscript: the teams of authors who helped to make our idea a reality and added to the strength of the book; and Professor Jim Lynch, who gave us his time to discuss the need for a policy statement and then agreed generously to write the Foreword.

Several anonymous reviewers endorsed our book proposal and informed some of our decisions, helping to shape the book into its current form. We were also helped by researchers who could not contribute to the manuscript, but pointed us in the direction of people who could.

We are grateful to Elsevier for giving us this opportunity to showcase our analysis of state-of-the-art biochar research, its future direction, and policy requirements. In particular, Marisa LaFleur is acknowledged gratefully for her friendly, task-oriented, and supportive professionalism.

Throughout the process we have been supported by our colleagues at Teesside University who: allowed us to pick their brains on the publishing process and shared their experiences of its inherent joys and pitfalls; made time to help develop our blog video clip; and, generally, provided us with a "Go for it!" attitude.

This book has also happened as a result of contributions from our student researchers Abayomi Olaifa, Pratima Rai, Daniel Dancsics, Emma Phillips, Stephen Anderson, Shaun Prior, Christopher Schroeter, Sean Lindsay, Joe Russell, Jodie Harris, and Paul Wilkinson. Their enthusiasm for our biochar research has helped focus our minds and kept the wheels going. Furthermore, their research would not have been possible without funding from the Teesside University Research Fund, Department for Learning Development Students as Researchers, and Society for Applied Microbiology Students into Work schemes.

Of course, there are many people who supported us academically before we had the idea for this book but whose encouragement has helped shape our careers. Specifically, Komang would like to thank Professor Eric Senior, Professor Henk van Verseveld (RIP), and Dr Wilfred Röling (RIP). Caroline would like to thank Professor Stephen Cummings, Dr Julia Cooper, Professor Jennifer Ames, Dr Lynn Dover, and Dr Andrew Nelson.

Thankfully, there is more to our lives than being academics and we are lucky enough to be surrounded by many fantastic people who remind us of, and nurture, this balance when we need it. We will never be able to communicate fully our gratitude to you, our family and friends, for your support this year, and always especially our parents: Malebitso and Alex (RIP), Jean and Michael.

Our final acknowledgment is reserved for our husbands, Eric Senior (and Our Two Ks) and Mark Dunkley. Thank you, mainly, for tolerating us, particularly when we come home late, get distracted with emails at weekends and holidays, and expect you to act as sounding boards during moments of ranting. Your unconditional love and support helps make us what and who we are.

Microbial Ecology Analysis of Biochar-Augmented Soils: Setting the Scene

T. Komang Ralebitso-Senior, C.H. Orr

Teesside University, Middlesbrough, United Kingdom

Biochar Application
http://dx.doi.org/10.1016/B978-0-12-803433-0.00001-1

1

OVERVIEW

Surpassing energy production and water provision, carbon sequestration, to slow the momentum of inimical climate change, is now the
single greatest challenge facing the scientific community. Of the various
sequestration options, charcoal production, to stabilize photosynthetically
fixed carbon, and its subsequent application to soil (biochar), is destined
to make a significant contribution particularly with its additionalities of
waste reduction and energy production. Together with carbon sequestration, significant economic benefits can be gained through ecosystem restoration, including contaminated land remediation and improved plant
productivity, by enhanced fertilizer efficacy, with small farmers in developing countries set to benefit most from this "climate-smart" agriculture
(Cernansky, 2015).

The importance of harnessing carbon can be gauged readily by recog-
nizing that between the 1850s and 2000 carbon loss from the soil organic
pool totaled 78 ± 12 gigatonnes (Gt) while natural fire biochar redressed
the balance by 35% at most.

According to the UK Environment Agency, contaminated land accounts
for >57,000 ha in England and Wales alone while the African continent is
blighted by 63 million hectares. Current bioremediation strategies, such
as land-farming/pump-and-treat/semipermeable and permeable reactive barriers, can be enhanced considerably by biochar as a vehicle for
biosupplementation, biostimulation, and bioaugmentation. With the
unprecedented global interest in the use of biochar as an environmental
management tool, this book aims to showcase the cutting-edge studies
and findings on the (molecular) microbial ecology of historical and contemporary biochar applications to soil.

BIOCHAR

Biochar, sometimes termed "pyrochar," is the carbon-rich, solid by-
product obtained from the carbonization of biomass, such as wood,
manure, or leaves, heated to temperatures between 300°C and 1000°C
under low (preferably zero) oxygen concentration. This process is known
as pyrolysis (Lehmann, 2007; Lehmann and Joseph, 2009; Verheijen
et al., 2010), which, typically, gives three products: liquid (bio-oil); solid
(biochar); and gas (syngas) with the yield of each varying dramatically
depending on the pyrolysis process (slow, fast, and flash) and conditions
(ie, feedstock, temperature, pressure, time, heating, and rate) (Fig. 1.1). In
particular, biochar production has been reported to decrease with increasing temperature (IEA, 2007 cited in Spokas et al., 2009).

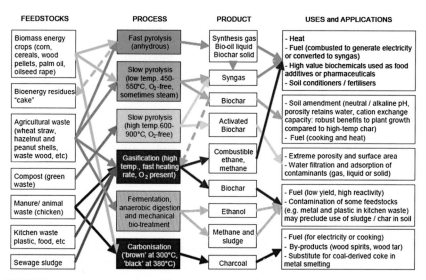

| FEEDSTOCKS | PROCESS | PRODUCT | USES and APPLICATIONS |

FIGURE 1.1 Summary of thermal conversion processes in relation to common feedstocks, typical products, and potential applications (Sohi et al., 2010).

Hydrochar differs from typical biochars as it is produced by liquefaction or hydrothermal carbonization, which results in an increased H/C ratio and decreased aromaticity (Lehman and Joseph, 2015). Furthermore, selections of feedstock and pyrolysis conditions have been considered as prime determinants of biochar properties and thus its efficacy in a wide range of contexts and/or ecosystems. Consequently, biochars vary widely and profoundly not only in their nutrient contents and pH but also in their organochemical and physical properties (Lehmann et al., 2011; Novak et al., 2009). Therefore it is crucial to consider the intended application prior to production to determine feedstock selection and pyrolysis protocol to produce bespoke biochar to address specific environmental concerns (Novak et al., 2009). According to Lehmann and Joseph (2009), the term "biochar" should not to be confused with "agrichar" as the latter could be derived from charred plastics or nonbiological materials. Also, although the production of biochar often mirrors that of charcoal, it is distinguished from it by its intended use in soil treatment, carbon storage, or filtration of percolating soil water (Lehmann and Joseph, 2009).

Biochar addition may affect soil biological community composition, as demonstrated for the biochar-rich *Terra Preta* soils in the Amazon (Grossman et al., 2010), and has been shown to increase soil microbial biomass (Liang et al., 2010, cited in Lehmann et al., 2011). Independent of possible microbial abundance increases, the net effects on soil physical properties depend on the interactions of the biochar with the physicochemical characteristics of the soil and other determinant factors such as the local climatic conditions and the biochar application regime (Verheijen et al., 2010). Also, parallel with soil enrichment, biochar may pose a direct risk to soil fauna and flora (Lehmann et al., 2011).

Terra Preta

The application of biochar to improve agricultural soils is not novel since it has been practiced for centuries. *Terra Preta*—Brazilian oxisols, anthropogenically modified by the inhabitants of the Amazonia region, also known as *"Terra Preta de Indio,"*—contains a high concentration of charcoal presumably through deliberate application by pre-Columbians and Amerindians (Kim et al., 2007) rather than the incidental presence of charred remains from forest clearing and burning (Novak et al., 2009; Verheijen et al., 2010). These anthrosols have been reported to persist over many centuries despite the prevailing humid tropical conditions and rapid mineralization rates (Lehmann et al., 2002), showing enhanced nutrient concentrations compared to surrounding soils. They have also been characterized to support a wider genetic agrobiodiversity relative to other soil types in the region (Atkinson et al., 2010).

These soils have attracted considerable interest with an extensive literature providing results on long-term carbon sequestration and microbial activity. The rich soils improve crop growth and are high in nutrients such as phosphorus and calcium (Warnock et al., 2007). Comparing microorganisms from *Terra Preta* fields to normal soils has facilitated diversity and abundance (Grossman et al., 2010) study although microbial community identification presents challenges with approximately only, as yet, 5% being cultured. In a study by Kim et al. (2007) a taxonomic cluster analysis was used to show the differences between two locations of *Terra Preta* and a preserved forest. The clusters highlighted site differences as well as species abundance changes with depth and distance. Both Kim et al. (2007) and Grossman et al. (2010) reported the dominance of *Verrucomicrobium* sp., Proteobacteria, and *Acidobacterium* sp. in the carbon-supplemented soil. In response to these results, Zhou et al. (2009) produced a table of microorganisms identified in Karst Forest by restriction fragment length polymorphism (RFLP) analysis and showed the widespread occurrence of these species.

Contemporary Biochar and Physico-/Biochemical Characteristics

It is well established that the substrates and pyrolysis parameters for producing different biochars determine their physico- and biochemical properties and thus their effects on the indigenous microbial populations (see also Chapter 2). Also the effects will be site and application ratio specific.

Feedstock determines the final chemical composition while process temperature, in particular, determines the surface area, pore size/volume/distribution, sorption, and partitioning of the biochar. Some of the substrates, production conditions, and properties of the generated biochars have been summarized in Table 2.1.

Generally, high-temperature pyrolysis (>550°C) produces biochars with high surface areas (>400 $m^2 g^{-1}$) (Downie et al., 2011; Keiluweit et al., 2010), increased aromaticity and therefore high recalcitrance (Singh and Cowie, 2008), and good adsorption properties (Lima and Marshall, 2005; Mizuta et al., 2004). In contrast, low-temperature pyrolysis produces higher yields but the products are, potentially, more phytotoxic although they may improve soil fertility because of the stability of the aromatic backbone and increased functional groups, which provide sites for nutrient exchange (Gai et al., 2014; Joseph et al., 2010). Together with temperature, pyrolysis time is an important variable since the two determine cation exchange capacity as has been deliberated widely in the literature (eg, Lee et al., 2010).

Although generally accepted, the underpinning mechanisms of the different responses of biochar application relative to feedstock, production parameters/conditions, receiving ecosystem, methods of physicochemical, biochemical, and microbial ecology analyses, and measurement of plant biomass, yield, or output, are yet to be elucidated fully.

BIOCHAR AND ITS APPLICATIONS

Together with the rapidly growing population, an increasing number of global threats, such as food security because of declining agricultural production, periodic fuel crises, water scarcity, and climate change, have motivated biochar applications with several ongoing research initiatives seeking solutions for immediate implementation. Magnitude and urgency dictate multi-/cross-disciplinary efforts of numerous technologies and approaches.

Climate Change Mitigation

The increasing atmospheric concentrations of carbon dioxide (CO_2), methane (CH_4), and nitrous oxide (N_2O) are of major concern when considering future climates. For all three, cycling, production, consumption, or storage are linked substantially with soils (Gärdenäs et al., 2010). Anthropogenic activities (eg, fossil fuel emission, industry) and natural phenomena (such as carbon cycle, organic matter mineralization) are the major contributors of these biogenic greenhouse gases and are considered to be the main influence on radiative forcing and perturbation of global climate.

Annually, plants are thought to harness 15–20 times the volume of CO_2 emitted by fossil fuel, although about half of this is returned immediately to the atmosphere through respiration with approximately 60 Gt retained in new plant growth (about 45% (w/w) of planet biomass is carbon) (Sohi and Shackley, 2010). It has been estimated that globally, soils hold more organic carbon (1–100 Gt) than the atmosphere (750 Gt) and the terrestrial biosphere (560 Gt). Additionally, the flux of CO_2 from soil to the atmosphere has been reported to be in the region of $60 \, GtC \, year^{-1}$, which results mainly through microbial respiration because of soil organic matter mineralization (Verheijen et al., 2010). Therefore a net sink of atmospheric CO_2 was a principal consideration for combining pyrolysis with soil biochar application to retain photosynthetically fixed CO_2 (Lehmann, 2007).

Biochar, returned to the soil, conveys both quality benefits and, potentially, carbon sequestration (Lehmann, 2007; Spokas et al., 2009). It has been estimated that about half of plant biomass carbon can be converted through pyrolysis into chemical forms that are biologically and chemically recalcitrant. Feedstock and pyrolysis temperature can, however, determine carbon retention (Lehmann, 2007). Consequently, carbon, trapped as biochar, increases its recalcitrance and has the potential to exist for hundreds to thousands of years or possibly longer (Spokas et al., 2009; Verheijen et al., 2010), thus increasing carbon input relative to microbial respiration output and thus climate change mitigation (Steinbeiss et al., 2009; Verheijen et al., 2010).

Several investigations have indicated that biochar additions not only lead to a net sequestration of CO_2 but also may decrease emissions of CH_4 and N_2O (Lehmann, 2007; Sohi and Shackley, 2010). Although the atmospheric concentrations of CH_4 (1.74–1.87 ppm) and N_2O (321–322 ppb) are substantially lower than that of CO_2 (385 ppm), their impacts on global warming on a mass basis are approximately 25 and 298 times greater, respectively, over a 100-year timeframe (US EPA, 2011 cited in Liebig et al., 2012). In a study by Rondon et al. (2005) it was observed that emission of N_2O from soybean plots was reduced by 50% and CH_4 was suppressed fully following acidic soil biochar supplementation ($20 \, mg \, ha^{-1}$) in the Eastern Colombian Plains (Spokas et al., 2009). Similarly, Yanai et al. (2007) observed an 85% reduction in N_2O production of rewetted soils with 10 wt% biochar compared to control soils possibly because of lower nitrification resulting from the higher C:N ratio or lower carbon quality (Lehmann, 2007). More work is, however, required to determine the effects of biochar on the soil nitrogen cycle and the associated greenhouse gas emissions.

The microbial ecology of biochar application in the context of climate change potential is explored in Chapters 7, 8, and 11 with a focus on greenhouse gas emissions or mitigation caused by soil application. In general the contradictory findings on whether biochar augmentation contributes to or alleviates NO_x gas production in agricultural contexts highlight the need for further global studies. To date, most published literature on this has focused on biochemical analyses with microecophysiology-based studies as indicated in Table 1.1.

Agriculture

There is real potential for biochar to close an agronomic circle through enhanced yields (microbial stimulation in the rhizosphere), reduced energy outputs (less fertilizer application), minimized carbon emissions (carbon sequestration), and decreased irrigation demands (improved soil moisture-holding capacity). The partial and/or full realization of these must be informed by elucidating, critically, the basic tenets of how soil microbial communities respond to biochar augmentation.

Biochar has been considered as a potential approach to develop more sustainable agricultural systems, together with addressing global food security and reducing greenhouse gas emissions as major concerns in agricultural management (Jones et al., 2012). The ability of biochar to store carbon and improve soil fertility depends on its physical and chemical properties, which vary according to pyrolysis conditions and feedstock choice (Atkinson et al., 2010; Novak et al., 2009). Also, since soil amendment requirements in different regions depend on specific quality issues, one biochar type is not always a solution for all. For example, biochar with a highly aromatic composition (pyrolyzed at 400–700°C) may be best suited for long-term C sequestration because of its recalcitrance. It has, however, a limited capacity to retain soil nutrients as a result of fewer ion exchange functional groups because of dehydration and decarboxylation (Novak et al., 2009). In contrast, lower temperature (250–400°C) biochars have higher yield recoveries because of more C=O and C—H functional groups that can serve as nutrient exchange sites after oxidation (Novak et al., 2009).

The central quality of biochar that makes it attractive as a soil supplement is its highly porous structure, which varies from <0.9 nm in nanopores to >50 nm in macropores, which improves water retention and increases soil surface area (Atkinson et al., 2010; Sohi et al., 2009). The study by Liang et al. (2006) on sandy soil biochar inclusion demonstrated specific surface area increase (×4.8) relative to adjacent soil. Typically, sandy soils have a limited specific surface area ($0.01–0.1\,m^2g^{-1}$) compared to clayey soils ($5–750\,m^2g^{-1}$) so their water and nutrient retention capacities are low (Atkinson et al., 2010). A crucial factor here is the ability of biochar to adsorb (and transport) nutrients, which are retained commonly in soil by adsorption to minerals and organic matter (Lehmann, 2007). As with biochar, the cation exchange capacity (CEC) of soil increases in proportion to the amount of organic matter. Since biochar has a higher surface area,

TABLE 1.1 Examples of microbial ecology techniques used to study biochar-impacted ecosystems as of January 2016

Method	Biochar type	Scale and duration	Context	Target communities/species	Target genes	References
ARISA	Corn stover	510-day Alfisol and Andisol microcosms with and without *Medicago sativa* L.	Impacts on organic matter chemical composition of two soil types	Total bacterial community	16S rRNA	Wang et al. (2015a)
ARISA, T-RFLP, and next-generation sequencing	Amazonian Dark Earth	Amazonian Dark Earth samples	Impacts of different Amazonian agricultural models	Bacterial, archaeal, and fungal communities	16S rRNA and 18S rRNA	Navarrete et al. (2010)
CLPP and T-RFLP	Jarrah (*Eucalyptus marginata*)	Agronomic field planted with wheat for 10 months	Carbon and nitrogen dynamics	Total and ammonia-oxidizing bacteria	*amoA* gene	Dempster et al. (2012)
DGGE	Broad-leaved konara oak (*Quercus serrata* Murray)	Addition to mixtures of poultry manure and rice husk/apple pomace wastes	Impact on compost quality re stabilization capacity, microbial biomass and enzyme activity	Bacterial and fungal communities	16S rRNA and 18S rRNA	Jindo et al. (2012)
	–	Field studies in north China brown soil planted with soybean	Impact on cultivable microbial abundance and community composition	Cultivable and total bacteria, and total fungi and actinomycetes	16S RNA and 18S rRNA	Sun et al. (2012)
	Bamboo	3-year soil mesocosms planted with maize; run-off and infiltration water used for aquatic mesocosms in third year	Impacts of biochar and organic fertilizers on soil and water microbial communities	Bacterial and viral communities	16S rRNA	Doan et al. (2014)
	Rice straw and garbage	Samples from paddy field	Paddy-upland rotation experiment	Total bacteria	16S rRNA	He et al. (2014)
	Mushroom medium, corn stalk, and rice straw	Impacts of fresh biochar-derived volatile organic compounds (VOCs) on the dynamics of *Bacillus mucilaginosus* and soil microbial communities	Inoculum survival and microbial community structure in response to biochar-based VOCs	*Bacillus mucilaginosus* and bacterial and fungal communities	16S rRNA and 18S rRNA	Sun et al. (2014)

Method	Feedstock	Experiment	Focus	Community	Marker	Reference
DGGE, R2A plating, RFLP, and sequencing	Miscanthus giganthus	Pot experiments using soil from "Cow-lands" experiment planted with Lolium perenne var. Malambo	Impacts on nutrient-mobilizing/plant growth promoting bacteria in the Lolium perenne rhizosphere	Total bacterial community structure, sulfur-/phosphorus-mobilizing rhizobacteria diversities and nematode counts	16S rRNA, asfA, and phnJ	Fox et al. (2014)
DGGE, T-RFLP, clone library, and sequencing	Amazonian Dark Earths	Amazonian Dark Earth samples	Amazonian Dark Earths or Terra Preta microbial community structure and composition	Total bacterial and archaeal communities	16S rRNA	Grossman et al. (2010)
PLFA	Maize straw	N-amendment on microbial communities	Microbial community structure	Bacterial, archaeal, and fungal communities	N/A	Lu et al. (2015a)
	Cotton straw	Field study on drip-irrigated desert soil planted to cotton	Soil microbial community composition	Bacteria and actinomycetes	N/A	Liao et al. (2016)
	Hard wood	2-year in situ experiment within vineyard topsoil	Biochar and compost amendments effects on copper immobilization	Total bacterial community	N/A	Mackie et al. (2015)
	Sugar maple wood	24-week incubation study	Shifts in microbial community composition and water-extractable organic matter with biochar amendment	Total bacterial community	N/A	Mitchell et al. (2015)
CLPP and PLFA	Sewage sludge, Miscanthus, and pine wood	Greenhouse experiments	Biochar interaction with earthworms	Total bacterial and fungal community	N/A	Paz-Ferreiro et al. (2015)

Continued

TABLE 1.1 Examples of microbial ecology techniques used to study biochar-impacted ecosystems as of January 2016—cont'd

Method	Biochar type	Scale and duration	Context	Target communities/species	Target genes	References
qPCR	Switchgrass	6-month pot study	Biochar impact on microbial nitrogen cycling	Nitrogen cycling microbial community	*nifH, amoA, nirS, nirK, nosZ* and 16S rRNA	Ducey et al. (2013) and Harter et al. (2013)
	Rice straw	Pot trials lasting six consecutive crop seasons	Impacts on nitrification potential, abundance, and community composition in intensively managed agricultural soil	Ammonia-oxidizing archaea and ammonia-oxidizing bacteria	16S rRNA	He et al. (2015)
	Pinewood	3-week pot trials containing cucumbers	Inoculum carrier in agricultural soil	Plant growth-promoting rhizobacteria	16S rRNA	Hale et al. (2014)
qPCR, DGGE, clone library, and sequencing	Corn stalk	Paddy soil microcosms transplanted with 22-day-old seedlings of "Wuxiangjing 14" (inbred *Japonica*) cultivar	Biochar impact on methane emissions from agronomic soil	Methanogenic archaea	Archaeal 16S rRNA and *pmoA*	Feng et al. (2012)
	–	84-day microcosms with coastal alkaline soil	Impact on nitrogen-fixing bacteria	*nifH* abundance and diversity	*nifH*	Song et al. (2014)
	Poultry manure	Field scale over 3 years	Biochar ability to improve crop production in saline soils	Total bacterial and fungal communities	16S rRNA and 18S rRNA	Lu et al. (2015b)

Method	Feedstock	Experimental setup	Study focus	Target community	Gene target	Reference
qPCR and T-RFLP	Rice straw and dairy manure	–	Biochars' ability to mitigate nitrous oxide emission	Ammonia-oxidizing bacteria and nitrite-oxidizing bacteria	*amoA* and *nirS*	Liu et al. (2014)
	Corn silage hydrochar and deciduous/coniferous wood chips pyrochar	80-day laboratory microcosms	Comparative effects of pyro- and hydrochar on bacterial and archaeal N_2O and CO_2 emissions	Bacterial and archaeal communities	16S rRNA	Andert and Mumme (2015)
	Sugar maple and white spruce sawdust	Forest soil over 2 years	Impact of biochar addition to acidic hardwood forest soils	Total bacterial and fungal communities	16S rRNA and 18S rRNA	Noyce et al. (2015)
qPCR and SIP	Wood pellets	Batch incubation experiments	Biochar impact on methane oxidation within landfill	Methanotrophs	*pmoA*	Reddy et al. (2014)
qPCR, DGGE, T-RFLP, and clone library	Wheat straw	Field experiment over 2 years	Biochar impact on microbial activity within cropland	Total bacterial and fungal communities	16S rRNA and 18S rRNA	Chen et al. (2013)
T-RFLP	Birch (*Betula pendula*) wood grown on trace metal contaminated and noncontaminated land	5-week pot experiments planted with *Lolium perenne* var. *Calibra* in growth chamber	Impact of feedstock groan on contaminated land on plant growth and metal accumulation	Total bacterial community	16S rRNA	Evangelou et al. (2014)
T-RFLP, qPCR, and next-generation sequencing	Monterey pine (*Pinus radiata*)	7-day field plots with and without bovine urine	Impacts on nitrogen-cycling microbial communities	Total, nitrifying, and denitrifying bacterial communities	16S rRNA, *nirS*, *nirK*, and *nosZ*	Anderson et al. (2014)
	Amazonian Dark Earths	Amazonian Dark Earth soil samples	Insights into catabolic microbial community structure, composition, and abundance in Amazonian Dark Earth	Total and functional community structure, composition, and abundance	16S rRNA and *bph*	Brossi et al. (2014)

Continued

TABLE 1.1   Examples of microbial ecology techniques used to study biochar-impacted ecosystems as of January 2016—cont'd

Method	Biochar type	Scale and duration	Context	Target communities/species	Target genes	References
T-RFLP and next-generation sequencing	Monterey pine (*P. radiata*)	Filed plot soil used subsequently for pot experiments	Impacts on rhizosphere and bulk soil communities	Total bacterial community	16S rRNA	Anderson et al. (2011)
	Wheat straw	Three field sites	Biochar impacts on bacterial diversity and activity within paddy fields	Total bacterial community	16S rRNA	Chen et al. (2015)
	Amazonian Dark Earths	Amazonian Dark Earth samples	Impacts of Amazonian Dark Earth on rhizosphere communities of *Mimosa debilis* and *Senna alata*	Rhizosphere bacterial community	16S rRNA	Barbosa et al. (2015)
Next-generation sequencing	Maize	180-day soil incubation	Soil priming effects with plant residue	Total bacterial and fungal communities	16S rRNA and 18S rRNA	Su et al. (2015)
Pyrosequencing	Citrus wood	Pot experiments with pepper plants	Impact on root-associated bacterial community structure	Total bacterial community	16S rRNA	Kolton et al. (2011)
	Anthrosols and wood	Two field sites	Comparison of anthropogenic biochar and Amazonian anthrosols	Total bacterial community	16S rRNA	Taketani et al. (2013)
	Amazonian Dark Earths	Comparison of four Amazonian Dark Earth sites	Fungal community composition across Amazonian Dark Earth sites	Total fungal community	18S rRNA	Lucheta et al. (2015)
	Acacia green waste	3.5-year field experiment	Coapplication of biochar and fertilizer to agricultural soil	Total bacterial and fungal communities	16S rRNA and 18S rRNA	Abujabhah et al. (2016)

Method	Biochar feedstock	Experimental setup	Purpose	Target community	Gene/marker	Reference
Sequencing	Soybean stover and pine needle	90-day soil incubation experiment	Biochar as a candidate for remediation of lead and copper	Total bacterial community	16S rRNA	Ahmad et al. (2016a,b)
	Corncob	34-month pot experiment	Microbial community growing on biochar pellets	Total bacterial and fungal communities	16S rRNA and 18S rRNA	Sun et al. (2016)
	Citrus wood	Greenhouse experiments with sweet pepper and tomato plants grown in soilless potting medium	Nutrient agar cultivable microbial strains from pepper experiments	Bacterial isolates with unique colony morphologies	16S rRNA gene	Graber et al. (2010)
Sequencing and DGGE	–	–	Effects of abiotic N, P, K, pH, and C:N ratio on microbial community composition	Bacterial, fungal, and actinomycete communities	16S rRNA and 18S rRNA	Sun et al. (2013)
	Wood chips	30-day laboratory column experiments	Combined impacts with volatile petroleum hydrocarbon addition	Bacterial community	16S rRNA	Meynet et al. (2014)
	Wheat straw	112-day pot macrocosms with NPK fertilizer incubated in the dark for 4 weeks and then seeded with Lolium multiflorum Lam with greenhouse maintenance	Heavy metal and PCB-contaminated soil	Total bacteria	16S rRNA	Liu et al. (2015)
	Peanut shell	70-day pot studies with acidic orchard soil	Impacts on nitrification and ammonia-oxidizing bacteria (AOB) community structure	AOB	16S rDNA of AOB	Wang et al. (2015b)
Stable isotope labeling/tracing	Grassland biomass	45-day soil pots planted with Trifolium pratense L. singly or in mixed cultures with Festuca rubra L. and Plantago lanceolata L.	Biological nitrogen fixation re biochar application rate and plant species	Nitrogen-fixing bacteria	N/A	Cayuela et al. (2013) and Mia et al. (2014)

ARISA, automated radioisotope substrate analysis; T-RFLP, terminal restriction fragment length polymorphism; CLPP, community-level physiological profiling; DGGE, denaturing gradient gel electrophoresis; qPCR, quantitative or real-time polymerase chain reaction; SIP, stable isotope probing; PLFA, phospholipid fatty acid; PCB, polychlorinated biphenyls.

greater negative surface charge, and greater charge density than other soil organic matter, it has the capacity to adsorb cations to a greater extent. The potential CEC of biochar has also been recorded to increase with increased temperature (Lehmann, 2007). Additionally, biochar has shown to contribute directly to nutrient adsorption, thereby decreasing nutrient leaching and, consequently, increasing nutrient use efficiency with resultant higher crop yield.

Unfortunately, very little is known about biochar impacts on soil physical properties. For example, while its mobility within the soil profile could be of vital interest with respect to enhanced plant production, its migration into ground- and surface waters has started to raise concern for potentially unintended negative impacts (Leifeld et al., 2007). It has been observed that biochar facilitates root penetration in soils with high impedance/compactness (Chan et al., 2007) and improves soil water permeability in upland rice-growing land (Asai et al., 2009). In contrast, Verheijen et al. (2010) proposed that small particle size biochar fractions may affect soil hydrology and, subsequently, decrease soil infiltration rate. Therefore evaluation of the extent and implications of biochar particle size distribution relative to its mobility and fates in soil processes and functions is essential.

So far, problematic soils such as those with extreme acidity or salinity, severe nutrient imbalances, and critically low soil organic matter have been the main focus of biochar supplementations often with dramatic impacts recorded (Jien and Wang, 2013; Spokas et al., 2012). These soils, however, do not represent fertile agricultural areas. Jones et al. (2012) made a 3-year field trial and observed that biochar addition to highly productive agricultural systems may not yield similar benefits as seen for poor soils. Importantly, no negative consequences were apparent in terms of general soil quality and crop growth and nutrition.

To reflect their importance, deliberations on these aspects are presented in Chapters 7 and 8, particularly for the nitrogen cycle. These, however, do not include biochar potential as inoculum carrier (Hale et al., 2014; Sun et al., 2014) and in plant disease mitigation (eg, Graber et al., 2014) so the possible applications and need for these new research areas are alluded to in Chapters 4 and 12. Briefly, for the latter, biochar can reduce the incidence of plant disease caused by the chemical (improved pH, cation exchange capacity, and P and S transformations), microbial (reduced number of fungal pathogens and promotion of mycorrhizal fungi), and physical (increased water-holding capacity and soil bulk density) improvements it makes to the soil (Elad et al., 2010). In this study, soil-applied biochar impacted positively on plant disease resistance to the foliar fungal pathogens *Botrytis cinerea* (gray mold) and *Leveillula taurica* (powdery mildew) on sweet pepper and tomato and *Podosphaera aphanis* (powdery mildew) on strawberry plants. This induced systemic resistance phenomenon was caused by the interactions between specific soil microorganisms, such as *Bacillus*, *Pseudomonas*, and *Trichoderma* spp., and plant roots (Kolton et al., 2011).

In summary, the roles and potential applications of biochar in agriculture contexts need to reflect biogeochemical cycling in general, but C and N in particular, and include an understanding of biochar-based greenhouse gas emission or mitigation. Naturally, fertilizer and water retention with respect to increased crop yield are well recognized and are being researched further to elucidate biochar- and soil-specific effects, but the exciting area of disease suppression is beginning to emerge and raise new questions and challenges for researchers and practitioners alike. Ultimately, the roles of micro- (bacteria, fungi, and archaea) and macroorganisms in all these specific agronomic subcontexts necessitate continued applications of established and novel microecophysiology tools.

Bioremediation

As introduced earlier, biochar has the potential to adsorb and sequester both organic and inorganic contaminants (eg, Shan et al., 2016; Zhang et al., 2013a,b) with their high microporosity and surface area and heterogeneous surface physicochemical properties as key factors (Beesley et al., 2010, 2011). Consequently, it has been investigated as a potential novel tool in polluted site remediation. In addition, it is being evaluated as a low-cost in situ remediation approach for leachates (such as heavy metal containing leonardite) and persistent pollutants such as polycyclic aromatic hydrocarbons (PAHs), polychlorinated biphenyls, pesticides, and metals (Bushnaf et al., 2011; Cheng et al., 2016; Li et al., 2016).

Activated charcoal or biochar, when mixed with contaminated soils and sediments, causes pollutants to repartition from the contaminated matrix to the added particles, leading to adsorption (Hale et al., 2012). The large surface area of biochar has been considered for its high affinity and capacity to sorb organic compounds such as PAHs and so reduce their availability to further leaching or organism assimilation (Beesley et al., 2010; Hale et al., 2012). Such occlusion may, however, reduce pollutant availability for microbial catabolism and so increase persistence. The efficacy of biochar in multimolecule-contaminated land remediation was highlighted by Beesley et al. (2010) who reported greatly reduced concentrations of PAHs, with more than a 50% decrease of the heavier, more toxicologically relevant PAHs, compared with a green waste compost. Also, unlike other soil supplements, biochar longevity obviates the need for repeated applications as required for heavy metal adsorption (Lehmann and Joseph, 2009).

Potentially there are constraints in using activated carbon and/or biochar as an effective tool in environmental management and these include inefficient mixing, which is required, for example, to enhance pollutant mass transfer from sediment particles to biochar. Similarly, biochar pore blockage by other anthropogenic contaminants or natural organic matter, either via a direct or a competitive sorption mechanism, can reduce treatment effectiveness (Hale et al., 2012).

Biochar can benefit soil microorganisms in many ways as it provides nutrients, such as carbon and minerals, which adsorb on its surface, and habitats in its pores, which afford protection from predators and grazers. Microorganisms, such as bacteria and fungi, together with other soil organisms, play important roles in nutrient cycle, catabolism, and soil health maintenance. Many studies have reported that biochar addition may stimulate the growth and activity of bacteria, such as *Pseudomonas* and *Bacillus* spp., which are capable of mineralizing pollutants such as PAHs and benzonitrile (Song et al., 2012). In contrast, decreased benzonitrile and atrazine degradation have also been reported in biochar-supplemented soil, possibly because of decreased molecule bioavailability via sorption (Song et al., 2012), and so commend contaminant bioavailability assessment.

Currently, magnetic biochar is being explored largely for its applicability in physicochemical adsorption of organic and inorganic contaminants from a range of industries such as biopharmaceuticals (carbamazepine, tetracycline—Shan et al., 2016; pentachlorophenol—Devi and Saroha, 2014), heavy metals (chromium—Han et al., 2015a; arsenic—Zhang et al., 2013b; lead—Yan et al., 2014), and pesticides (Chikezie, 2011). A comprehensive literature search up to January 2016 (Scopus) revealed no record of published studies on microbial ecological analyses following ecosystem augmentation with magnetic biochar. Notwithstanding this, potential experimental designs for future work with this are suggested in Chapter 4. Thus combinations of culture-based and molecular approaches can be applied in relatively simple and controlled investigative models to address the complexity of biochar effects when magnetism is added to the already growing list of biochar variables.

In general, the strengths, limitations, potentials, possible challenges, etc. of biochar application to effect contaminant attenuation are therefore presented in Chapters 9 and 10. These deliberations include knowledge that has been gained relative to different molecular microbial ecology techniques that have been applied to date, key knowledge gaps, and some complementary physicochemical tools.

BIOCHAR AS HABITAT FOR SOIL ORGANISMS

During pyrolysis, biochar porosity and therefore specific surface area increase considerably. On the one hand, plant-based feedstocks provide anatomical structures through xylem and phloem vessels; on the other hand, the pyrolysis process effects the loss of volatile matter that may fill micropores and so increase pore space availability by several orders of magnitude (Mukherjee et al., 2011). Since the early days of biochar research, it has been hypothesized that micropores serve as habitats for microorganisms where they are protected from grazers and/or desiccation (Steinbeiss et al., 2009).

The importance of micropores depends to a large extent on biochar type and pore size distribution. Quilliam et al. (2013) observed that about 17% of the pores of a wood-based biochar were uninhabitable for most microorganisms because of their diameters of <1 μm, although the larger pores were less densely populated than hypothesized and sometimes even blocked by mineral aggregations. Depending on the availability of organic carbon, the soil pores seemed more colonized by microbial species than the biochar pores, which had lower mineralizable carbon at both the external and internal surfaces. This observation might have resulted from the researchers analyzing relatively large pieces of biochar (1–2 cm), which, even after 3 years of field soil exposure, had not been populated as much as the smaller particles. In biochar field applications, farmers tend to use their usual fertilizer distribution equipment optimized for particle sizes typical for mineral fertilizers (1–3 mm). The authors expected that in the long term, microbial colonization would increase significantly after progressive physical biochar weathering, accompanied by higher availability of metabolically labile carbon.

Studies by Jaafar et al. (2015) confirmed the habitability of different wood-based biochars for soil microorganisms. For example, the chars were populated by fungal hyphae within an incubation period of 56 days with differences in porosity and pore structure of different chars having little importance. Rather, the effects of soil clumping on biochar surfaces and pore closure by soil particles tended to reduce the surface area and pore availability for microbial colonization. As the characteristics of such interactions are dependent on the respective soil and biochar, this study, which focused on one soil type, highlighted the need for further investigations with multiple biochars and soils. One could expect that a less acidic soil, for example, with a different composition of soil organic matter, might interact differently with the same biochars as used in the study. Also the particle size insignificance might prove more relevant if the interactions of the soil mineral phase with the biochar surfaces do not dominate pore accessibility.

Evidence of fungal colonization of biochar was also found by Hammer et al. (2014) who observed mycorrhizal growth and phosphorus recovery from biochar surfaces. Increased phosphorus availability may also be mediated by bacterial phosphatases that contribute to nutrient provision (Ennis et al., 2012). As these bacteria can be found in the rhizosphere of many soils (Cattaneo et al., 2014), they should also colonize readily biochar surfaces. According to Warnock et al. (2007), mycorrhizae and plant roots interact by means of signaling molecules. Also, although biochar may interfere with these allelochemicals, which are important for signaling during root colonization, it can still contribute to the detoxification of fungicidal compounds in soil. Therefore the authors listed some additional mechanisms by which mycorrhizae may benefit from biochar addition to soil and these are: nutrient availability; indirect effects on other soil microbes that may be beneficial or detrimental; and provision of refuge from fungal grazers.

Example studies, such as that of Masiello et al. (2013), have highlighted the need for detailed and fundamental investigations, if initially in simplified controlled systems, on the effects of different temperature biochars on cellular communication. The researchers indicated that their findings must be extended to consider different feedstock biochars and production parameters other than temperature, together with other biochemical signaling molecules. The ultimate aim is to establish how the adsorption capacity of the material for cell signaling molecules influences gene expression and therefore potentially dictates ecosystem function, for example, biofilm formation, carbon sequestration, respiration, nitrogen fixation, and disease suppression.

Microbial abundance, diversity, and activity in soil are determined mainly by pH, redox potential (E_h), soil humidity, water activity, and temperature (Thies et al., 2015). Biochar as a soil supplement affects all of these. If not specially adjusted, fresh biochars are alkaline so their addition increases pH particularly in acidic soils (Kloss et al., 2014). Bacterial diversity is highest in pH-neutral soils (Fierer and Jackson, 2006) whereas fungi are tolerant of more acidic and more alkaline conditions. Therefore biochar addition to acidic soils may increase bacterial abundance whereas addition to neutral soils may increase the ratio of fungi to bacteria (Chen et al., 2013). Ultimately, a shift in microbial association composition is accompanied frequently by a change in bacterial and fungal grazers as well as by their predators and so alters the soil faunal communities (Domene et al., 2015; McCormack et al., 2013).

The surfaces and water soluble compounds of biochar are redox active and may therefore modify the soil redox potential (Husson, 2013). Together with pH, E_h determines the metabolic types of microbial communities and involves the transfer of electrons. Different microorganisms are adapted to specific E_h conditions whereby anaerobic bacteria require a narrow range of low E_h values (Husson, 2013). Under highly reducing conditions ($E_h < 0\,mV$), bacteria are more abundant than fungi whereas the latter dominate moderately reducing E_h conditions (Seo and DeLaune, 2010). Given the predominantly negative charge of biochar and its low E_h, the microbiological reduction of sulfate to sulfide can be supported by electron transfer from biochar to sulfate. Easton et al. (2015) reported that the addition of amorphous $Fe(OH)_3$ mitigated this reduction, which is important to prevent methyl mercury generation in Hg-contaminated soils.

The porosity of biochar contributes to an increase in the water-holding capacity and provides plants with available water during periods of precipitation deficit (Basso et al., 2013). The higher water availability benefits both the rhizosphere bacteria and other soil-dwelling organisms. In years with extreme precipitation, however, the permanently wet soil may benefit phytopathogens such as *Fusarium* that cause root rot in several crops (Elmer and Pignatello, 2011). The dark color of biochar decreases soil albedo, thereby, potentially, increasing soil temperature and affecting radiative forcing and the negative carbon balance of biochar production (Meyer et al., 2012; Verheijen et al., 2013). Existing results are, however, based mainly on either soil surface studies disregarding plant cover or unrealistically high addition rates. Following field application it has been shown that the effects on albedo disappear in the second year (Genesio and Miglietta, 2012). For soil organisms in temperate climates, spring soil warming may benefit microbial activity. When, during progressive crop development, the vegetation cover reaches leaf area indices of >1, the effect of soil color becomes less important. So during summer excessive warming of vegetated soil surfaces constitutes a limited risk factor.

SOIL BIOTA RESPONSE TO BIOCHAR

Soil is very complex with its varied biological communities (such as bacteria, archaea, fungi, algae, etc.) and specifically their structures and functions. The health and diversity of soil micro-/macroorganisms are critical to soil function and ecosystem services, which, in turn, have implications for soil structure and stability, nutrient cycling, aeration, water use efficiency, disease resistance, and C storage capacity (Atkinson et al., 2010; Lehmann et al., 2011). It is proposed that organic supplements are perhaps the most important means of managing soil diversity.

Microfauna

To understand, critically, the impacts of biochar on soil microbial communities and therefore ecosystem functions it is important to use different and multiple techniques. Several molecular techniques have been applied to elucidate microbial community responses to biochar and they include denaturing gradient gel electrophoresis (DGGE), phospholipid fatty acid (PLFA), terminal restriction fragment length polymorphism (T-RFLP), quantitative or real-time polymerase chain reaction (qPCR), and next-generation sequencing.

As previously stated, Amazon soils have been found to contain diverse ranges of microorganisms adapted to the soil's biochemistry and ecology. In taxonomic studies based on molecular approaches, Kim et al. (2007)

recorded higher numbers of operational taxonomic units (OTU) in *Terra Preta* soils (396 OTU) compared with pristine forests soils (291 OTU). Similarly, more bacterial richness, ie, about 25% richer than forest soils, was found with 14 phylogenetic groups compared to nine in the forest soils (Kim et al., 2007). The quality, quantity, and distribution of organic supplements all affect the soil food web trophic structure so each must be considered when biochar is used as a soil management tool (Lehmann et al., 2011).

Although biochar induces changes in microbial community structure, composition, and diversity, it is extremely unlikely to effect the same outcomes across different phylotypes or functional groups. Instead, the altered soil environment, in terms of resource base (eg, available C, nutrients, water), abiotic factor shifts (eg, pH, toxic elements), or different habitats, may increase the competitiveness of some microbial groups and so lead to community composition and structure changes (Lehmann et al., 2011). Consequently, studies on different types of soils (*Terra Preta*, biochar supplemented soil, and soil rich in char from vegetation fires) have shown substantial changes in community composition and diversity of fungal, bacterial, and archaeal populations (Kim et al., 2007).

Archaea were once thought to be synonymous with extreme environments although data suggests their presence is ubiquitous and diverse (Grossman et al., 2010). Although members of soil archaea are difficult to culture, molecular studies have demonstrated the presence of archaeal 16S rRNA genes in a wide variety of soils. The major group observed was Crenarchaeota, which is the dominant phylum in mesophilic environments (Navarrete et al., 2011; Taketani and Tsai, 2010). In contrast, molecular studies have revealed the abundance of archaeal ammonia oxidizers (nitrifiers) in soils compared to their well-known bacterial counterparts indicating their importance in the nitrogen cycle (Grossman et al., 2010; Navarrete et al., 2011). Similarly, the dominance of certain groups of archaea in acidic soil has suggested that they play important roles in low-pH soils (Taketani and Tsai, 2010).

Several studies have compared archaea communities of anthrosols to adjacent soils and have recorded lower diversity, with striking differences in their band patterns by >90% (Grossman et al., 2010). When compared to bacteria, only fewer OTUs were resolved in the DGGE profiles of the archaea 16S rRNA genes. The total quantity and diversity of archaeal communities are known to decrease by soil depth (Grossman et al., 2010) but increase with lower pH indicating higher richness and diversity in adjacent soils compared to anthrosols (Taketani and Tsai, 2010). The structure comparison by T-RFLP for archaea revealed that the soil variables were more important than the soil type. DNA sequencing showed that *Candidatus* was the most abundant genus in both soil types. Furthermore, the role of methanogenic archaea in methane gas emission from paddy fields, which is now considered to be the largest anthropogenic source of methane, is under scrutiny. In particular, biochar addition can lower methane emission by promoting methanotrophic proteobacteria and so reduce the ratio of methanogenic to methanotrophic species (Feng et al., 2012). Hence this knowledge can be used to develop an effective greenhouse gas mitigation process.

The soil fungi, such as arbuscular mycorrhizal fungi (AMF), ectomycorrhizal fungi (EMF), and ericoid mycorrhizal fungi, are known to have well-organized roles in terrestrial ecosystems (Warnock et al., 2007) with biochar additions to AMF and EMF often seen as positive. Mycorrhizal response to the host plant is most commonly assessed by root colonization measurement to determine the abundance of host fungal tissue. In a study by Makoto et al. (2010) (cited in Lehmann et al., 2011) it was observed that biochar addition increased by 19–157% the formation rate and tip number of larch seedling EMF infection. In contrast, decreases in AMF abundance after biochar addition have also been reported (Warnock et al., 2010) possibly because of increased nutrients, particularly phosphorus, and water. Other possibilities include changed soil conditions or direct negative effects of high mineral or organic compound concentrations of, for example, salts or heavy metals (Lehmann et al., 2011).

Specific functional microbial strains and communities respond differently to biochar type and application regime. To exemplify this, detailed discussions are made in Chapters 7 and 8 for nitrogen cycling communities. Most chapters include a section on the impacts of biochar on soil micro- and macropopulations as relevant to the specific context under consideration. Therefore, while there is some degree of overlap between individual chapters, we feel that this is justified as it affords continuity and cross-referencing, while ensuring that each chapter has the requisite degree of autonomy.

In Chapter 3, Ng and Cavagnaro explore the use of PLFA and identify: "These results are yet another indicator that the effect of biochar on the soil food web and soil processes are dependent on biochar properties and biochar–soil interactions." Although only addressed briefly in Chapter 12, biochar addition could interfere with the chemical signals involved in "communication" both between microorganisms and between plants and microorganisms (Warnock et al., 2007). A study by Marsiello et al. (2013) highlighted this key knowledge gap on understanding biochar effects on microbial communities at the cellular/single-cell level, ie, cell communication.

The impacts of biochar additions on soil microbial communities can be explored by culture-based analyses (Chapter 4). Since, however, 95–99% of microorganisms cannot, as yet, be cultured the development and application of advanced molecular biological techniques has enabled researchers to explore and study complex microbial genetic diversities and identify several uncultured microorganisms from different ecosystems (Cenciani et al., 2010; Muyzer et al., 1993; Muyzer, 1999). Genetic fingerprinting is one such approach that is applied to determine microorganism diversity in natural ecosystems and monitor temporal microbial community behavior. These techniques provide patterns or profiles of community diversity on the basis of physical separation of unique nucleic acid sequences (Muyzer, 1999). For example, He et al. (2014) supplemented paddy-upland soils with rice straw and garbage biochars with and without chemical NPK fertilizer. Their DGGE profiles showed that biochar influenced the soil bacterial community structure and diversity as measured by the Shannon index. Another popular fingerprinting technique is T-RFLP, with both techniques complemented by sequencing information. Grossman et al. (2010) used a combination of DGGE, T-RFLP, cloning, and sequencing to make ecogenomic comparisons of Brazilian anthrosols and adjacent nonanthrosol soils. Conventional soil respiration, PLFA, and culturing to probe microbial response to contemporary biochar application are also possible options (Steinbeiss et al., 2009).

Regardless of the molecular approach used, none of these techniques gives a full picture when used alone. Since all techniques have biases it is not always possible to compare directly the results when two different methods are used. In recent years the trend has been to use next-generation sequencing of bar-coded 16S rRNA amplicons for the comprehensive analysis of bacterial community composition in heterogeneous environments, including soil and the rhizosphere (eg, Kolton et al., 2011). Typically, the results found, regardless of the approach, are comparable with higher bacterial, fungal, and archaeal numbers and diversity reported with biochar addition, independent of the soil type and land use.

Overall, a critical mining of the literature repository Scopus revealed that most publications on the microbiological analysis of biochar-impacted ecosystems applied DNA- and lipids or fatty acid-based techniques as exemplified in Table 1.1. The applicability, potential for knowledge furtherance, and limitations of these techniques are discussed in the relevant chapters. For example, Chapter 3 aims to illustrate how biomolecules other than DNA can be targeted to determine how biochar affects the structural, compositional, and functional capacities of microbial communities in different soil ecosystems. Furthermore, understanding the physiological properties of populations in response to environmental perturbations remains central to effective implementation of different environmental biotechnologies, including those that are biochar driven. Therefore, despite ongoing debates of their values compared to advanced high-throughput techniques, Chapter 4 provides an illustration of how culture-based and ecogenomic analyses can be applied for fundamental elucidation of biochar-augmented soils.

Meso-/Macrofauna

As with soil microorganisms, soil fauna also play a role in the soil ecosystem, particularly in the exchange of nutrients between trophic levels and the distribution of organic material around the soil (Bardgett, 2005; Wilkinson et al., 2009). This capacity to alter soil physicochemical structures is brought about by processes such as burrowing and fecal deposition (Hardie et al., 2014). Macrofauna can also have negative impacts such as those that are disease causal agents (Fry, 2012).

As is explored in Chapter 11, soil macrofauna are also affected by biochar addition. Although the literature is relatively limited, there is emerging evidence of contradictory impacts of biochar additions on macrofauna such as the positive impact on earthworm communities, and in turn on rice growth as observed by Noguera et al. (2010), contrasting the zero impact on heavy metal reduction and earthworm populations reported by Beesley and Dickinson (2011). As for other communities, it is likely that these effects depend on: soil type; pH; biochar application ratio; biochar type *re* source, pyrolysis conditions, age, contaminant occurrence, and concentration; and time after application.

Fauna are also affected by other environmental perturbations. Like plant roots, soil fauna contact contaminants directly so their uptake is a measure of bioavailability and absorbability through the digestion tract. To study sediments, benthic organisms such as chironomid larvae are reliable indicators of ecotoxicity. Shen et al. (2012) observed a sharp decrease in the biota-sediment accumulation factor for PAH following biochar additions of <1.5%. For higher applications the accumulation factor fell apparently because of increased particle ingestion by the larvae. A study of blackworms (*Lumbriculus variegatus*) in sediments supplemented with 8% magnetic activated carbon or biochar for PAH remediation revealed a continuing ecotoxic effect even when 77% of the magnetic sorbent had been removed and aqueous PAH concentrations had decreased by 98% (Han et al., 2015b).

Earthworms are effective bioindicators of ecotoxicity in agricultural soils. Tammeorg et al. (2014), for example, did not observe earthworm avoidance in soil supplemented with 1.6% biochar and found the highest earthworm densities and biomasses in a field soil where $30\,t\,ha^{-1}$ had been added. When Gomez-Eyles et al. (2011) examined a PAH-contaminated soil the bioavailable (cyclodextrin-extractable) PAH concentrations decreased both in the soil and in the earthworms following a 10% (w/w) addition of biochar. The earthworms themselves increased the PAH bioavailability apparently through ingested soil digestion. Biochar also reduced hexachlorobenzene uptake by earthworms in the studies of Song et al. (2012), even when the application was as low as 0.1% (w/w). As a result, the need to explore the responses of this important soil community is considered in Chapter 10, in part, and Chapter 11, in detail.

POLICY GUIDELINES AND REQUIREMENTS FOR BIOCHAR APPLICATION

A debate that is not new to scientific research can probably be encapsulated by the mirror question: Science before policy, where science is pursued to inform/underpin policy directly and indirectly or policy before science where policy development directs scientific investigation? Notwithstanding the perspective or where emphasis is placed on this conundrum by different stakeholders, the fact remains that biochar production and subsequent application, including in soils, must be linked to appropriately and robustly deliberated policy guidelines. For example, guidelines, directives, policies, and legislation on pollutant abatement, control, containment, and (bio)remediation are established and implemented globally, although to different degrees of adherence, management, regulation, and (legislative/constitutional) rigor. Thus well-recognized and lesser known regulatory bodies have been established worldwide and they include:

1. Australia—Commonwealth Scientific and Industrial Research Organisation (CSIRO);
2. Brazil—UK & Brazil Science and Innovation Network;
3. Canada—Environment and Climate Change Canada; Canadian Environment Assessment Agency;
4. China—Chinese Agency for Environmental Planning;
5. Europe—European Union Waste Framework, Organisation for Economic Co-operation and Development (OECD), Registration, Evaluation and Authorization of Chemicals;
6. Republic of South Africa—Council for Scientific and Industrial Research (CSIR);
7. United States of America—Environment Protection Agency (US EPA);
8. United Kingdom—Environment Agency, Department for Environment Food & Rural Affairs; and
9. New Zealand—Environmental Protection Authority.

Interestingly, to our knowledge, with the exception of the Scottish Environmental Protection Agency's *Interim Regulatory Position Statement on Biochar*, none of these bodies has firm guidelines on biochar (Table 1.2). Hence there is considerable paucity of policy statements on its application despite multiple acclaimed benefits, recognized potential limitations, and apparent ongoing uses, particularly at medium to large agronomic scales. Some of biochar key characteristics, which are underpinned by its unique physicochemical architecture, are its capacity to: sequester carbon and, consequently, mitigate climate change; act as a soil modifier, to then enhance agricultural output while managing fertilizer and irrigation water requirements; and facilitate (bio) remediation efficacy. Current research findings on these three elements can be summarized as follows:

1. Mitigate climate change. Contradictory and/or different reports regarding the short- and long-term recalcitrance of biochar-C, and different emission concentrations of greenhouse gases, following biochar application, especially from agronomic soils.
2. Agricultural output. Contradictory trends with some biochars in some agronomic soils increasing plant/crop yield; different biochar capacities for pesticide and fertilizer adsorption with various implications for crop output; and different impacts on soil microbial communities, including mycorrhizal fungi, in bulk and rhizosphere soils, again with different outcomes for agronomic yields.
3. (Bio)remediation potential. Balancing contaminant sequestration, for physicochemical attenuation, to mitigate migration within and beyond the impacted ecosystems, and bioavailability, for microbial catabolism, is essential.

Unfortunately, these three specific but interrelated thrusts/foci are exacerbated by disparate studies where a wide array of different feedstocks, production conditions, application ratios, soils, climatic conditions, and plant species are researched. Nonetheless, several deliberations, as exemplified by Ortega-Calvo (2013, 2015) on the importance of addressing contaminant bioavailability relative to managing its potential risk to (soil) ecosystem health, must be transferred and translated to biochar application, which could intensify such risk. Other rigorous debates consider the need for consistency, including the use of "gold standard" biochar(s) (eg, Lehman and Joseph, 2015; Schloz et al., 2014), to ensure comparability between and across studies/soils/biochars. It is, however, doubtful whether this strategy can be adopted realistically by researchers and practitioners nationally, regionally, and globally.

Overall, the need for concerted effort, including for policy development and implementation, is evidenced by the emergence of bodies (established as of Jan. 2016) such as the International Biochar Initiative (IBI), the British Biochar Foundation (BBF), the UK Biochar Research Centre (UKBRC), the European Cooperation on Science and Technology Action on Biochar/European Biochar Research Network, the United States Biochar Initiative (USBI), the Australia and New Zealand Biochar Researchers Network, and the China Biochar Network (Table 1.2). Each has different approaches with formalized and nonformalized strategies and declarations.

The IBI was founded in 2002 with three key objectives: supporting biochar research, providing clear, nonbiased information around all aspects of biochar, and creating standards and policies to guide the public. Subsequently, the organization formulated the *IBI Biochar Standards*, which provides guidelines/deliberations on feedstock, biochar production, handling

TABLE 1.2   Examples of biochar "Policy" bodies and/or initiatives and their key statements prior to February 2016

Body	Policy document	Key "Policy" statement(s)/parameters	Comments	Document location
International Biochar Initiative (IBI)	IBI Biochar Standards—Standardized Product Definition and Product Testing Guidelines for Biochar That is Used in Soil	Category A (Basic Utility Properties, required for all biochars); Category B (Toxicant Assessment, required for all biochars); Category C (Advanced Analysis and Soil Enhancement Properties, optional)	Developed by consulting a range of frameworks/bodies/methods, eg, OECD, US EPA, ASTM, AOAC. Policy revisions inform the IBI Biochar Standards	http://www.biochar-international.org/sites/default/files/IBI_Biochar_Standards_V2%200_final_2014.pdf
British Biochar Foundation (BBF)	Biochar Quality Mandate		Focuses specifically on soil biochar addition; mandate is informed by expertise from UK funding, research, and policy bodies	http://www.britishbiocharfoundation.org/
European Biochar Foundation/Certificate (EBC)	European Biochar Certificate—Guidelines for a Sustainable Production of Biochar	Biochars graded into "basic" and "premium"; certification is dependent on feedstock, method of production, physicochemical characteristics, and application methods/aims; nine key certification requirements stipulated for biochar properties	–	http://www.european-biochar.org/en

UK Biochar Research Centre (UKBRC)	N/A	N/A	Produces the UKBRC Standard Biochar; Repository for commissioned reports and the Scottish Environmental Protection Agency *Interim Regulatory Position Statement on Biochar* (http://www.sepa.org.uk/media/156613/wst-ps-031-manufacture-and-use-of-biochar-from-waste.pdf)	http://www.biochar.ac.uk/index.php
European Commission Environment, eg, Waste Framework Directive	N/A. Repository of funded project reports and research articles	Biochar must comply with, for example, Directives 2001/95/EC and 2009/31/EC on product safety and "geological storage of carbon dioxide," respectively	Biochar classified as waste, hence implications for safety and use on agricultural land and other pristine ecosystems	http://ec.europa.eu/environment/index_en.htm
UK Environment Agency	N/A. Example deliberations include: *Product comparators for materials applied to land: non-waste biochar Report—SC130040/R6*	Nonwaste biochar feedstock listed; biochar applied to sequester carbon and improve soil conditions; requisite consideration for biochar-derived N re "nitrate vulnerable zone"; analytical and certifiable physicochemical parameters identified; limits for heavy metals/metalloids, PAHs, dioxins/furans and PCBs stipulated	Document informed, in part, by published scientific literature	https://www.gov.uk/government/organisations/environment-agency

PAHs, Polycyclic aromatic hydrocarbons; *PCBs*, polychlorinated biphenyls; *OECD*, Organisation for Economic Co-operation and Development; *US EPA*, United States Environmental Protection Agency; *ASTM*, America Standard Test Method; *AOAC*, Association of Analytical Communities.

and storage, and categories (test category A, B, and C) for physicochemical characterization. The European Biochar Foundation/Certificate (EBC) aims to ensure biochar safety and consistency/assurance relative to production technology. The guidelines developed are similar to IBI's and are informed by the latest research findings and practices with the objective to afford knowledge transfer "as a sound basis for future legislation." The BBF has developed the *Biochar Quality Mandate* (v.1.0 published in 2014) that covers the full scope of biochar technology from feedstock, production, analysis/characterization, application relative to quality and good practices, potential markets, health and safety, risks management, and legislative compliance. Like the IBI and EBF/C, the BBF document identifies the basic physicochemical properties that must be determined for every biochar. For these three bodies, guidelines and recommendations on substrates, parameters to be analyzed for biochar physicochemical property identification (see also Chapter 12), and protocols to be used for each parameter/property are underpinned by established methods of registered bodies such the British and European Standards, the International Organization for Standardization, the America Standard Test Method (ASTM), and the US Composting Council. Where possible the analyses should be made by accredited laboratories to then determine classification and, therefore, recommended applications. As a result, Behrens et al. (Chapter 7) propose, for example, that, for the European context, biochar should (as a minimum) be characterized according to the standards that have been stipulated by the EBC or the IBI.

Other bodies include the USBI, which posts updates on key legislation deliberations where biochar is identified as one of the likely strategies that can inform specific bills or Acts in relation to water efficiency, carbon sequestration, greenhouse gas emission, and clean energy provision. Thus it engenders discussions on policy, ethics, and sustainability of biochar as exemplified also by its USBI Conference of 2013. Although the South Africa CSIR has no apparent policy statement, it documents dialogs and reports deliberating the potential applications of biochar. For example, a team of South African researchers (Konz et al., 2015) produced a comprehensive report for the country's Department of Environmental Affairs entitled *Assessment of the Potential to Produce Biochar and its Application to South African soils as a Mitigation Measure*, which is also available in the IBI repository. While not a policy statement, the authors presented a comprehensive proposal that was similar to the BBF biochar compilation. Thus they aimed to cover a wide remit for biochar scope in South Africa and presented a discourse on its recognized global potential, feedstock including "alien vegetation," production and application considerations, (techno)economics *re* different value chains, socioeconomic impacts including prospective stakeholders, legislation, and research opportunities.

The US EPA has no explicit statement on biochar policy or legislation but the agency seemingly sponsors relevant research especially regarding climate change mitigation. Australia CSIRO has no specific formal policy on biochar. Notwithstanding this the country's Carbon Farming Initiative was developed in 2011 to "allow landholders to generate offset credits from activities that reduce emissions or sequester carbon, including biochar application" (see IBI site). This was amended subsequently in 2015 to establish the Emissions Reduction Fund with particular focus on carbon sequestration to satisfy carbon offsetting criteria as also certified against "Kyoto and non-Kyoto" guidelines. Currently, discourse on biochar is seemingly championed by academic and nonacademic stakeholders under the Australia and New Zealand Biochar Researchers Network, which is aligned with the IBI. In contrast to the IBI, BBF, and EBC, the Network has no documented guidelines on biochar. Instead it identifies cation exchange- and water-holding capacity as parameters that have critical significance relative to long-term biochar impacts and soil stability implications for climate change mitigation potential.

Similar to Australia and New Zealand, there is no publicly available formal biochar legislation for China although the China Biochar Network was established in 2010 also in alignment with IBI. The Canada Biochar Initiative (CBI; est. 2008) aligns with the IBI and "…promotes biochar as a sustainable vehicle to sequester carbon, improve soil fertility, and enhance food security by advancing the research, development, deployment, investment and commercialization of biochar in Canada" (http://www.biochar.ca/). Consequently, "potential benefits" and "needed research" statements are provided by the CBI and include the requirements to analyze changes in soil characteristics, plants, microbial communities, and ecosystems following soil biochar augmentation. The country also has Biochar-Ontario (Canada), which is separate from, but a close affiliate of, the CBI.

Generally, the Australia, New Zealand, Canada, China, and South Africa models are, perhaps, examples of local/regional initiatives where research findings and deliberations create critical mass with the potential to help formulate and/or inform biochar policy and legislative frameworks, even if through the IBI, initially. Thus national/regional groups, as members of the IBI, can use this conduit to enter global discussions such as negotiating and drafting of the relevant United Nations Framework Convention on Climate Change documents (http://www.biochar-international.org/policy/international).

Despite being central to considerable research and application interest with its Amazonian Dark Earths or *Terra Preta*, Brazil seemingly has no formal policy on biochar. This is reflected probably by publications such as Rittl et al. (2015) who explored the different stakeholders of the material, including where (decision-making) power and the focus of debates on the topic presently lie. Nevertheless, it is highly likely that historical *Terra Preta* inspired and has direct impact on concepts and strategies such as "Climate-Smart Agriculture," which espouses "sustainable agricultural practices that enhance productivity and are adaptable toward reduced greenhouse gas emissions to mitigate climate change." These contemporary initiatives must therefore be informed and feed back into relevant policy and legislation.

According to Stone et al. (2016), "The establishment of the range of soil biodiversity found within European soils is needed to guide EU policy development regarding the protection of soil. Such a base-line should be collated from a wide-ranging sampling campaign to ensure that soil biodiversity from the majority of soil types, land-use or management systems, and European climatic (bio-geographical zones) were included." This statement must, however, be extended to, and superimposed with, soil biochar application. Two likely questions could then be presented to, potentially, provide impetus and focus on biochar policy/legislation deliberations: (1) are existing proposals and future suggestions mere wish lists or achievable targets; and (2) how would they be informed by existing and ratified policies such as the United Nations Environment Programme Agenda-21, Kyoto Protocol and Paris Agreement?

Also complete realization of the potential of ecologically and economically dynamic "win–win–win" biochar scenarios, as discussed in Chapter 7, for example, will require an underpinning by clear policy guidelines. So the impacts on human/animal gut microbiota following biochar-augmented supplementation regimes, and the implications of these for economic gains in soils to which the produced manures are applied, would need close analysis. Similar detailed investigations would be necessary for digestates from anaerobic digesters such as upflow anaerobic sludge blankets where biochar may be used as the support for biofilm formation.

While not a likely preoccupation for researchers, the relationship between the "in-draft" or ratified policy statements (if any) and laboratory-/pilot-/full-scale investigations must be deliberated by all stakeholders. Indeed, as has now become the norm for health, agriculture, and contaminated ecosystems, specific policy developers, managers, and regulators could fund biochar investigations to be directed by academics either individually or jointly with practitioners. Similarly, open dialogs between scholars, practitioners, and policymakers could ensure that scientific data and trends begin to inform biochar policy drafting and revision. For example, some regulatory policies have recognized the importance of pollution assessments that entail bioavailability analyses. Thus, according to Ortega-Calvo et al. (2015), the practice of conducting risk assessments solely by measuring total or apparent, directly extractable pollutant concentrations is being replaced gradually by incorporation of bioavailability data into both risk evaluations and environmental management regulations (see also Chapter 10).

Together with scientific and humanitarian benefits, economic benefits can accrue from biochar directly, and its value-added by-products, and this is evidenced by the emergence of registered independent or university spin-off companies. Therefore requisite impartiality by funding bodies and policymakers would have to be adopted and implemented rigorously to avoid and/or minimize conflicts of interest. Fortunately, research funding schemes, including Horizon 2020 and its predecessor the European Commission Framework Programme, have stringent criteria for this together with risk management and ethical adherence to be declared and mitigated at the grant bid submission stage.

As with any national, geopolitical region and global policies, it is undeniable that, once existent and endorsed, actual management, implementation, and policing of biochar-specific statements by the different signatories/stakeholders would probably be a challenge. These, however, do not preempt the policy formulation.

SUMMARY

As introduced earlier, an increasing number of global threats require sustainable mitigating strategies. To this end, deliberations and research are ongoing to find realistic and achievable solutions that can be implemented immediately and in the future. The magnitude and urgency for these require joint multi- and cross-disciplinary initiatives that are underpinned by multiple technologies and approaches. Biochar is increasingly considered as one such viable approach, including for environmental management. Thus, because of its unique properties, it has been identified as both a valuable soil supplement to increase sustainably soil fertility and an appropriate tool for protracted (atmospheric) carbon sequestration in soil, thereby playing an important role in climate change mitigation. Furthermore, the large surface area and cation exchange capacity of biochar facilitate sorption of both organic and inorganic contaminants, thus, possibly, reducing pollutant mobility and increasing bioavailability/biodegradability. Consequently, biochar is potentially a timely tool for physicochemical and biological amelioration of contaminated ecosystems, including soils. It does, however, have limitations and, possibly, negative impacts, which are acknowledged and so drive future investigations.

To date, most published textbooks and/or book chapters have focused mainly on the historical, economic, agricultural, and practical applications of biochar. This publication addresses a paucity and is relevant for a scientific audience including microbial ecologists, environmental biotechnologists, and research-informed practitioners in biochar exploitation. The book consolidates emerging studies, which elucidate ecosystem structural and functional dynamics in response to biochar application with a focus on microbial ecology-based analysis.

Independent of the biochar application context, its different and multiple potentials are realizable only through an underpinning with comprehensive cutting-edge microecophysiology tools and relevant complementary physicochemical analyses. Thus the narrative considers the effects of biochar on bacterial, archaeal, fungal, and mesofaunal populations in pristine ecosystems, agronomic contexts, and contaminated sites. The thrust is on the application of culture-based and molecular profiling and functional microbial community analyses that target whole communities and their different biomolecules including DNA, RNA, fatty acids or lipids, and proteins/enzymes. We have attempted to minimize and eliminate repetition between chapters. Instead, overlaps in content and application have been identified and linked in our synthesis by cross-referencing between chapters.

In summary, this compilation is designed to highlight the existing body of knowledge that has been developed through use of different microbial ecology tools. Parallel to this is the identification of paucities that, consequently, mandate further research as well as the need for appropriately informed global deliberations on biochar policy. The ultimate objective is to ensure a progressively robust understanding of the impacts of biochar application on soil ecosystem function with an underpinning of informed decision making and cost–benefit analyses both in the immediate future and the long term. The contributing authors/teams have ensured that the book content truly reflects the global biochar + microbial ecology research requirements and impetus.

References

Abujabhah, I.S., Bound, S.A., Doyle, R., Bowman, J.P., 2016. Effects of biochar and compost amendments on soil physico-chemical properties and the total community within a temperate agricultural soil. Appl. Soil Ecol. 98, 243–253.

Ahmad, M., Ok, Y.S., Kim, B.-Y., Ahn, J.-H., Lee, Y.H., Zhang, M., Moon, D.H., Al-Wabel, M.I., Lee, S.S., 2016a. Impact of soybean stover- and pine needle-derived biochars on Pb and as mobility, microbial community, and carbon stability in a contaminated agricultural soil. J. Environ. Manag. 166, 131–139.

Ahmad, M., Ok, Y.S., Rajapaksha, A.U., Lim, J.E., Kim, B.-Y., Ahn, J.-H., Lee, Y.H., Al-Wabel, M.I., Lee, S.-E., Lee, S.S., 2016b. Lead and copper immobilization in a shooting range soil using soybean stover- and pine needle-derived biochars: chemical, microbial and spectroscopic assessments. J. Hazard. Mater. 301, 179–186.

Anderson, C.R., Condron, L.M., Clough, T.J., Fiers, M., Stewart, A., Hill, R.A., Sherlock, R.R., 2011. Biochar induced soil microbial community change: implications for biogeochemical cycling of carbon, nitrogen and phosphorus. Pedobiologia 54, 309–320.

Anderson, C.R., Hamouts, K., Clough, T.J., Condron, L.M., 2014. Biochar does not affect soil N-transformations or microbial community structure under ruminant urine patches but does not alter relative proportions of nitrogen cycling bacteria. Agric. Ecosyst. Environ. 191, 63–72.

Andert, J., Mumme, J., 2015. Impact of pyrolysis and hydrothermal biochar on gas-emitting activity of soil microorganisms and bacterial and archaeal community composition. Appl. Soil Ecol. 96, 225–239.

Asai, H., Samson, B.K., Stephan, H.M., Songyikhangsuthor, K., Homma, K., Kiyono, Y., Inoue, Y., Shiraiwa, T., Horie, T., 2009. Biochar amendment techniques for upland rice production in northern Laos: soil physical properties, leaf SPAD and grain yield. Field Crop Res. 111, 81–84.

Atkinson, C.J., Fitzgerald, J.D., Hipps, N.A., 2010. Potential mechanism for achieving agricultural benefits from biochar application to temperate soils: a review. Plant Soil 337, 1–18.

Barbosa Lima, A., Cannavan, F.S., Navarrete, A.A., Teixeira, W.G., Kuramae, E.E., Tsai, S.M., 2015. Amazonian Dark Earth and plant species from the Amazon region contribute to shape rhizosphere bacterial communities. Microb. Ecol. 69, 855–866.

Basso, A.S., Miguez, F.E., Laird, D.A., Horton, R., Westgate, M., 2013. Assessing potential of biochar for increasing water-holding capacity of sandy soils. Glob. Change Biol. Bioenergy 5, 132–143.

Bardgett, E., 2005. The Biology of Soil: A Community and Ecosystem Approach. Oxford University Press.

Beesley, L., Moreno-Jimenez, E., Gomez-Eyles, J.L., 2010. Effects of biochar and greenwaste compost amendments on mobility, bioavailability and toxicity of inorganic and organic contaminants in a multi-element polluted soil. Environ. Pollut. 158, 2282–2287.

Beesley, L., Dickinson, N., 2011. Carbon and trace element fluxes in the pore water of an urban soil following greenwaste compost, woody and biochar amendments, inoculated with the earthworm *Lumbricus terrestris*. Soil Biol. Biochem. 43, 188–196.

Beesley, L., Moreno-Jimenez, E., Gomez-Eyles, J.L., Harris, E., Robinson, B., Sizmur, T., 2011. A review of biochars' potential role in the remediation, revegetation, and restoration of contaminated soils. Environ. Pollut. 159, 3269–3282.

Brossi, M.J.D.L., Mendes, L.W., Germano, M.G., Lima, A.B., Tsai, S.M., 2014. Assessment of bacterial *bph* gene in Amazonian Dark Earth and their adjacent soils. PLoS One 9 (6), e99597.

Bushnaf, K.M., Puricelli, S., Saponaro, S., Werner, D., 2011. Effect of biochar on the fate of volatile petroleum hydrocarbon in an aerobic sandy soil. J. Contam. Hydrol. 126, 208–215.

Cattaneo, F., Di Gennaro, P., Barbanti, L., Giovannini, C., Labra, M., Moreno, B., Benitez, E., Marzadori, C., 2014. Perennial energy cropping systems affect soil enzyme activities and bacterial community structure in a South European agricultural area. Appl. Soil Ecol. 84, 213–222.

Cayuela, M.L., Sanchez-Monedero, M.A., Roig, A., Hanley, K., Enders, A., Lehmann, J., 2013. Biochar and denitrification in soils: when, how much and why does biochar reduce N_2O emissions. Scientific Reports 3. http://dx.doi.org/10.1038/srep01732.

Cenciani, K., Mazzetto, A.M., Lammel, D.R., Fracetto, F.J., Fracetto, G.G.M., Frazao, L., Cerri, C., Feigl, B., 2010. Genetic and functional diversities of microbial communities in Amazonian soils under different land uses and cultivation. InTechOpen 125–141.

Cernansky, R., 2015. Agriculture: state-of-the-art soil. Nature 517, 258–260.

Chan, K.Y., Van Zwieten, L., Meszaros, I., Downie, A., Joseph, S., 2007. Agronomic values of green waste biochar as a soil amendment. Aus. J. Soil Res. 45, 629–634.

Chen, J., Liu, X., Li, L., Zheng, J., Qu, J., Zheng, J., Zhang, X., Pan, G., 2015. Consistent increase in abundance and diversity but variable change in community composition of bacteria in topsoil of rice paddy under short term biochar treatment across three sites from South China. Appl. Soil Ecol. 91, 68–79.

Chen, J., Liu, X., Zheng, J., Zhang, B., Lu, H., Chi, Z., Pan, G., Li, L., Zheng, J., Zhang, X., Wang, J., Yu, X., 2013. Biochar soil amendment increased bacterial but decreased fungal gene abundance with shifts in community structure in a slightly acid rice paddy from Southwest China. Appl. Soil Ecol. 71, 33–44.

Cheng, Q., Huang, Q., Khan, S., Liu, Y., Liao, Z., Li, G., Ok, Y.S., 2016. Adsorption of Cd by peanut husks and peanut husk biochar from aqueous solutions. Ecol. Eng. 87, 240–245.

Chikezie, V.O., 2011. Formation and Properties of Magnetic Biochar (MSc thesis). Teesside University.

Dempster, D.N., Gleeson, D.B., Solaiman, Z.M., Jones, D.L., Murphy, D.V., 2012. Decreased soil microbial biomass and nitrogen mineralization with Eucalyptus biochar addition to a coarse textured soil. Plant Soil 354, 311–324.

Devi, P., Saroha, S., 2014. Synthesis of the magnetic biochar composites for use as an adsorbent for the removal of pentachlorophenol from the effluent. Bioresour. Technol. 169, 525–531.

Doan, T.T., Bouvier, C., Bettarel, Y., Bouvier, T., Henry-des-Tureaux, T., Janeau, J.L., Lamballe, P., Nguyen, B.V., Jouquet, P., 2014. Influence of buffalo manure, compost, vermicompost and biochar amendments on bacterial and viral communities in soil and adjacent aquatic systems. Appl. Soil Ecol. 73, 78–86.

Domene, X., Hanley, K., Enders, A., Lehmann, J., 2015. Short-term mesofauna responses to soil additions of corn stover biochar and the role of microbial biomass. Appl. Soil Ecol. 89, 10–17.

Downie, A.E., Van Zwieten, L., Smernik, R.J., Morris, S., Munroe, P.R., 2011. *Terra preta australis*: reassessing the carbon storage capacity of temperate soils. Agric. Ecosyst. Environ. 140, 137–147.

Ducey, T.F., Ippolito, J.A., Cantrell, K.B., Novak, J.M., Lentz, R.D., 2013. Addition of activated switchgrass biochar to an aridic subsoil increases microbial nitrogen cycling gene abundances. Appl. Soil Ecol. 65, 65–72.

Easton, Z.M., Rogers, M., Davis, M., Wade, J., Eick, M., Bock, E., 2015. Mitigation of sulfate reduction and nitrous oxide emission in denitrifying environments with amorphous iron oxide and biochar. Ecol. Eng. 82, 605–613.

Elad, Y., David, D.R., Harel, Y.M., Borenshtein, M., Kalifa, H.B., Silber, A., Graber, E.R., 2010. Induction of systemic resistance in plants by biochar, a soil-applied carbon sequestering agent. Phytopathology 100, 913–921.

Elmer, W.H., Pignatello, J.J., 2011. Effect of biochar amendments on mycorrhizal associations and Fusarium crown and root rot of asparagus in replant soils. Plant Dis. 95, 960–966.

Ennis, C.J., Evans, G.A., Islam, M., Ralebitso-Senior, T.K., Senior, E., 2012. Biochar: carbon sequestration, land remediation and impacts on soil microbiology. Crit. Rev. Environ. Sci. Technol. 42, 2311–2364.

Evangelou, M.W.H., Brem, A., Ugolini, F., Abiven, S., Schulin, R., 2014. Soil application of biochar produced from biomass grown on trace element contaminated land. J. Environ. Manag. 146, 100–106.

Feng, Y., Xu, Y., Yu, Y., Xie, Z., Lin, X., 2012. Mechanisms of biochar decreasing methane emission from Chinese paddy soils. Soil Biol. Biochem. 46, 80–88.

Fierer, N., Jackson, R.B., 2006. The diversity and biogeography of soil bacterial communities. PNAS 103, 626–631.

Fox, A., Kwapinski, W., Griffiths, B.S., Schmalenberger, A., 2014. The role of sulfur- and phosphorus-mobilizing bacteria in biochar-induced growth promotion of *Lolium perenne*. FEMS Microbiol. Ecol. 90, 78–91.

Fry, W.E., 2012. Principles of Plant Disease Management. Academic Press, London, UK.

Gai, X., Wang, H., Liu, J., Zhai, L., Liu, S., Ren, T., Liu, H., 2014. Effects of feedstock and pyrolysis temperature on biochar adsorption of Ammonium and nitrate. PLoS One 9, 12.

Gärdenäs, A.I., Ågren, G.I., Bird, J.A., Clarholm, M., Hallin, S., Ineson, P., Kätterer, T., Knicker, H., Nilsson, I.S., Näsholm, T., Ogle, S., Paustian, K., Persson, T., Stendahl, J., 2010. Knowledge gaps in soil carbon and nitrogen interactions – from molecular to global scale. Soil Biol. Biochem. 43, 702–717.

Genesio, L., Miglietta, F., 2012. Surface albedo following biochar application in durum wheat. Environ. Res. Lett. 7, 014025.

Gomez-Eyles, J.L., Sizmur, T., Collins, C.D., Hodson, M.E., 2011. Effects of biochar and the earthworm *Eisenia fetida* on the bioavailability of polycyclic aromatic hydrocarbons and potentially toxic elements. Environ. Pollut. 159, 616–622.

Graber, E.R., Frenkel, O., Jaiswal, A.K., Elad, Y., 2014. How may biochar influence the severity of diseases caused by soilborne pathogens? Carbon Manag. 5, 169–183.

Graber, E.R., Harel, Y.M., Kolton, M., Cytryn, E., Silber, A., David, D.R., Tsechansky, L., Borenshtein, M., Elad, Y., 2010. Biochar impact on development and productivity of pepper and tomato grown in fertigated soilless media. Plant Soil 337, 481–496.

Grossman, J.M., O'Neill, B.E., Tsai, S.M., Liang, B., Neves, E., Lehmann, J., Thies, J.E., 2010. Amazonian anthrosols support similar microbial communities that differ distinctly from those extant in adjacent, unmodified soils of the same mineralogy. Microb. Ecol. 60, 192–205.

Hale, S.E., Elmquist, N., Brandlis, R., Hartnik, J., Jakob, L., Henriksen, T., Werner, D., Cornelissen, G., 2012. Activated carbon amendment to sequester PAHs in contaminated soil: a lysimeter field trial. Chemosphere 87, 177–184.

Hale, L., Luth, M., Kenny, R., Crowley, D., 2014. Evaluation of pinewood biochar as a carrier of bacterial strain *Enterocbacter cloacae* UW5 for soil inoculation. Appl. Soil Ecol. 84, 192–199.

Hammer, E.C., Balogh-Brunstad, Z., Jakobsen, I., Olsson, P.A., Stipp, S.L.S., Rillig, M.C., 2014. A mycorrhizal fungus grows on biochar and captures phosphorus from its surfaces. Soil Biol. Biochem. 77, 252–260.

Han, Y., Cao, X., Ouyang, X., Sohi, S., Chen, J., 2015a. Adsorption kinetics of magnetic biochar derived from peanut hull on removal of Cr (VI) from aqueous solution: effects of production conditions and particle size. Chemosphere 145, 336–341.

Han, Z., Sani, B., Akkanen, J., Abel, S., Nybom, I., Karapanagioti, H.K., Werner, D., 2015b. A critical evaluation of magnetic activated carbon's potential for the remediation of sediment impacted by polycyclic aromatic hydrocarbons. J. Hazard. Mater. 286, 41–47.

Hardie, M., Clothier, B., Bound, S., Oliver, G., Close, D., 2014. Does biochar influence soil physical properties and soil water availability? Plant Soil 376, 347–361.

Harter, J., Krause, H.-M., Schuettler, S., Ruser, R., Fromme, M., Scholten, T., Kappler, A., Behrens, S., 2013. Linking N_2O emissions from biochar-amended soil to the structure and function of the N-cycling microbial community. ISME J. 8, 660–674.

He, L., Yang, H., Zhong, Z., Gong, P., Liu, Y., Lü, H., Yang, S., 2014. PCR-DGGE analysis of soil bacterium community diversity in farmland influenced by biochar. Acta Ecol. Sin. 34, 4288–4294.

He, L., Liu, Y., Zhao, J., Bi, Y., Zhao, X., Wang, S., Xing, G., 2015. Comparison of straw-biochar-mediated changes in nitrification and ammonia oxidizers in agricultural oxisols and cambosols. Biol. Fert. Soils 13.

Husson, O., 2013. Redox potential (E_h) and pH as drivers of soil/plant/microorganism systems: a transdisciplinary overview pointing to integrative opportunities for agronomy. Plant Soil 362, 389–417.

Jaafar, N.M., Clode, P.L., Abbott, L.K., 2015. Soil microbial responses to biochars varying in particle size, surface and pore properties. Pedosphere 25, 770–780.

Jien, S.H., Wang, C.S., 2013. Effects of biochar on soil properties and erosion potential in a highly weathered soil. Catena 110, 225–233.

Jindo, K., Suto, K., Matsumoto, K., García, C., Sonoki, T., Sanchez-Monedero, M.A., 2012. Chemical and biochemical characterisation of biochar-blended composts prepared from poultry manure. Bioresour. Technol. 110, 396–404.

Jones, D.L., Rousk, J., Edwards-Jones, G., DeLuca, T.H., Murphy, D.V., 2012. Biochar-mediated changes in soil quality and plant growth in a three year field trial. Soil Biol. Biochem. 45, 113–124.

Joseph, S.D., Camps-Arbestain, M., Lin, Y., Munroe, P., Chia, C.H., Hook, J., van Zwieten, L., Kimber, S., Cowie, A., Singh, B.P., Lehmann, J., Foidl, N., Smernik, R.J., Amonette, J.E., 2010. An investigation into the reactions of biochar in soil. Aus. J. Soil Res. 48, 501–515.

Keiluweit, M., Nico, P.S., Johnson, M.G., Kleber, M., 2010. Dynamic molecular structure of plant biomass-derived black carbon (biochar). Environ. Sci. Technol. 44, 1247–1253.

Kim, J.S., Sparovek, G., Longo, R.M., De Melo, W.J., Crowley, D., 2007. Bacterial diversity of terra preta and pristine forest soil from the Western Amazon. Soil Biol. Biochem. 39, 684–690.

Klos, P.Z., Link, T.E., Abatzoglou, J.T., 2014. Extent of the rain-snow transition zone in the western US under historic and projected climate. Geophys. Res. Lett. 41, 4560–4568.

Kolton, M., Harel, Y.M., Pasternak, Z., Graber, E.R., Elad, Y., Cytryn, E., 2011. Impact of biochar application to soil on the root-associated bacterial community structure of fully developed greenhouse pepper plants. Appl. Environ. Microbiol. 14, 4924–4930.

Konz, J., Brett, C., van der Merwe, A.B., 2015. Assessment of the Potential to Produce Biochar and Its Application to South African Soils as a Mitigation Measure. Available at: https://www.environment.gov.za/sites/default/files/reports/biocharreport2015.pdf.

Lee, J.W., Kidder, M., Evans, B.R., Paik, S., Ill, A.C.B., Garten, C.T., Brown, R.C., 2010. Characterization of biochars produced from cornstovers for soil amendment. Environ. Sci. Technol. 44, 7970–7974.

Lehmann, J., 2007. Bio-energy in the black. Front. Ecol. Environ. 5, 381–387.

Lehmann, J., Joseph, S. (Eds.), 2009. Biochar for Environmental Management Science and Technology. Earthscan, London.

Lehmann, J., Joseph, S., 2015. Biochar for environmental management: an introduction. In: Lehmann, J., Joseph, S. (Eds.), Biochar for Environmental Management: Science, Technology and Implementation, second ed. Routledge, New York, pp. 1–14.

Lehmann, J., Rilling, M.C., Thies, J., Masiello, C.A., Hockaday, W.C., Crowley, D., 2011. Biochar effects on soil biota—a review. Soil Biol. Biochem. 43, 1812–1836.

Lehmann, J., da Silva, J.P., Rondon, M., Cravo, M.S., Greenwood, J., Nehls, T., 2002. Slash-and-char-a feasible alternative for soil fertility management in the central Amazon. In: Proceedings of the 17th World Congress of Soil Science, pp. 1–12.

Leibig, M.A., Franzluebbers, A.J., Follett, R.F., 2012. Managing Agricultural Greenhouse Gases: Coordinated Agricultural Research through GRACEnet to Address Our Changing Climate. Elsevier Inc, UK and USA.

Leifeld, J., Fenner, S., Muller, M., 2007. Mobility of black carbon in grained peatland soils. Biogeosciences 4, 425–432.

Li, H., Ye, X., Geng, Z., Zhou, H., Guo, X., Zhang, Y., Zhao, H., Wang, G., 2016. The influence of biochar type on long-term stabilization for Cd and Cu in contaminated paddy soils. J. Hazard. Mater. 304, 40–48.

Liang, B., Lehmann, J., Solomon, D., Kinyangi, J., Grossman, J., O'Neill, B., Skjemstad, J.O., Theis, J., Luizão, F.J., Peterson, J., Neves, E.G., 2006. Black carbon increases cation exchange capacity in soils. Soil Sci. Soc. Am. J. 70, 1719–1730.

Liao, N., Li, Q., Zhang, W., Zhou, G., Ma, L., Min, W., Ye, J., Hou, Z., 2016. Effects of biochar on soil microbial community composition and activity in drip-irrigated desert soil. Eur. J. Soil Biol. 72, 27–34.

Lima, I.M., Marshall, W.E., 2005. Granular activated carbons from broiler manure: physical, chemical and adsorptive properties. Bioresour. Technol. 96, 699–706.

Liu, L., Shen, G., Sun, M., Cao, X., Shang, G., Chen, P., 2014. Effect of biochar on nitrous oxide emission and its potential mechanisms. J. Air Waste Manag. Assoc. 64, 894–902.

Liu, W., Wang, S., Lin, P., Sun, H., Hou, J., Zuo, Q., Huo, R., 2015. Response of $CaCl_2$-extractable heavy metals, polychlorinated biphenyls, and microbial communities to biochar amendment in naturally contaminated soils. J. Soils Sediments 3, 1–10.

Lu, W., Ding, W., Zhang, J., Zhang, H., Luo, J., Bolan, N., 2015a. Nitrogen amendment stimulated decomposition of maize straw-derived biochar in a sandy loam soil: a short-term study. PLoS One 10 (7), e0133131.

Lu, H., Lashari, M.S., Liu, X., Ji, H., Li, L., Zheng, J., Kibue, G.W., Joseph, S., Pan, G., 2015b. Changes in soil microbial community structure and enzyme activity with amendment of biochar-manure compost and pyroligneous solution in a saline soil from central China. Eur. J. Soil Biol. 70, 67–76.

Lucheta, A.R., de Souza Cannavan, F., Roesch, L.F.W., Tsai, S.M., Kuramee, E.E., 2015. Fungal community assembly in the Amazonian Dark Earth. Microb. Ecol. 1–12.

Mackie, K.A., Marhan, S., Ditterich, F., Schmidt, H.P., Kandeler, E., 2015. The effects of biochar and compost amendments on copper immobilization and soil microorganisms in a temperate vineyard. Agric. Ecosyst. Environ. 201, 58–69.

Makotu, K., Tamai, Y., Kim, Y.S., Koike, T., 2010. Buried charcoal layer and ectomycorrhizae cooperatively promote the growth of *Larix gmelinii* seedlings. Plant Soil 327, 143–152.

Masiello, C.A., Chen, Y., Gao, X., Liu, S., Cheng, H.-Y., Bennett, M.R., Rudgers, J.A., Wagner, D.S., Zygourakis, K., Silberg, J.J., 2013. Biochar and microbial signalling: production conditions determine effects on microbial communication. Environ. Sci. Technol. 47, 11496–11503.

McCormack, S., Ostle, N., Bardgett, R.D., Hopkin, D.W., VanBergen, A.J., 2013. Biochar in bioenergy cropping systems: impacts on soil faunal communities and linked ecosystem processes. GCB Bioenergy 5, 81–95.

Meyer, S., Bright, R.M., Fischer, D., Schulz, H., Glaser, B., 2012. Albedo impact on the suitability of biochar systems to mitigate global warming. Environ. Sci. Technol. 46, 12726–12734.

Meynet, P., Moliterni, E., Davenport, R.J., Sloan, W.T., Camacho, J.V., Werner, D., 2014. Predicting the effects of biochar on volatile petroleum hydrocarbon biodegradation and emanation from soil: a bacterial community finger-print analysis inferred modelling approach. Soil Biol. Biochem. 68, 20–30.

Mia, S., van Groenigen, J.W., van de Voorde, T.F.J., Oram, N.J., Bezemer, T.M., Mommer, L., Jeffery, S., 2014. Biochar application rate affects biological nitrogen fixation in red clover conditional on potassium availability. Agric. Ecosyst. Environ. 191, 83–91.

Mitchell, P.J., Simpson, A.J., Soong, R., Simpson, M.J., 2015. Shifts in microbial community and water-extractable organic matter composition with biochar amendment in a temperate forest soil. Soil Biol. Biochem. 81, 244–254.

Mizuta, K., Matsumoto, T., Hatate, Y., Nishihara, K., Nakanishi, T., 2004. Removal of nitrate nitrogen from drinking water using bamboo powder charcoal. Bioresour. Technol. 95, 255–257.

Mukherjee, A., Zimmerman, A.R., Harris, W., 2011. Surface chemistry variations among a series of laboratory-produced biochars. Geoderma 163, 155–247.

Muyzer, G., 1999. DGGE/TGGE a method for identifying genes from natural ecosystems. Curr. Opin. Microbiol. 2, 317–322.

Muyzer, G., de Waal, E.C., Uitterlinden, A.G., 1993. Profiling of complex microbial populations by denaturing gradient gel electrophoresis analysis of polymerase chain reaction-amplified genes coding for 16S rRNA. Appl. Environ. Microbiol. 59, 695–700.

Navarrete, A.A., Cannavan, F.S., Taketani, R.G., Tsai, S.M., 2010. A molecular survey of the diversity of microbial communities in different Amazonian agricultural model systems. Diversity 2 (5), 787–809.

Navarrete, A.A., Taketani, R.G., Mendes, L.W., de Souza Cannavan, F., de Souza Moreira, F.M., Tsai, S.M., 2011. Land-use systems affect Archaeal community structure and functional diversity in western Amazon soils. Rev. Bras. Ciên. Solo 35, 1527–1540.

Noguera, D., Rondón, M., Laossi, K.R., Hoyos, V., Lavelle, P., de Carvalho, M.H.C., Barot, S., 2010. Contrasted effect of biochar and earthworms on rice growth and resource allocation in different soils. Soil Biol. Biochem. 42, 1017–1027.

Novak, J.M., Lima, I., Xing, B., Gaskin, J.W., Steinera, C., Das, K.C., Ahmedna, M., Rehrah, D., Watts, D.W., Busscher, W.J., Schomberg, H., 2009. Characterization of designer biochar produced at different temperatures and their effects on a loamy sand. Ann. Environ. Sci. 3, 195–206.

Noyce, G.L., Basiliko, N., Fultorpe, R., Sackett, T.E., Thomas, S.C., 2015. Soil Microbial responses over 2 years following biochar addition to a north temperate forest. Biol. Fert. Soils 51, 649–659.

Ortega-Calvo, J., Tejeda-Agredano, M., Jimenez-Sanchez, C., Congiu, E., Sungthong, R., Niqui-Arroyo, J., Cantos, M., 2013. Is it possible to increase bioavailability but not environmental risk of PAHs in bioremediation? J. Hazard. Mater. 261, 733–745.

Ortega-Calvo, J.J., Harmsen, J., Parsons, J.R., Semple, K.T., Aitken, M.D., Ajao, C., Eadsforth, C., Galay-Burgos, M., Naidu, R., Oliver, R., Peijnenburg, W.J., Roembke, J., Streck, G., Versonnen, B., 2015. From bioavailability science to regulation of organic chemicals. Environ. Sci. Technol. 49, 10255–10264.

Paz-Ferreiro, J., Liang, C., Fu, S., Mendez, A., Gasco, G., 2015. The effect of biochar and its interaction with the earthworm *Pontoscolex corethrurus* on soil microbial community structure in tropical soils. PLoS One 10.

Quilliam, R.S., Deluca, T.H., Jones, D.L., 2013. Biochar application reduces nodulation but increases nitrogenase activity in clover. Plant Soil 366, 83–92.

Reddy, K.R., Yargicoglu, E.N., Yue, D., Yaghoubi, P., 2014. Enhanced microbial methane oxidation in landfill cover soil amended with biochar. J. Geotech. Geoenviron. 140, 9.

Rittl, T.F., Arts, B., Kuyper, T.W., 2015. Biochar: an emerging policy arrangement in Brazil? Environ. Sci. Policy 51, 45–55.

Rondon, M., Ramirez, J.A., Lehmann, J., 2005. Charcoal additions reduce net emissions of greenhouse gases to the atmosphere. In: Proceedings of the 3rd USDA Symposium on Greenhouse Gases and Carbon Sequestration, Baltimore, USA, March 21–24, 2005, p. 208.

Seo, D.C., DeLaune, R.D., 2010. Effect of redox conditions on bacterial and fungal biomass and carbon dioxide production in Louisiana coastal swamp forest sediment. Sci. Total Environ. 408, 3623–3631.

Scholz, S.M., Sebres, T., Roberts, K., Whitman, T., Wilson, K., Lehmann, J., 2014. Biochar Systems for Smallholders in Developing Countries. World Bank Publications, Washington DC.

Shan, D., Deng, S., Zhao, T., Wang, B., Wang, Y., Huang, J., Yu, G., Winglee, J., Wiesner, M.R., 2016. Preparation of ultrafine magnetic biochar and activated carbon for pharmaceutical adsorption and subsequent degradation by ball milling. J. Hazard. Mater. 305, 156–163.

Shen, M., Xia, X., Wang, F., Zhang, P., Zhao, X., 2012. Influences of multiwalled carbon nanotubes and plant residue chars on bioaccumulation of polycyclic aromatic hydrocarbons by *Chironomus plumosus* larvae in sediment. Environ. Toxicol. Chem. 31, 202–209.

Singh, B.P., Cowie, A.L., 2008. A novel approach, using ^{13}C natural abundance, for measuring decomposition of biochars in soil. In: LD Currie, Yates, L.J. (Eds.), Carbon and Nutrient Management in Agriculture, Fertilizer and Lime Research Centre Workshop Proceedings. Massey University, Palmerston North, New Zealand, p. 549.

Sohi, S.P., Krull, E., Lopez-Capel, E., Bol, R., 2010. A review of biochar and its use and function in soil. In: Sparks, D. (Ed.), Advances in Agronomy, vol. 105. Elsevier, pp. 47–82.

Sohi, S., Lopez-Capel, E., Krull, E., Bol, R., 2009. Biochar, Climate Change and Soil: A Review to Guide Future Research. CSIRO Land and Water Science Report.

Sohi, S., Shackley, S., 2010. An Assessment of the Benefits and Issues Associated with the Application of Biochar to Soil. UK Biochar Research Centre.

Song, Y.-J., Zhang, X.-L., Gong, J., 2014. Effects of biochar amendment on the abundance and community structure of nitrogen-fixing microbes in a coastal alkaline soil. Chinese J. Ecol. 33 (8), 2168–2175.

Spokas, K.A., Cantrell, K.B., Novak, J.M., Archer, D.W., Ippolito, J.A., Collins, H.P., Boateng, A.A., Lima, I.M., Lamb, M.C., McAloon, A.J., Lentz, R.D., Nichols, K.Z., 2012. Biochar: a synthesis of its agronomic impact beyond carbon sequestration. J. Environ. Qual. 41, 973–989.

Spokas, S.A., Koskinen, W.C., Baker, J.M., Reicosky, D.C., 2009. Impacts of woodchip biochar addition on greenhouse gas production and sorption/degradation of two herbicides in a Minnesota soil. Chemosphere 77, 574–581.

Song, Y., Wang, F., Bian, Y., Kengara, F.O., Jia, M., Xie, Z., Jiang, X., 2012. Bioavailability assessment of hexachlorobenzene in soil as affected by wheat straw biochar. J. Hazard. Mater. 217, 391–397.

Steinbeiss, S., Gleixner, G., Antonietti, M., 2009. Effect of biochar amendment on soil carbon balance and soil microbial activity. Soil Biol. Biochem. 41, 1301–1310.

Stone, D., Blomkvist, P., Hendriksen, N.B., Bonkowski, M., Jørgensen, H.B., Carvalho, F., Dunbar, M.B., Gardi, C., Geisen, S., Griffiths, R., Hug, A.S., Jensen, J., Laudon, H., Mendes, S., Morais, P.V., Orgiazzi, A., Plassart, P., Römbke, J., Rutgers, M., Schmelz, R.M., Sousa, J.P., Steenbergen, E., Suhadolc, M., Winding, A., Zupan, M., Lemanceau, P., Creamer, R.E., 2016. A method of establishing a transect for biodiversity and ecosystem function monitoring across. Eur. Appl. Soil Ecol. 97, 3–11.

Su, P., Lou, J., Brookes, P.C., Lou, Y., He, Y., Xu, J., 2015. Taxon-specific responses of soil microbial communities to different soil priming effects induced by addition of plant residues and their biochars. J. Soils Sediments 1–11. http://dx.doi.org/10.1007/s11368-015-1238-8.

Sun, D., Jun, M., Zhang, W., Guan, X., Huang, Y., Lan, Y., Gao, J., Chen, W., 2012. Implication of temporal dynamics of microbial abundance and nutrients to soil fertility under biochar application – field experiments conducted in a brown soil cultivated with soybean, north China. Adv. Mat. Res. 518–523, 384–394.

Sun, D., Meng, J., Chen, W., 2013. Effects of abiotic components induced by biochar on microbial communities. Acta Agric. Scand. Sec. B Soil Plant Sci. 63, 633–641.

Sun, D., Meng, J., Liang, H., Yang, E., Huang, Y., Chen, W., Jiang, L., Lan, Y., Zhang, W., Gao, J., 2014. Effect of volatile organic compounds absorbed to fresh biochar on survival of *Bacillus mucilaginosus* and structure of soil microbial communities. J. Soils Sediments 15 (2), 271–281.

Sun, D., Meng, J., Xu, E.G., Chen, W., 2016. Microbial community structure and predicted bacterial metabolic functions in biochar pellets aged in soil after 34 months. Appl. Soil Ecol. 100, 135–143.

Taketani, R.G., Lima, A.B., Da Conceição Jesus, E., Teixeira, W.G., Tiedje, J.M., Tsai, S.M., 2013. Bacterial community composition of anthropogenic biochar and Amazonian anthrosols assessed by 16S rRNA gene 454 pyrosequencing. Antonie van Leeuwenhoek 104, 233–242.

Taketani, R.G., Tsai, S.M., 2010. The influence of different land uses on the structure of archaeal communities in Amazonian anthrosols based on 16S rRNA and *amoA* genes. Microb. Ecol. 59, 734–743.

Tammeorg, P., Parviainen, T., Nuutinen, V., Simojoki, A., Vaara, E., Helenius, J., 2014. Effects of biochar on earthworms in arable soil: avoidance test and field trial in boreal loamy sand. Agric. Ecosyst. Environ 191, 150–157.

Thies, J.E., Rillig, M.C., Graber, E.R., 2015. Biochar effects on the abundance, activity and diversity of the soil biota. In: Lehmann, J., Joseph, S. (Eds.), Biochar for Environmental Management: Science, Technology and Implementation, second ed. Earthscan Publications Ltd, London, UK.

Verheijen, F.G.A., Jeffery, S., van der Velde, M., Penizek, V., Beland, M., Bastos, A.C., Keizer, J.J., 2013. Reductions in soil surface albedo as a function of biochar application rate: implications for global radiative forcing. Environ. Res. Lett. 8, 44008.

Verheijen, F., Jeffery, S., Bastos, A.C., van der Velde, M., Diafas, I., 2010. Biochar Application to Soils: A Critical Scientific Review of Effects on Soil Properties, Process and Functions. JRC Scientific and Technical Reports. European Communities, Luxembourg.

Wang, C., Anderson, C., Suárez-Abelenda, M., Wang, T., Camps-Arbestain, M., Ahmad, R., Herath, H.M.S.K., 2015a. The chemical composition of native organic matter influences the response of bacterial community to input of biochar and fresh plant material. Plant Soil 395, 87–104.

Wang, Z., Zong, H., Zheng, H., Liu, G., Chen, L., Xing, B., 2015b. Reduced nitrification and abundance of ammonia-oxidizing bacteria in acidic soil amended with biochar. Chemosphere 138, 576–583.

Warnock, D.D., Lehmann, J., Kuyper, T.W., Rillig, M.C., 2007. Mycorrhizal responses to biochar in soil – concepts and mechanisms. Plant Soil 300, 9–20.

Warnock, D.D., Mummey, D.L., McBride, B., Major, J., Lehmann, J., Rillig, M.C., 2010. Influences of non-herbaceous biochar on arbuscular mycorrhizal fungal abundances in roots and soils: Results from growth-chamber and field experiments. Appl. Soil Ecol 46, 450–456.

Wilkinson, M.T., Richards, P.J., Humphreys, G.S., 2009. Breaking ground: pedological, geological and ecological implications of soil bioturbation. Earth-Science Rev. 97, 257–272.

Yan, L., Kong, L., Qu, Z., Li, L., Shen, G., 2014. Magnetic biochar decorated with ZnS nanocrystals for Pb (II) removal. ACS Sustain. Chem. Eng. 3, 125–132.

Yanai, Y., Toyota, K., Okazaki, M., 2007. Effects of charcoal addition on N_2O emissions from soil resulting from rewetting air-dried soil in short-term laboratory experiments. Soil Sci. Plant Nutr. 53, 181–188.

Zhang, X., Wang, H., He, L., Lu, K., Sarmah, A., Li, J., Bolan, N.S., Pei, J., Huang, H., 2013a. Using biochar for remediation of soils contaminated with heavy metals and organic pollutants. Environ. Sci. Pollut. Res. Int. 20, 8472–8483.

Zhang, M., Gao, B., Varnoosfarderani, S., Hebard, A., Yao, Y., Inyang, M., 2013b. Preparation and characterization of a novel magnetic biochar for arsenic removal. Bioresour. Technol. 130, 457–462.

Zhou, J., Huang, Y., Mo, M., 2009. Phylogenetic analysis on the soil bacteria distributed in Karst forest. Braz. J. Microbiol. 40, 827–837.

Feedstock and Production Parameters: Effects on Biochar Properties and Microbial Communities

C. Steiner[1], A.O. Bayode[2], T. Komang Ralebitso-Senior[2]

[1]University of Kassel, Witzenhausen, Germany; [2]Teesside University, Middlesbrough, United Kingdom

BIOCHAR CHARACTERISTICS AND KEY DETERMINING PARAMETERS

Biochar can be produced by pyrolysis or gasification of biomass. Unlike pyrolysis, gasification does not exclude oxygen (O_2) completely.

Gasification is optimized to yield burnable gases while pyrolysis yields more biochar (solid carbonized biomass residues) and potentially pyrolysis oil. Slow and fast pyrolysis can be distinguished, whereas fast pyrolysis is optimized for oil production (Antal and Gronli, 2003; Brewer et al., 2009).

The potential feedstocks for biochar production, used either singly or as blends, are almost unlimited. Theoretically, all woody and nonwoody biomass can be converted to biochar. Depending on the technology used, the practical implementation is frequently limited by the moisture or mineral content of the feedstock. For instance, the presence of chlorine and alkali metals (potassium, K) can cause corrosion and ash sintering problems (Steenari et al., 2009).

As a consequence of different production technologies and feedstock, the properties of the produced biochar can range widely. Most important are the feedstock characteristics. While elements such as C, hydrogen (H), O, nitrogen (N), and sulfur (S) are volatilized during pyrolysis, minerals such as phosphorus (P), K, calcium (Ca), magnesium (Mg), and silicon (Si) remain and their concentrations increase in the resultant biochar (Gaskin et al., 2008). Therefore biochar produced from feedstock rich in minerals (ash) has a high ash content (Steiner, 2015). The ash content is mainly responsible for the biochar's alkalinity, which is an important characteristic. Other properties such as the C content, degree of carbonization, amount of elements volatilized, and remaining volatile matter are determined by several production conditions including temperature, heating rate (fast or slow pyrolysis), and O_2 supply (gasification). The presence of toxic compounds or elements in biochar can either be a consequence of contaminated feedstock (eg, heavy metals) or formation during pyrolysis/gasification (Steiner, 2015). Therefore careful selection of feedstock can prevent contamination with heavy metals, while the formation of polycyclic aromatic hydrocarbons (PAHs) needs to be avoided during biochar production.

The resulting diversity of biochars with distinct characteristics has different direct and indirect effects on soil microbial communities. Many characteristics such as pH and ash content have well-known effects on soil biology. In contrast, others such as porosity and surface area are more biochar specific and to delineate their effects on soil, microbial populations are objectives of recent research.

Biochar Carbon

The key characteristic of biochar C is its stability (recalcitrance) in comparison to the uncarbonized feedstock. The stability enables long-term C sequestration but also has important consequences for the microbial community. Kuzyakov et al. (2009) estimated that approximately 0.5% of biochar decomposes in soil under optimal conditions per year. This is very low in comparison to uncarbonized biomass. Thus carbonization withdraws a source of feed for the resident microbial associations. Generally the degree of aromaticity can be used as an indicator to predict biochar's stability where the content of aliphatic compounds correlates with stability (Harris et al., 2013). Additional predictive indicators for this characteristic are the molar oxygen-to-carbon ($O:C_{org}$) and hydrogen-to-carbon ratios ($H:C_{org}$) (Spokas, 2010).

Both feedstock selection and the technology used for biochar production determine the quantity and quality of C remaining in the biochar. The total C content can be reported on an ash-free basis by subtracting the minerals (ash content) in the feedstock and biochar from the total mass. The C consists of relatively stable aromatic C and relatively labile aliphatic (volatile) C. The proportion of aliphatic C decreases with increasing carbonization temperature (Gaskin et al., 2008; Novak et al., 2009). Ameloot et al. (2013) produced biochar from willow (*Salix dasyclados*) wood and swine manure at 350°C and 700°C and found similar volatile matter content for both feedstocks at the same temperature but only half the amount (10–15%) in biochar produced at 700°C compared with a production temperature of 350° C. Also biochars produced from fast pyrolysis are characterized by relatively high volatile matter contents (Brewer et al., 2011, 2009). Volatiles often undergo secondary reactions with the solid C especially when the gases do not escape quickly (Antal and Grønli, 2003). A third type of C possibly present in biochar is uncarbonized or only partially pyrolyzed (torrefied biomass). This type of carbon remains if pyrolysis temperature and/or residence time are insufficient. Both torrefied biomass and volatile matter do not have the same stability as the stable aromatic C and can therefore be utilized by the microbial community as a C source.

Biochar Structure

The porous structure and high surface area of biochar is a further specific characteristic. The surface area is maximized by physical and chemical activation processes (Ioannidou and Zabaniotou, 2007). Both the surface area and surface characteristics are responsible for retention of water, nutrients, and organic compounds. The porous structure may provide a preferred habitat for microbes, hence it is hypothesized that the pores protect microbes from predators. Biochar has a relatively low bulk density and, when added to soil, changes soil physical properties such as the bulk density (Mukherjee and Lal, 2013). Thus biochar structure has direct and indirect effects on the microbial community.

Surface characteristics and porous structure are influenced by pyrolysis conditions such as temperature where surface area typically increases with high pyrolysis temperatures (Mukherjee et al., 2011). In addition, the density and pore size distribution of the feedstock determines the surface area. The pore size distribution is also influence by the lignin or cellulose content of the feedstock. While lignin favors the formation of macropores, the formation of micropores is favored by cellulose (Ioannidou and Zabaniotou, 2007). Further activation steps such as elevated pressure, steam, and acid treatments maximize the formation of a large surface area (activated carbon).

Mineral (Ash) Content of Biochar

The mineral content of biochar depends on the feedstock used for its production and the degree of volatiles, such as C, O, H, N, and S, lost during conversion (Steiner, 2015). Typically, the minerals contained in the feedstock remain in the biochar and are concentrated with increasing losses of volatile elements (Gaskin et al., 2008). Also the amount of volatiles lost during the carbonization process depends on pyrolysis temperature and oxygen supply (gasification). Only the ashes remain after complete gasification or combustion of biomass. Therefore the ash content can range widely depending on feedstock and the conversion technology applied (Steiner, 2015). Ultimately the minerals affect various soil properties such as pH and mineral nutrition for plants and microbes. As a result the effects of biochar can be advantageous or detrimental depending on the original soil properties and application rates.

INFLUENCE OF BIOCHAR ON MICROBIAL COMMUNITIES

Biochar additions to soil may directly and indirectly influence the microbial community. Microbial communities might be supported by the provision of habitat, feed (labile C source), and nutrients (minerals). On the other hand, the conversion of crop residues into biochar may reduce the available C for microbial utilization. Furthermore, biochar's influence on retention of dissolved organic matter and nutrients may indirectly influence soil biology. Changes in soil physical and chemical properties consequently influence microbial communities. Depending on soil properties, relatively small additions of biochar (1–2% w/w) are sufficient to reduce bulk density and increase water-holding capacity (Mukherjee and Lal, 2013). Increased water-holding capacity may mitigate the adverse effects of drying on the microbial community. The soil's pH has been identified as a key parameter influencing microbial community composition and activity (Wakelin et al., 2008). Thus, while fungi dominate at lower pH, the bacteria are more abundant at higher pH (Bååth and Anderson, 2003).

The anthropogenic *Terra Preta* soils in the Amazon Basin provide evidence that the effects of biochar on the soil microbial community could be long lasting (see also Chapter 5). They show a higher bacterial species richness as well as total numbers (Kim et al., 2007). However, *Terra Preta* contains other ingredients such as bone apatite (Lima et al., 2002) in addition to pyrogenic carbon. Therefore, the extent to which pyrogenic carbon is responsible for the observed changes remains uncertain. This is exemplified by some contrasting findings in the literature as follows and also as summarized in Table 2.1. Here, effects of different biochars on microbial community dynamics, relative to feedstock, production conditions, and char characteristics, are provided. Briefly, Khodadad et al. (2011) found a decrease in microbial diversity after biochar addition in forest soils collected in Florida, USA. Short-term changes in soil properties such as availability of labile C, minerals, and increased pH might disappear because of mineralization, leaching, and plant uptake. Single-dosed biochar addition in short-term experiments cannot delineate the long-term effects. Castaldi et al. (2011) found increased soil pH, higher rates of N mineralization, soil respiration, and denitrification 3 months after biochar application but no effects after 14 months. Quilliam et al. (2012) did not find any differences in soils amended with biochar (25 and 50 Mg ha^{-1}) 3 years after application, but a reapplication of biochar increased microbial growth and colonization by mycorrhizal fungi.

The Role of Biochar Carbon for Interacting Microbial Consortia

Higher CO_2 emissions and a larger microbial biomass were found after the addition of biochar produced by fast pyrolysis compared to that produced by slow pyrolysis (Bruun et al., 2012). The authors suggested that the higher content of relatively labile C accounted for this increase. The mineralization of biochar C depends strongly on the temperature at which biochar was produced and thus on the relative amount of more aliphatic and volatile components (Luo et al., 2013). Therefore increased microbial biomass and activity is likely to be short-lived and dependent on the availability of labile C (Castaldi et al., 2011; Rutigliano et al., 2014). Also biochar consists of relatively labile and relatively stable compounds, explaining the frequently observed biphasic decomposition characterized by a fast mineralization of volatiles and subsequent slow mineralization of relatively stable aromatic C (Ameloot et al., 2013).

However, intrinsic properties of biochar C have also been used to explain reductions in nitrous oxide emissions after biochar application. Biochar may act as an electron shuttle, facilitating the transfer of electrons to denitrifying microorganisms (Cayuela et al., 2013) (see also Chapters 7 and 8). Biochar has sorptive properties and can reduce the bioavailability of toxic elements such as PAHs (Gomez-Eyles et al., 2011), and thus may

TABLE 2.1 Example of substrates/feedstocks, production conditions, physicochemical properties of the resultant biochars and impacts on microbial communities

Feedstock	Production conditions	Biochar physicochemical properties	Context/ecosystem	Target microbial clade	Microbial community response	References
Willow (*Salix dasyclados*) wood, swine manure	350 and 700°C	pH=11.1±0.42 C/N ratio=74.9±0.06 C %=80.3%±0.02 Ash=1.28%±0.15	Laboratory of thermochemical conversion of biomass	–	Increased microbial activity	Ameloot et al. (2013)
Pine (*Pinus radiata*)			Ryegrass (*Lolium perenne*) planted soil—bulk soil and rhizosphere		Abundance increases (eg, Bradyrhizobiaceae, Hyphomicrobiaceae) and decreases (eg, Streptomycetaceae, Micromonosporaceae)	Anderson et al. (2011)
Poultry litter	600°C O_2^- nil	pH=10.33 C/N ratio=25 C=23.6% Ash=55.8% CEC=58.7 mmolc kg^{-1}		Rhizobia and other bacteria	Increased response	Lehmann et al. (2011)
Corn stalk	300, 400 and 500°C		Paddy soils: Inceptisol and Ultisol	Methanogens and methanotrophs	Abundance increase for methanotrophic Proteobacteria	Feng et al. (2012)
Algae (*Cladophora coelothrix*)	512±5°C O_2^- nil	pH 8.72 C:N ratio=10.4 C=34.6% Ash=32.1% CEC=19 mmolc kg^{-1}	Saline environment	–	–	Bird et al. (2011)

Feedstock	Production conditions	Properties	Soil/application	Microorganisms	Influence	Reference
Hardwood tree (*Quercus serrate* Murray)	Between 400 and 600°C	pH (H_2O)=7.23 C=791.5 g kg^{-1} O=91.5 mg kg^{-1} H=18.9 mg kg^{-1} Ash=78.7 g kg^{-1} N=37.6 g kg^{-1} P=2.3 g kg^{-1} K=14.1 g kg^{-1} Surface area=255.0 m^2 g^{-1}	Composting of poultry and cow manure		Increased Gram-positive to Gram-negative bacteria ratio; higher fungal biodiversity	Jindo et al. (2012a,b)
Corn stover	600°C O_2- nil	pH=9.42 C/N ratio=66 C=70.6% Ash=16.7% CEC=252.1 mmolc kg^{-1}	–	Rhizobia and other bacteria		Lehmann et al. (2011)
Chipped trunks and large branches of *Fraxinus excelsior* L. and *Fagus sylvatica* L., and *Quercus robur* L.	450°C for 48h		Agriculture soil planted with maize and grass		Bacterial-dominated community shift in year 2	Jones et al. (2012)
Oak wood	600°C O_2- nil	pH=6.38 C/N ratio=489 C=87.5% Ash=1.3% CEC=75.7 mmolc kg^{-1}	–	Rhizobia and other bacteria		Lehmann et al. (2011)
Laurel oak heartwood (*Quercus laurifolia* Michx.); Eastern gamagrass (*Tripsacum dactyloides*)			Fire-impacted and non-impacted soil		Abundance of bacterial members of the Actinobacteria and Gemmatimonadetes phyla	Khodadad et al. (2011)

Continued

TABLE 2.1 Example of substrates/feedstocks, production conditions, physicochemical properties of the resultant biochars and impacts on microbial communities—cont'd

Feedstock	Production conditions	Biochar physicochemical properties	Context/ ecosystem	Target microbial clade	Microbial community response	References
Sludge	700°C O_2^- nil	pH=12 C/N ratio=489 C=20.4% Ash=72.5%	Pilot scale	–	–	Hossain et al. (2011)
Pine wood (*Pinus ponderosa*)	450°C under N_2 for 5 h	C/N ratio=110 H/C ratio=0.5 O/C ratio=0.1 C=779 g kg^{-1} N=7.1 g kg^{-1}	Temperate forest subsoil		Utilized preferentially by Gram-positive bacteria; no significant changes in the microbial community composition	Santos et al. (2012)
Sugar cane bagasse	600°C for 1h 30min O_2 <0.5%	pH=7.7 C/N ratio=489 C=76.5% Ash=72.5% CEC=4.19 mmolc kg^{-1}	Laboratory scale	–	–	Inyang et al. (2010)
Glucose Yeast	Hydrothermal carbonization; 180°C for 24 h	GB: C=64.6%; YB: C=67.4%; N = 5%	Arable soil and forest soil		Glucose biochar utilized preferentially by bacteria	Steinbeiss et al. (2009)

Feedstock	Pyrolysis conditions	Properties	Soil/scale	Microbial groups	Effects	Reference
Ripped vines' wood (*Vitis vinifera*)	550°C O₂ free	pH = 10 C/N ratio = 83.7 C = 40.2% Ash = 13.4% CEC = 11.6 dS m⁻¹	Pilot scale	–	–	Rosas et al. (2015)
Peanut hulls (*Arachis hypogaea*)	400 and 500°C; with and without steam activation	At 500°C and steam activated: pH = 9.96 ± 0.01 C = 806 g kg⁻¹ CEC = 4.46 cmol/ kg ± 0.13	Agricultural soils			Gaskin et al. (2008)
Soybean stover; pine needles	300°C for 3 h O₂ nil	For soybean stover: pH = 7.27 C/N ratio = 489 C = 68.8% Ash = 10.4%	Agricultural land adjacent an abandoned mine	Bacteria, actinomycetes, and arbuscular mycorrhizal fungi	Abundance changes of Gram-positive and negative bacteria, fungi, actinomycetes, and arbuscular mycorrhizal fungi were feedstock and pyrolysis temperature dependent	Ahmad et al. (2016)
Maize stover	400°C for 13 h	pH = 10.7 C/N ratio = 489 C = 66% CEC = 63.5 cmol kg⁻¹	Hyperthermic Typic Haplustept farm soil	Bacteria, fungi, and protozoa	Decreased microbial biomass C and dehydrogenase activity	Purakayastha et al. (2015)

CEC, Cation exchange capacity.

Biochar carbon is largely unaffected by microbial decomposition. Therefore only the labile fraction such as the volatile solids are rapidly utilized by the microbial metabolism and thus contribute to an effective C/N ratio. The reported total C/N ratios have little agronomic relevance.

improve the habitat for microbial populations. The sorption of enzymes would be a further mechanism influencing microbial processes in soil (Lehmann et al., 2011).

Effects of Biochar Structure on Microbial Community Dynamics

The large surface area and porous structure of biochar is thought to provide a preferred and protective habitat for microbial colonization. Nevertheless, many biochar pores are too small to be inhabited by microbes; hence Quilliam et al. (2013) reported rather low microbial colonization. In general it seems that colonization depends largely on the availability of labile C utilized for microbial metabolism. This C can be provided either by the biochar itself or as adsorbed carbohydrates originating from somewhere else. This is strengthened by the findings of Luo et al. (2013) who found significantly more microbial colonization on biochar that provided more labile C (produced at 350°C) than biochar with less labile C (produced at 750°C).

Microbial Impacts of Biochar Mineral (Ash) Content

Applications of wood ash to soils increase pH and nutrient availability and has been linked to the observed increases in microbial population and biomass (Zimmermann and Frey, 2002). However, ash (coal fly ash) can also reduce these microbial population properties because of its alkalinity (El-Mogazi et al., 1988). Other negative factors include salinity (Wichern et al., 2006), toxicity of boron, and other trace elements (Pandey and Singh, 2010). Therefore the effect of ash on microbial communities depends on the initial acidity of the soil, the alkalinity of the biochar, and its application rate. Ash is only one component of biochar. Hence increases in pH and the availability of minerals were observed frequently after biochar additions (Biederman and Harpole, 2013) where they were presumed to be responsible for increases in soil microbial activity (Ameloot et al., 2013; Steiner et al., 2008; Watzinger et al., 2014). However, decreased microbial biomass and N mineralization because of biochar amendments were also observed (Dempster et al., 2012). The latter authors used a biochar with relatively low ash content produced at a relatively high temperature (600°C). Luo et al. (2013) found a marked decrease in microbial biomass after the addition of biochar produced at 700°C to a soil with high pH (7.37) but an increase if added to soil with low pH (3.73). *Terra Preta* soils are characterized by a higher pH than the adjacent unmodified soils. A decrease in acidity was also linked to an increase in biological nitrogen fixation. Similarly, the application of wood ash has been beneficial to enhance arbuscular mycorrhizae and has also been observed after biochar application (Atkinson et al., 2010; Biederman and Harpole, 2013).

CONCLUSIONS AND OUTLOOK

The functions of biochar in soil are manifold while biochar properties are influenced by feedstock and production technology. This stresses the importance of assessing feedstock, biochar, and soil properties for a beneficial use. Influences on soil pH, aeration, bulk density, water-holding capacity, availability of nutrients, and sorption of organic and inorganic compounds because of biochar additions to soil can have profound effects on microbial activity. The microbial community is crucial for soil functions such as the provision of plant nutrients, breakdown of agrochemicals, and emissions of greenhouse gases such as methane and nitrous oxide. Overall, small changes in soil chemical and physical characteristics can alter the microbial community and thus important soil functions. Therefore gaining comprehensive knowledge about biochar properties, using the relevant physicochemical techniques, and elucidating its interactions with microbial communities are crucial to optimize biochar utilization.

References

Ahmad, M., Ok, Y.S., Kim, B.-Y., Ahn, J.-H., Lee, Y.H., Zhang, M., Moon, D.H., Al-Wabel, M.I., Lee, S.S., 2016. Impact of soybean stover- and pine needle-derived biochars on Pb and as mobility, microbial community, and carbon stability in a contaminated agricultural soil. J. Environ. Manage. 166, 131–139.

Ameloot, N., De Neve, S., Jegajeevagan, K., Yildiz, G., Buchan, D., Funkuin, Y.N., Prins, W., Bouckaert, L., Sleutel, S., 2013. Short-term CO_2 and N_2O emissions and microbial properties of biochar amended sandy loam soils. Soil Biol. Biochem. 57, 401–410. http://dx.doi.org/10.1016/j.soilbio.2012.10.025.

Anderson, C.R., Condron, L.M., Clough, T.J., Fiers, M., Stewart, A., Hill, R.A., Sherlock, R.R., 2011. Biochar induced soil microbial community change: implications for biogeochemical cycling of carbon, nitrogen and phosphorus. Pedobiologia 54, 309–320.

Antal, M.J., Grønli, M., 2003. The art, science, and technology of charcoal production. Industrial Eng. Chem. Res. 42, 1619–1640.

Atkinson, C., Fitzgerald, J., Hipps, N., 2010. Potential mechanisms for achieving agricultural benefits from biochar application to temperate soils: a review. Plant Soil 337, 1–18. http://dx.doi.org/10.1007/s11104-010-0464-5.

Bååth, E., Anderson, T.H., 2003. Comparison of soil fungal/bacterial ratios in a pH gradient using physiological and PLFA-based techniques. Soil Biol. Biochem. 35, 955–963. http://dx.doi.org/10.1016/S0038-0717(03)00154-8.

Biederman, L.A., Harpole, W.S., 2013. Biochar and its effects on plant productivity and nutrient cycling: a meta-analysis. GCB Bioenergy 5, 202–214. http://dx.doi.org/10.1111/gcbb.12037.

Bird, M., Wurster, C., de Paula Silva, P., Bass, A., de Nys, R., 2011. Algal biochar production and properties. Bioresour. Technol. 102, 1886–1891.

Brewer, C., Unger, R., Schmidt-Rohr, K., Brown, R., 2011. Criteria to select biochars for field studies based on biochar chemical properties. Bioenerg. Res. 4, 312–323. http://dx.doi.org/10.1007/s12155-011-9133-7.

Brewer, C.E., Schmidt-Rohr, K., Satrio, J.A., Brown, R.C., 2009. Characterization of biochar from fast pyrolysis and gasification systems. Environ. Prog. Sustain. Energy 28, 386–396. http://dx.doi.org/10.1002/ep.10378.

Bruun, E.W., Ambus, P., Egsgaard, H., Hauggaard-Nielsen, H., 2012. Effects of slow and fast pyrolysis biochar on soil C and N turnover dynamics. Soil Biol. Biochem. 46, 73–79. http://dx.doi.org/10.1016/j.soilbio.2011.11.019.

Castaldi, S., Riondino, M., Baronti, S., Esposito, F.R., Marzaioli, R., Rutigliano, F.A., Vaccari, F.P., Miglietta, F., 2011. Impact of biochar application to a Mediterranean wheat crop on soil microbial activity and greenhouse gas fluxes. Chemosphere 85, 1464–1471. http://dx.doi.org/10.1016/j.chemosphere.2011.08.031.

Cayuela, M.L., Sánchez-Monedero, M.A., Roig, A., Hanley, K., Enders, A., Lehmann, J., 2013. Biochar and denitrification in soils: when, how much and why does biochar reduce N_2O emissions? Sci. Rep. 3. http://dx.doi.org/10.1038/srep01732.

Dempster, D.N., Gleeson, D.B., Solaiman, Z.M., Jones, D.L., Murphy, D.V., 2012. Decreased soil microbial biomass and nitrogen mineralisation with *Eucalyptus* biochar addition to a coarse textured soil. Plant Soil 354, 311–324. http://dx.doi.org/10.1007/s11104-011-1067-5.

El-Mogazi, D., Lisk, D.J., Weinstein, L.H., 1988. A review of physical, chemical, and biological properties of fly ash and effects on agricultural ecosystems. Sci. Total Environ. 74, 1–37. http://dx.doi.org/10.1016/0048-9697(88)90127-1.

Feng, Y., Xu, Y., Yu, Y., Xie, Z., Lin, X., 2012. Mechanisms of biochar decreasing methane emission from Chinese paddy soils. Soil Biol. Biochem. 46, 80–88.

Gaskin, J.W., Steiner, C., Harris, K., Das, K.C., Bibens, B., 2008. Effect of low-temperature pyrolysis conditions on biochar for agricultural use. Trans. ASABE 51, 2061–2069.

Gomez-Eyles, J.L., Sizmur, T., Collins, C.D., Hodson, M.E., 2011. Effects of biochar and the earthworm *Eisenia fetida* on the bioavailability of polycyclic aromatic hydrocarbons and potentially toxic elements. Environ. Pollut. 159, 616–622. http://dx.doi.org/10.1016/j.envpol.2010.09.037.

Harris, K., Gaskin, J., Cabrera, M., Miller, W., Das, K.C., 2013. Characterization and mineralization rates of low temperature peanut hull and pine chip biochars. Agronomy 3, 294–312. http://dx.doi.org/10.3390/agronomy3020294.

Hossain, M., Strezov, V., Chan, K., Ziolkowski, A., Nelson, P., 2011. Influence of pyrolysis temperature on production and nutrient properties of wastewater sludge biochar. J. Environ. Manage. 92, 223–228.

Inyang, M., Gao, B., Pullammanappallil, P., Ding, W., Zimmerman, A., 2010. Biochar from anaerobically digested sugarcane bagasse. Bioresour. Technol. 101, 8868–8872.

Ioannidou, O., Zabaniotou, A., 2007. Agricultural residues as precursors for activated carbon production - a review. Renew. Sustain. Energy Rev. 11, 1966–2005.

Jindo, K., Sánchez-Monedero, M.A., Hernández, T., García, C., Furukawa, T., Matsumoto, K., Sonoki, T., Bastida, F., 2012a. Biochar influences the microbial community structure during manure composting with agricultural wastes. Sci. Total Environ. 416, 476–481.

Jindo, K., Suto, K., Matsumoto, K., García, C., Sonoki, T., Sanchez-Monedero, M.A., 2012b. Chemical and biochemical characterisation of biochar-blended composts prepared from poultry manure. Bioresour. Technol. 36, 396–404.

Jones, D.L., Rousk, J., Edwards-Jones, G., DeLuca, T.H., Murphy, D.V., 2012. Biochar-mediated changes in soil quality and plant growth in a three year field trial. Soil Biol. Biochem. 45, 113–124.

Khodadad, C.L.M., Zimmerman, A.R., Green, S.J., Uthandi, S., Foster, J.S., 2011. Taxa-specific changes in soil microbial community composition induced by pyrogenic carbon amendments. Soil Biol. Biochem. 43, 385–392. http://dx.doi.org/10.1016/j.soilbio.2010.11.005.

Kim, J.-S., Sparovek, G., Longo, R.M., Melo, W.J.D., Crowley, D., 2007. Bacterial diversity of terra preta and pristine forest soil from the Western Amazon. Soil Biol. Biochem. 39, 684–690. http://dx.doi.org/10.1016/j.soilbio.2006.08.010.

Kuzyakov, Y., Subbotina, I., Chen, H., Bogomolova, I., Xu, X., 2009. Black carbon decomposition and incorporation into soil microbial biomass estimated by ^{14}C labeling. Soil Biol. Biochem. 41, 210–219. http://dx.doi.org/10.1016/j.soilbio.2008.10.016.

Lehmann, J., Rillig, M.C., Thies, J., Masiello, C.A., Hockaday, W.C., Crowley, D., 2011. Biochar effects on soil biota - a review. Soil Biol. Biochem. http://dx.doi.org/10.1016/j.soilbio.2011.04.022.

Lima, H.N., Schaefer, C.E.R., Mello, J.W.V., Gilkes, R.J., Ker, J.C., 2002. Pedogenesis and Pre-Colombian land use of "terra preta anthrosols" ("Indian black earth") of Western Amazonia. Geoderma 110, 1–17.

Luo, Y., Durenkamp, M., De Nobili, M., Lin, Q., Devonshire, B.J., Brookes, P.C., 2013. Microbial biomass growth, following incorporation of biochars produced at 350°C or 700°C, in a silty-clay loam soil of high and low pH. Soil Biol. Biochem. 57, 513–523. http://dx.doi.org/10.1016/j.soilbio.2012.10.033.

Mukherjee, A., Lal, R., 2013. Biochar impacts on soil physical properties and greenhouse gas emissions. Agronomy 3, 313–339. http://dx.doi.org/10.3390/agronomy3020313.

Mukherjee, A., Zimmerman, A.R., Harris, W., 2011. Surface chemistry variations among a series of laboratory-produced biochars. Geoderma 163, 247–255. http://dx.doi.org/10.1016/j.geoderma.2011.04.021.

Novak, J.M., Lima, I., Xing, B., Gaskin, J.W., Steiner, C., Das, K.C., Ahmedna, M., Rehrah, D., Watts, D.W., Busscher, W.J., Schomberg, H., 2009. Characterization of designer biochar produced at different temperatures and their effects on a loamy sand. Ann. Environ. Sci. 3, 195–206.

Pandey, V.C., Singh, N., 2010. Impact of fly ash incorporation in soil systems. Agric. Ecosyst. Environ. 136, 16–27. http://dx.doi.org/10.1016/j.agee.2009.11.013.

Purakayastha, T.J., Kumari, S., Pathak, H., 2015. Characterisation, stability, and microbial effects of four biochars produced from crop residues. Geoderma 239-240, 293–303.

Quilliam, R.S., Glanville, H.C., Wade, S.C., Jones, D.L., 2013. Life in the 'charosphere': does biochar in agricultural soil provide a significant habitat for microorganisms? Soil Biol. Biochem. 65, 287–293.

Quilliam, R.S., Marsden, K.A., Gertler, C., Rousk, J., DeLuca, T.H., Jones, D.L., 2012. Nutrient dynamics, microbial growth and weed emergence in biochar amended soil are influenced by time since application and reapplication rate. Agric. Ecosyst. Environ. 158, 192–199. http://dx.doi.org/10.1016/j.agee.2012.06.011.

Rosas, J., Gamez, N., Cara, J., Ubalde, J., Sort, X., Sanchez, M., 2015. Assessment of sustainable biochar production for carbon abatement from vineyard residues. J. Anal. Appl. Pyrol. 113, 239–247.

Rutigliano, F.A., Romano, M., Marzaioli, R., Baglivo, I., Baronti, S., Miglietta, F., Castaldi, S., 2014. Effect of biochar addition on soil microbial community in a wheat crop. Eur. J. Soil Biol. 60, 9–15.

Santos, F., Torn, M.S., Bird, J.A., 2012. Biological degradation of pyrogenic organic matter in temperate forest soils. Soil Biol. Biochem. 51, 115–124.

Spokas, K.A., 2010. Review of the stability of biochar in soils: predictability of O:C molar ratios. Carbon Manag. 1, 289–303.

Steenari, B.-M., Lundberg, A., Pettersson, H., Wilewska-Bien, M., Andersson, D., 2009. Investigation of ash sintering during combustion of agricultural residues and the effect of additives. Energy Fuels 23, 5655–5662. http://dx.doi.org/10.1021/ef900471u.

Steinbeiss, S., Gleixner, G., Antonietti, M., 2009. Effect of biochar amendment on soil carbon balance and soil microbial activity. Soil Biol. Biochem. 41, 1301–1310.

Steiner, C., 2015. Considerations in Biochar Characterization. Agricultural and Environmental Applications of Biochar: Advances and Barriers. Soil Science Society of America, Inc., Madison, WI.

Steiner, C., Das, K.C., Garcia, M., Förster, B., Zech, W., 2008. Charcoal and smoke extract stimulate the soil microbial community in a highly weathered Xanthic Ferralsol. Pedobiologia 51, 359–366. http://dx.doi.org/10.1016/j.pedobi.2007.08.002.

Wakelin, S.A., Macdonald, L.M., Rogers, S.L., Gregg, A.L., Bolger, T.P., Baldock, J.A., 2008. Habitat selective factors influencing the structural composition and functional capacity of microbial communities in agricultural soils. Soil Biol. Biochem. 40, 803–813. http://dx.doi.org/10.1016/j.soilbio.2007.10.015.

Watzinger, A., Feichtmair, S., Kitzler, B., Zehetner, F., Kloss, S., Wimmer, B., Zechmeister-Boltenstern, S., Soja, G., 2014. Soil microbial communities responded to biochar application in temperate soils and slowly metabolized ^{13}C-labelled biochar as revealed by ^{13}C PLFA analyses: results from a short-term incubation and pot experiment. Eur. J. Soil Sci. 65, 40–51. http://dx.doi.org/10.1111/ejss.12100.

Wichern, J., Wichern, F., Joergensen, R.G., 2006. Impact of salinity on soil microbial communities and the decomposition of maize in acidic soils. Geoderma 137, 100–108. http://dx.doi.org/10.1016/j.geoderma.2006.08.001.

Zimmermann, S., Frey, B., 2002. Soil respiration and microbial properties in an acid forest soil: effects of wood ash. Soil Biol. Biochem. 34, 1727–1737. http://dx.doi.org/10.1016/S0038-0717(02)00160-8.

3

Biochar Effects on Ecosystems: Insights From Lipid-Based Analysis

E.-L. Ng[1], T.R. Cavagnaro[2]

[1]Future Soils Laboratory, Melbourne, VIC, Australia; [2]The Waite Research Institute, The University of Adelaide, SA, Australia

OUTLINE

Biochar Application
http://dx.doi.org/10.1016/B978-0-12-803433-0.00003-5

BACKGROUND

Soils contain the most diverse terrestrial communities on the planet (Bardgett and Wardle, 2010; Fierer et al., 2009). The vast majority of this diversity (abundance and richness) can be attributed to soil microbes. They, in turn, provide essential regulating, provisioning, and supporting ecosystem services, such as C and nutrient cycling, that underpin global agricultural production (Jackson et al., 2008; Wagg et al., 2011). Despite their tremendous importance the vast majority of soil microbes remain undescribed.

Biochars are products of elevated thermal decomposition of organic matter in the absence of oxygen. When waste material is pyrolyzed to produce bioenergy, the application of the biochar by-product into soil may be a feasible way to sequester carbon (C) (Lehmann, 2007a). Biochar is increasingly being considered, and used, as a soil ameliorant, be it to increase soil C stocks, supply nutrients, or to provide habitat for soil microbes. The provision of habitat is particularly interesting in the context of soil microbial ecology, as the porous nature of biochars gives not only a large surface area, but also provides microsites where soil microbes can reside (Ascough et al., 2010; Ennis et al., 2012; Pietikäinen et al., 2000). The colonization of biochar by soil microbes may also be important in terms of helping to bind the biochar to soil particles, thereby contributing to soil structure (see also Chapter 11). The large surface area of biochars, their cation exchange capacity, and nutrient content (Biederman and Harpole, 2013; Lehmann, 2007b; Muhammad et al., 2014; Singh et al., 2012) also make them important sites of nutrient cycling and release in soils. Given that soil microbes play a central role in soil nutrient cycling, and the increasing use of biochars in agriculture, understanding biochar–microbe–soil interactions is of high priority.

One of the great challenges in soil ecology is linking soil microbes, be it individual taxonomic groups through to whole communities, to the ecosystem services that they provide. In order to do this, a number of different approaches have been developed. These methods include those that measure microbial biomass, activity, and diversity (Drenovsky et al., 2008; Hirsch et al., 2010; Schinner et al., 1996). In this chapter we focus on one such method, that is, the use of phospholipid fatty acids (PLFA), which can be used to assess changes in microbial biomass, community composition, and (to a lesser extent) specific groups of the microbial community (Frostegård and Bååth, 1996; Frostegård et al., 2011).

Before we discuss the use of PLFA in microbial ecology, it is first necessary to consider what PLFA are. PLFA are a vital structural component of the membranes of cells. These lipids, which are comprised of C chains of varying length and can be saturated or unsaturated, depend on the identity of the organism and its physiological status (Zelles, 1999). Importantly, different cells contain different types and amounts of PLFA, and as such can be used to characterize a whole community, or (with due caution, see later) components of microbial communities (Frostegård et al., 2011; Yao et al., 2015).

A key feature of the PLFA approach is that it is a fairly rapid and inexpensive method to obtain a snapshot of the community at the time of sampling (Spiegelman et al., 2005; see Table 3.1 for comparison with DNA- and RNA-based methods). Another feature of PLFA is that they can be correlated with other measures such as microbial biomass carbon by fumigation extraction (Leckie et al., 2004; Potthoff et al., 2006), thus gaining the advantage of providing information on microbial biomass and community composition from a single analysis. It is culture independent and therefore avoids culturing bias. When used in conjunction with isotopically labeled organic matter sources, it is possible to trace C flow into the microbial biomass (Yao et al., 2015). Finally, certain PLFA have been associated with specific groups of microbes at a coarse level, for example, fungi versus bacteria. Specific PLFA have also been associated with specific groups of soil microbes (eg, Gram-positive vs. Gram-negative

TABLE 3.1 Comparison of PLFA, DNA- and RNA-based approaches to soil microbial community determination

Method	Measure of biomass	Measure of abundance of specific organisms	Provides measure of diversity	Cost	Ease of use and need for specialized equipment
PLFA	Yes	No	Yes	Intermediate	Relatively easy to use, but requires access to gas chromatography and mass spectroscopy
DNA-based methods	No	Yes, with downstream analysis	Yes, with downstream analysis	Intermediate	Relatively easy to use, but requires specific equipment depending on downstream analysis
RNA-based methods	No	Yes, with downstream analysis	Yes, with downstream analysis	High	Can be difficult to use, and requires specific equipment depending on downstream analysis

PLFA, Phospholipid fatty acids.

bacteria; but see later). The characterization of PLFA to corresponding specific microbial groups has been done by various studies, and such details can be found in reviews by Frostegård et al. (2011), Zelles (1999), and White (1993), and for recent discussion of when and/or where such specific characterization are best applied.

In this chapter we focus on PLFA because they can be used to provide a measure of microbial biomass and reveal changes in whole microbial community composition. As mentioned earlier, they can also be used to make inferences about specific groups of organisms, but only with due caution (see later). Our goal is to provide an overview of methods for extracting and analyzing PLFA, examine current insights into biochar effects on soil microbial composition, structure and function, and present some thoughts on potential future research avenues.

EXTRACTING LIPIDS AND PRODUCTION OF FATTY ACID METHYL ESTERS

Various modifications of the method first proposed by White et al. (eg, Vestal and White, 1989; White et al., 1979) are used routinely for the extraction and analysis of PLFA. Briefly the methods involve multiple rounds of extraction of lipids from the soil samples, followed by separation of phospholipids and converting them to fatty acid methyl esters (Frostegård et al., 1991; Zelles, 1997). The extracted lipids are then analyzed using gas chromatography and/or mass spectrometry to separate, identify, and quantify the presence of fatty acid methyl esters.

The specific steps in the methodology have been well examined and improved over the years. Combinations of extraction mixture, organic solvents, and extraction method and duration have been used and their extraction efficiency examined (Frostegård et al., 1991; Nielsen and Petersen, 2000; Papadopoulou et al., 2011; Wu et al., 2009). The effects of storage and soil mass used have also been investigated (Olsson et al., 1995; Wu et al., 2009). We use the following method in our research, with slight modification from Ng et al. (2014a):

"PLFA are extracted following a modified procedure of Bossio and Scow (1998). Lipids are extracted from 4 g of lyophilized soil using 15.6 mL of 0.8:1:2 (v/v/v) citrate buffer (0.15 M, pH 4.0):CHCl$_3$:methanol mixture. Samples are shaken at room temperature for 1 h and then centrifuged at 1900 × g for 10 min. The supernatant is transferred into a clean glass tube. A further 11.7 mL of 0.9:1:2 (v/v/v) citrate:CHCl$_3$:methanol mixture is added to the soil pellet. The samples are shaken and centrifuged again as before, and the supernatant is isolated and combined with the first supernatant. A further 13.3 mL of 0.9:1 (v/v) citrate:CHCl$_3$ mixture is added to the pooled supernatant. Samples are left overnight for phase separation, after which the aqueous layer is removed and the $CHCl_3$ layer evaporated under a stream of N_2. Each sample is re-dissolved in 2 mL of $CHCl_3$ and transferred to solid phase extraction cartridges for separation of lipid classes. An aliquot of 3 mL of $CHCl_3$ is added followed by 2 aliquots of 5 mL of acetone. These extracts are discarded. The phospholipid fraction is collected by extracting the cartridges with 5 mL of methanol; the methanol is then evaporated under N_2. For methanolysis, the phospholipid fraction is incubated at 37°C for 20 min with 1 mL of a 1:1 mixture of methanol and toluene and 1 mL of methanolic KOH (0.2 M). The samples are neutralized with 0.3 mL acetic acid (1 M) and 2 mL of ultrapure H_2O. Two extractions are carried out with a mixture of 2 mL of 4:1 (v/v) hexane:$CHCl_3$ and the organic phases combined. The organic layer is collected and evaporated again under a stream of N_2. Each sample is resuspended in 200 mL of hexane containing methyl decanoate (0.005 mg mL^{-1}) and analyzed using gas chromatography. The chromatography is conducted with a 30 m (5%-phenyl)-methylpolysiloxane column (Varian CP 3800), using He as a carrier gas, an FID detector, and a temperature program of 120°C initial temperature, ramped to 220°C at 4°C min^{-1}, ramped to 325°C at 20°C min^{-1}, and held at 325°C for 8 min. Bacterial phospholipid markers used are i15:0, a15:0, 15:0, i16:0, 16:1ω7, i17:0, a17:0, 17:0cy, 17:0, and 19:0cy (cf. with Frostegård and Bååth, 1996, and references therein). Linoleic acid (18:2ω6,9) is used as an indicator of fungal biomass (Frostegård and Bååth, 1996)."

MAKING SENSE OF PLFA: ANALYTICAL APPROACH AND INTERPRETATION OF PLFA DATA

The use of multivariate analysis in community ecology has been increasing but such tools have not been applied extensively in the interpretation of PLFA data from biochar studies. The appropriate tools provide means to visualize and explore the patterns in the community composition, and to compare and evaluate statistically the PLFA profiles across treatments. This can include a combination of univariate and multivariate approaches. Boxplots, scatterplots, bar plots, or 3D plots can be used to view summarized data such as total PLFA or bacterial-to-fungal (B:F) ratios. For multivariate analysis, cluster analysis, principal component analysis, principal coordinate analysis, nonnumeric multidimensional scaling (NMDS), and correspondence analysis can be used to explore and visualize multidimensional data. Also redundancy analysis (RDA), constrained canonical analysis, linear discriminant analysis, and variation partitioning have been used to reveal the relationship between the observed pattern and the measured environmental variables.

Here we present an example of the use of a combination of univariate and multivariate analyses to explore and explain patterns in soil microbial community composition and structure in two soils amended with organic inputs. This example draws specifically on our previous work on PLFA and biochars (Ng et al., 2014a,b). The soil PLFA composition was examined in the presence of applied municipal raw green waste (green waste, Gw), composted green waste (compost, Co) and charred green waste (biochar, Ch) over a 12-week experiment.

Using NMDS, it was observed that the microbial community composition in biochar-treated soils changed over 12 weeks (Fig. 3.1A). In the Cranbourne sandy soil (Cr), the soil microbial composition in biochar-amended soil became more similar to the unamended soil while in the Werribee clayey soil (We), the microbial composition in biochar-treated

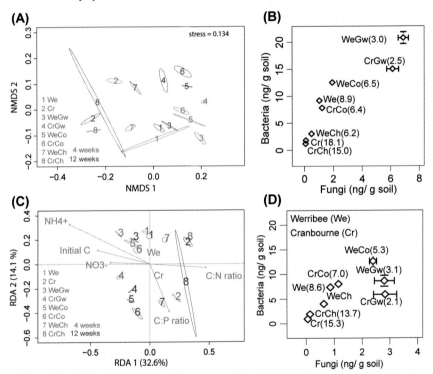

FIGURE 3.1 (A) Microbial phospholipid fatty acids (PLFA) composition changed in Cranbourne sandy soil (Cr) and Werribee clayey soil (We) at 4 and 12 weeks under different organic inputs, (B) bacteria-to-fungal biomass (mean ± SE) at 4 weeks, (C) the environmental variables contributing to the variations in microbial PLFA composition, and (D) bacteria-to-fungal biomass (mean ± SE) at 12 weeks. 1 = *We*, Werribee unamended soil; 2 = *Cr*, Cranbourne unamended soil; 3 = *WeGw*, Werribee soil + green waste; 4 = *CrGw*, Cranbourne soil + green waste; 5 = *WeCo*, Werribee soil + compost; 6 = *CrCo*, Cranbourne soil + compost; 7 = *WeCh*, Werribee soil + biochar; 8 = *CrCh*, Cranbourne soil + biochar. *Figures adapted from Ng, E.L., Patti, A.F., Rose, M.T., Schefe, C.R., Wilkinson, K., Cavagnaro, T.R., 2014a. Functional stoichiometry of soil microbial communities after amendment with stabilised organic matter. Soil Biol. Biochem. 76, 170–178.* http://dx.doi.org/10.1016/j.soilbio.2014.05.016.

soil became more dissimilar from the unamended equivalent over the same period of time. Nevertheless, both biochar-amended soils had consistently lower B:F ratios compared to unamended soils (Fig. 3.1B and D). The influence of the input on soil available nitrogen in the form of ammonium was found to be strongly correlated with the B:F ratio and total PLFA throughout the experiment. Furthermore, the soil C:nutrient ratios were identified by redundancy analysis to be important in explaining these variations in soil microbial community composition (Fig. 3.1C).

At 12 weeks, combining cluster analysis of PLFA data and NMDS of soil microbial activity showed that the soil microbial activity of each amended and unamended soil remained distinct. Also the microbial community composition in biochar-amended Werribee clayey soil was more similar to that of biochar-amended Cranbourne sandy soil than its unamended counterpart (Fig. 3.2A). Further analysis with multivariate regression tree (MRT) and redundancy analysis indicated that the C forms, as determined by ^{13}C-NMR, were able to explain between 46% and 86% of total variations in soil microbial community composition (Fig. 3.2B and C). The MRT further indicated that 60% of the treatments were mostly distinguished by their differences in bacterial y19:0 and fungi 18:2ω6 (Table 3.2). In particular, aryl-C, O-aryl-C, carbonyl-C, and alkyl-C were the four carbon forms identified to be responsible for the distinction between soil microbial community compositions. The former two are associated with more recalcitrant C while the latter two are associated with more labile C. Interestingly these C forms had greater influence on the microbial community composition than the soil microbial activity, where between 57% and 79% of total variations were explained by the four C forms. Considering the other findings from the experiment, microbial activity may be faced with greater constraints from nutrient availability and functional stoichiometry.

In addition to the insights gained previously, utilizing multiple approaches also allowed comparisons to be made with the various data analytical tools. Where results are corroborated by different methods, it provides a strong support that the observed patterns are indeed robust. Additionally, if constrained and unconstrained analysis yields similar ordination plots, that provides a strong indication that the observed patterns were mostly explained by the environmental variables incorporated into the constrained analysis.

In multivariate analysis, the measure used to quantify the similarity or differences between objects is called the association coefficient. The choice of appropriate and ecologically meaningful association coefficients should be chosen based on the data. Dierksen et al. (2002) compared three variations of discriminant analysis (linear discriminant analysis, quadratic discriminant analysis, and nonparametric density estimation) and found that the choice of data analysis matters, with less assumption on the distribution improving the classification error and therefore the

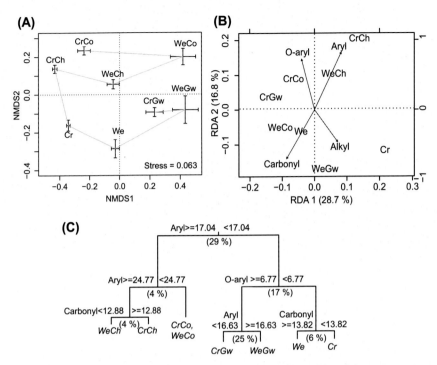

FIGURE 3.2 (A) Nonnumeric multidimensional scaling (NMDS) plot (showing mean ± SE) shows distinct microbial activity at 12 weeks. Phospholipid fatty acids (PLFA) composition, shown as line of minimum spanning tree, indicates that PLFA composition of biochar-amended Werribee clayey soil (WeCh) is more similar to that of Cranbourne sandy soil (CrCh) than its unamended control soil (We), (B) redundancy analysis (RDA) indicates aryl, O-aryl, carbonyl-C, and alkyl-C as the most important carbon forms in explaining the variations in PLFA composition, and (C) multivariate regression tree of PLFA composition shows variations explained by each carbon form at each branch (in parentheses). We, Werribee soil; Cr, Cranbourne soil; Gw, green waste; Co, composted green waste; Ch, charred green waste (eg, CrGw refers to Cranbourne soil amended with green waste). Figures adapted from Ng, E.L., Patti, A.F., Rose, M.T., Schefe, C.R., Wilkinson, K., Smernik, R.J., Cavagnaro, T.R., 2014b. Does the chemical nature of soil carbon drive the structure and functioning of soil microbial communities? Soil Biol. Biochem. 70, 54–61. http://dx.doi.org/10.1016/j.soilbio.2013.12.004.

correct identification of contrasting land management and field sites. As the number of PLFA may need to be reduced to provide sufficient degrees of freedom in the analysis (Legendre and Legendre, 2012), they employed a Bayesian approach to select a subset of fatty acids for the discriminant analyses. Based on that approach, they reduced the number of fatty acids from 73 to 6. A further test with stepwise discriminant analysis reduced the number to three fatty acids, 16:0 10 methyl (marker for Gram-positive bacteria), 17:1ω7c (marker for Gram-negative bacteria), and 18:0 (marker for total community biomass), which can accurately classify the soils to their contrasting land management and field sites. Additionally, Dierksen and colleagues

TABLE 3.2   Variance of soil microbial PLFA composition explained by the tree analysis

PLFA	PLFA variance (%) explained by tree splits and whole tree							
	Aryl <17.04	O-aryl <6.77	Aryl <24.77	Aryl <16.63	Carbonyl <13.82	Carbonyl <12.88	Tree total	PLFA total
Bacterial 14:0	0.0	0.0	0.0	0.0	0.0	0.0	0.1	0.1
Bacterial i15:0	0.6	1.3	0.2	0.0	0.1	0.7	2.8	3.3
Bacterial a15:0	0.1	0.6	0.1	0.0	0.0	0.2	1.0	1.2
Bacterial 15:0	0.0	0.0	0.0	0.0	0.0	0.0	0.1	0.1
2-OH 14:0	0.0	0.0	0.0	0.0	0.0	0.0	0.0	0.0
3-OH 14:0	0.0	0.0	0.0	0.0	0.0	0.0	0.0	0.0
Bacterial i16:0	0.2	1.7	0.0	0.0	0.0	0.1	2.1	2.3
Bacterial 16:1ω7c	0.6	0.5	0.0	0.2	0.0	0.0	1.4	1.6
Bacterial 16:1ω7t	0.6	0.1	0.0	0.2	0.0	0.0	0.9	1.0
Bacterial 16:0	2.4	1.0	0.7	2.3	0.2	0.5	7.0	7.5
Bacterial i17:0	1.0	0.8	0.1	0.1	0.0	0.0	2.0	2.2
Bacterial a17:0	0.4	0.4	0.0	0.0	0.0	0.0	0.8	1.0
Bacterial 17:0cy	0.0	0.1	0.0	0.2	0.0	0.1	0.4	0.6
Bacterial 17:0	0.0	0.1	0.0	0.0	0.0	0.0	0.1	0.2
2-OH 16:0	0.0	0.0	0.0	0.0	0.0	0.0	0.0	0.0
Fungi 18:2ω6,9	4.4	1.8	0.7	17.8	1.5	0.6	26.6	27.9
18:1ω9c	0.1	2.4	0.2	0.6	0.5	0.0	3.9	4.6
18:1ω9t	0.2	5.3	0.1	0.1	0.3	0.0	6.0	7.4
18:0	0.1	0.6	0.9	1.7	0.0	1.6	5.0	5.5

Continued

TABLE 3.2   Variance of soil microbial PLFA composition explained by the tree analysis —cont'd

PLFA	PLFA variance (%) explained by tree splits and whole tree							
	Aryl <17.04	O-aryl <6.77	Aryl <24.77	Aryl <16.63	Carbonyl <13.82	Carbonyl <12.88	Tree total	PLFA total
bacterial cy19:0	18.2	0.2	0.7	1.6	3.0	0.0	23.8	31.8
20:0	0.2	0.1	0.1	0.7	0.0	0.4	1.5	1.6
Total PLFA variance	29.3	17.0	3.8	25.4	5.7	4.3	85.6	100.0

PLFA, Phospholipid fatty acids.
Obtained from Ng, E.L., Patti, A.F., Rose, M.T., Schefe, C.R., Wilkinson, K., Smernik, R.J., Cavagnaro, T.R., 2014b. Does the chemical nature of soil carbon drive the structure and functioning of soil microbial communities? Soil Biol. Biochem. 70, 54–61. http://dx.doi.org/10.1016/j.soilbio.2013.12.004.

compared the PLFA approach to the 16rDNA length heterogeneity–polymerase chain reaction. They found that the PLFA approach was able to discriminate land management in an adjacent field site whereas the latter was able to resolve to field site level. Similarly a study by Noble et al. (2000) utilized the neural network method to compare PLFA profiles of marine sediments with the results cross-validated using linear discriminant analysis. They found that the neural network method indicated 2.7% incorrect classification versus 8.4% in the latter. These results indicate that there is no magic bullet approach to PLFA data analysis. Instead there is a variety of possible approaches to making sense of PLFA data, depending on the question of interest. Further details of numerical tools have been reviewed and described by Ramette (2007) and references therein. They have also provided some guidelines for data type and preparation, to choice of appropriate methods.

COMPOSITIONAL AND STRUCTURAL INSIGHTS FROM PLFA

Because of its relative affordability, suitability for rapidly processing a large number of samples, and informativeness, PLFA has been widely used to compare effects of land use and management (eg, Böhme et al., 2005; Bowles et al., 2014; Calderón et al., 2000; Steenwerth et al., 2002) and change over time (eg, Balser and Firestone, 2005; Waldrop and Firestone, 2006; Williams and Rice, 2007). Although this approach has been used less extensively in biochar studies thus far, some valuable insights have been gained. Therefore we summarize here some of the findings on the effects of biochar on soil microbes elucidated using PLFA (Table 3.3).

TABLE 3.3   Effects of biochar on the soil microbial community composition and the soil microbes that are breaking it down based on PLFA analysis

Organization level	Effect of biochar
SPECIFIC GROUPS	
Gram-positive bacteria	Little increase over 100 days after addition of wheat husk biochar at 3% w:w and not involved in its degradation (Watzinger et al., 2014); decreased with yeast biochar and enriched by glucose biochar added at a rate of 30% of initial soil organic carbon content in an arable and a forest soil (Steinbeiss et al., 2009); preferentially used pine biochar compared to precursor pine wood C and native soil organic matter, the primary consumer of pine biochar (Santos et al., 2012); dominate first 3 days' uptake of biochar (Farrell et al., 2013); increased with addition of biochar produced from swine manure, fruit peels, *Phragmites australis*, or *Brassica rapa* after 90-day incubation applied at rate of 1% and 3% w:w (Muhammad et al., 2014)
Gram-negative bacteria	Little uptake of pine biochar carbon applied at rate of 7.5% of soil C (Santos et al., 2012); increased after wheat husk biochar addition at rate of 3% w:w, but uptake of biochar carbon small and occurred earliest after 5 weeks (Watzinger et al., 2014); increased with addition of biochar produced from swine manure, fruit peels, *Phragmites australis*, or *Brassica rapa* after 90-day incubation applied at a rate of 1% and 3% w:w (Muhammad et al., 2014); enriched by glucose biochar and decreased with yeast biochar added at a rate of 30% of initial soil organic carbon content in an arable and a forest soil (Steinbeiss et al., 2009)
Fungi	Little uptake of biochar carbon applied at rate of 7.5% of soil C (Santos et al., 2012); little increase over 100 incubations after wheat husk biochar addition at a rate of 3% w:w and not involved in its degradation (Watzinger et al., 2014); increased with addition of biochar produced from swine manure, fruit peels, *Phragmites australis*, or *Brassica rapa* after 90-day incubation applied at rate of 1% and 3% w:w (Muhammad et al., 2014); increased with yeast biochar and unaffected by glucose biochar added at a rate of 30% of initial soil organic carbon content in an arable and a forest soil (Steinbeiss et al., 2009); decreased after 2 years with wood-derived biochar applied at 49 t ha^{-1} and unaffected by wood-derived biochar applied at 30 t ha^{-1} (Ameloot et al., 2014)
Actinomycetes	Little uptake of biochar C applied at a rate of 7.5% of soil C (Santos et al., 2012); increased with wheat husk biochar applied at 3% w:w (Watzinger et al., 2014); increased with addition of biochar produced from swine manure, fruit peels, *Phragmites australis*, or *Brassica rapa* after 90-day incubation (Muhammad et al., 2014).

Continued

TABLE 3.3   Effects of biochar on the soil microbial community composition and the soil microbes that are breaking it down based on PLFA analysis—cont'd

Organization level	Effect of biochar
ECOSYSTEM	
Microbial PLFA composition	**Unchanged**: No response to biochar in 180-day incubation with two temperate subsoils amended with pine biochar at a rate of 7.5% of soil C (Santos et al., 2012); composition unchanged by glucose biochar and changed with yeast biochar in two soils added at rate of 30% of initial soil organic carbon content (Steinbeiss et al., 2009); no response to willow biochar applied at 3% w:w in two temperate agricultural soils (Watzinger et al., 2014). **Changed**: Composition of two soils amended with green waste biochar at a rate to increase total soil C by 1% distinctly different from unamended soils and continued to change and differ from the control over 12 weeks' incubation (Ng et al., 2014a).
Microbial PLFA biomass	**Unchanged or reduced**: Similar to control over 180-day incubation in two temperate subsoils amended with pine biochar at a rate of 7.5% of soil C (Santos et al., 2012); reduced by glucose biochar but no change with yeast biochar in two soils added at a rate of 30% of initial soil organic carbon content (Steinbeiss et al., 2009); short-term (weeks) suppression by or no response to green waste biochar applied at a rate to increase total soil C by 1% in two agricultural soils (Ng et al., 2014a). **Increased or unchanged**: Increased over 90 days in biochar produced from swine manure, fruit peels, or *Phragmites australis* but no effect from *Brassica rapa* biochar after 90-day incubation at application rate of 1% and 3% w:w (Muhammad et al., 2014)
B:F ratio	**Unchanged**: Unchanged in three sites aged 7 months to 4 years that received wood-derived biochar up to 30 t ha^{-1} (Ameloot et al., 2014). **Increased**: Increased in site aged 2 years that received wood-derived biochar at 49 t ha^{-1} (Ameloot et al., 2014). **Lowered**: Lowered with green waste biochar compared to control applied at a rate to increase total soil C by 1% in two agricultural soils (Ng et al., 2014a)

PLFA, Phospholipid fatty acids; *B:F ratio,* bacterial-to-fungal ratio.

Of the abiotic and biotic pathways of decomposition, studies so far suggest that the decomposition of biochar is primarily biological (Farrell et al., 2013; Santos et al., 2012; Watzinger et al., 2014). While all major groups of heterotrophic microbes (eg, fungi and bacteria) have been reported to utilize biochar C, studies differed on which group actually has the biggest role in the degradation of biochar. For example, Santos et al. (2012) found that Gram-positive bacteria were primary users of pine biochar C.

Gram-positive bacteria have previously been identified to preferentially utilize older organic C compared to new rhizodeposits or fresh litter (Bird et al., 2011; Fierer et al., 2003; Kramer and Gleixner, 2008). On the other hand, Watzinger et al. (2014) found that actinomycetes and Gram-negative bacteria were the primary users of wheat husk biochar C. Similarly, Ng et al. (2014b) have reported that Gram-negative bacteria were more abundant in soils with higher relative aryl-C content, which are associated with stabilized organic matter. Overall, these studies reported minimal use of biochar C as a C source by soil microbes in the short term. However, a recurring theme observed here is the absence of long-term studies. As pyrolysis stabilizes the biochar C into a more "durable" form, long-term study is truly needed to understand the transformation of biochar in soil, its legacy on soil microbes, and their processes.

Given the small numbers of studies currently available on microbial PLFA response to biochar, it is not yet possible to determine with a high level of confidence the patterns and mechanisms by which biochar influences microbial community composition across ecosystems. However, the observed variations in responses of soil microbial communities to biochar indicate the complex interplay of biochar with the soil's physicochemical environment and their resident microbial community. Several factors have emerged as key variables that should be measured in future studies if we are to determine the mechanisms underlying soil microbial community response to biochar.

First, there is a strong influence of the ecosystem properties on microbial PLFA dynamics in the presence of biochars. These include soil pH, soil C:N:P, soil texture, and mineral surfaces (Marschner et al., 2008; Ng et al., 2014a; Watzinger et al., 2014). Soil pH affects nutrient availability, and where labile C and nutrients are present, they are potentially the preferred microbial food source compared to biochar. Additionally, stoichiometric constraints suggest that while soil microbial composition varies from ecosystem to ecosystem in response to organic inputs, the overall stoichiometry of the soil organic matter and functional stoichiometry of the microbial community proceed toward a constant ratio (Kirkby et al., 2011; Ng et al., 2014a; Sinsabaugh et al., 2009). On the other hand, soil texture and mineral surfaces influence physicochemical protection and stabilization of the biochar, thereby determining its availability to soil microbes (Marschner et al., 2008).

Second, biochars are highly variable in their chemical and physical properties because of differences in the parent material and the pyrolysis conditions (Singh et al., 2012; Zhang et al., 2015). These differences influence biochar properties such as cation exchange capacity, nutrient availability, water-holding capacity, pore structure, surface area, mineral matter, and biochar C stability (Lehmann et al., 2011; Muhammad et al.,

2014; Singh et al., 2012). These characteristics have important implications for the interaction of biochar with other soil compounds and their subsequent stabilization in the soil matrix (Marschner et al., 2008). It has also been reported that the ability of biochar to raise soil pH results in more pronounced effects in highly acidic soils (Muhammad et al., 2014; Watzinger et al., 2014). Measures of these variables from multiple studies will enable us to establish mechanisms through which biochar affects soil physicochemical properties.

Third, each habitat has its own microbial community, whose ability and/or need to utilize biochar as a resource may vary widely. In addition, biochar may arrive with its own resident microbes. As such, it is important to examine microbial functional changes corresponding to the microbial community compositional response to biochar. This is addressed in the following section.

BIOCHAR AS HABITAT

While examining the compositional and structural insights, responses to, and impact from biochar, it is important to recognize that biochars, with their distinctive chemical and physical properties combined, such as their water-holding capacity and large range of surface volume and pore sizes, meet four fundamental requirements to qualify as a habitat for soil microbes—water, food, air, and refuge (Lehmann et al., 2011; Pietikäinen et al., 2000; Warnock et al., 2007). When the microbial communities of biochar, activated charcoal, and pumice were compared, higher microbial biomass C on biochar-treated samples were associated with their physicochemical properties such as total surface area availability for microbial attachment, food resource adsorption capacity, and water retention capacity (Pietikäinen et al., 2000).

Ascough et al. (2010) carefully recorded charcoal colonization by two species of saprophytic fungi. They found that both fungi colonized mainly the surface and, to a lesser extent, the interiors through physical cracks present in the charcoal, but hyphae growth forms indicated preferential search for readily available nutrient sources rather than use of the charcoal structure itself as a source of nutrients.

While it seems clear that biochar can function as a habitat, it is unclear how its favorability as a habitat differs from organism to organism, or if other environmental factors, such as predation and food resources, strongly influence how microbes adapt to biochar as a habitat. There is also a need for longer-term studies that investigate biochar as a habitat for microbes and the successional dynamics that may take place. Ideally this information would also be related to measures of the ecosystem services that soil microbes provide.

LINKING WHO TO WHAT: FUNCTIONAL INSIGHTS FROM PLFA

Few approaches allow direct yet relatively inexpensive assessment of the linkage between soil microbial community structure and C dynamics as PLFA. This is particularly useful in examining the effects of biochar on the decomposition of soil organic C, soil C stocks, and CO_2 emissions. In the context of climate change and increasing global interest in the use of biochar for carbon sequestration, it is important to determine biochar contribution and retention in soil and its influence on the rate of soil C cycling through potential priming of soil organic matter turnover. Positive priming occurs when biochar addition stimulates the increase in mineralization of native soil organic C while negative priming occurs when the introduction of biochar inhibits the mineralization of native soil organic C (Cross and Sohi, 2011; Lu et al., 2014), although some studies consider that priming can also occur vice versa, where mineralization of biochar increases in the presence of soil organic C (Zimmerman et al., 2011). Here we consider both possibilities as priming.

Contrasting results on the priming effect of biochar have been obtained so far where studies have reported positive, negative, and/or zero priming effects of biochar (Cross and Sohi, 2011; Lu et al., 2014; Zimmerman et al., 2011). This has been associated with the variable nature of biochar and its production, and the soil to which the biochar is applied (Santos et al., 2012; Ventura et al., 2014; Zimmerman et al., 2011). For example, Cross and Sohi (2011) found that the labile C component in biochar was mainly responsible for higher soil CO_2 respiration in soils amended with biochar. Also the material had a stabilizing effect on soil labile organic C in soils with higher organic C. Similarly, Lu et al. (2014) found that corn straw-derived biochar, even when combined with inorganic nitrogen amendment, reduced CO_2 emission from native soil organic C and reduced dissolved organic C from native soil organic matter. On the other hand, Zimmerman et al. (2011) found a general pattern of positive priming for soils amended with grass-derived biochars produced at lower temperatures (250 and 400°C) compared to negative priming for soils amended with hard wood-derived biochars produced at higher temperatures (525–650°C). Positive priming generally occurred in the first 90 days while negative priming dominated later at 250–500 days. They concluded that over the long term, biochar–soil interaction will improve soil C stock.

These insights came largely from the application of stable isotope probing. The use of natural abundance of stable isotope is somewhat restricted to quantifying substrate uptake by total microbial biomass rather than individual PLFA because of microbial-driven isotope fractionation and inhomogeneous distribution of ^{13}C in substrates (Watzinger, 2015). Various studies have mitigated this limitation through artificial labeling.

Labeled substrates, when combined with PLFA analysis (eg, Kuzyakov et al., 2009; Yao et al., 2015), can help identify microbial-mediated mechanisms that control the magnitude and direction of priming effects from biochar addition to soil.

Artificial labeling mitigates biases related to fractionation and, as such, allows exploration of substrate utilization by individual PLFA, construction of food webs and identification of unknown microbial functions (Watzinger, 2015). This approach was utilized by Santos et al. (2012), where dual ^{13}C and ^{15}N labeling gave insights into the biological mechanism underlying carbon and nitrogen mineralization of biochar derived from the wood of *Pinus ponderosa* compared to its precursor. The researchers found that pine biochar was utilized readily by all heterotrophic microbial groups in two soil types. However, the native soil C and N turnovers were unaffected by addition of the pine biochar.

Using ^{13}C-labeled biochars derived from wheat or eucalypt shoots, Farrell et al. (2013) also observed immediate use of biochar C upon addition to soil, with subsequent contribution to as much as 40–60% of total respiration in the first 3 days while total microbial biomass remained similar between biochar amended and unamended soils. Additionally, both biochars stimulated the breakdown of native soil organic matter over the first 9 days. In the first 3 days, Gram-positive bacteria were responsible for most uptake of biochar C but this was not maintained over time, whereas actinomycetes and fungi uptake of biochar C occurred more gradually and was sustained over time. Nevertheless, the authors were not able to determine if the biochar ^{13}C content in the actinomycetes and fungi were caused by direct uptake or as an indirect result of turnover of the Gram-positive bacteria.

The use of radiocarbon labeling further reduces ambiguity of the contribution of biochar to total CO_2 efflux and its transformation in soil over time. For example, Kuzyakov et al. (2014) used ^{14}C-labeled *Lolium* residue to produce biochar and were able to determine its transformation into microbial biomass and dissolved organic C over 3.5 years, and loss as CO_2 over 8.5 years. They found that only about 6% of the added biochar C was lost as CO_2, with decomposition rate estimated to be less than 0.3% per year under optimal conditions between the 5th and 8th year. At 3.5 years, biochar-derived carbon only constituted less than 1% of all lipid fractions (neutral lipids, glycolipids, PLFA). Based on their observations, the mean residence time of the biochar was estimated to be 400 years under optimal decomposition conditions to 4000 years under field condition in a temperate climate. This clearly indicates the inadequacy of short-term studies to address biochar effects on soil, and more importantly that biochar is unlikely an important carbon source for microbes. In fact, Kuzyakov et al. (2009) observed that the addition of glucose stimulated biochar decomposition for 2 weeks to 3 months in different soils, suggesting that cometabolism was the main process for biochar decomposition. Interestingly, Kuzyakov et al. (2014) found that "you are what you eat" since microbial carbon of biochar origin had a slower turnover within microbes compared to other carbon sources. This is a positive sign for carbon retention in the ecosystem.

Slower C and nutrient cycling is associated with a fungal-based energy pathway in below-ground decomposition. Thus the ecosystem's main food resource is characterized by high contents of phenolic, lignin, and structural cellulose and, correspondingly, low nitrogen content. In a bacterial-based energy pathway, the ecosystem's main food resource is rich in labile C, low in phenolic, lignin, and structural cellulose and, correspondingly, high nitrogen content. These, in turn, are reflected in the C and nutrient cycling. In a fungal-based soil food web, slow nutrient cycles are coupled with high soil C sequestration whereas in a bacteria-based soil food web, rapid and potentially leaky nutrient cycles are coupled with low carbon sequestration (Wardle et al., 2004). Based on available studies, we obtained (eg, Ameloot et al., 2014; Ng et al., 2014a) or estimated the B:F ratio in biochar-amended soils based on microbial biomass or proportion of PLFA provided (eg, Muhammad et al., 2014; Steinbeiss et al., 2009). The B:F ratios obtained were variable, that is, biochar addition increased, decreased, or had no effect on the B:F ratio. In Ng et al. (2014a) we observed that B:F ratio was positively correlated with the biochar C:nutrient at 4 weeks, and negatively correlated with soil C:N at 12 weeks. These results are yet another indicator that the effect of biochar on the soil food web and soil processes are dependent on biochar properties and biochar–soil interactions.

In Ng et al. (2014a) and Steinbeiss et al. (2009), microbial structural and functional responses were strongly influenced by biochar addition. Specifically, Ng et al. (2014a) found two different arable soils became more similar in their microbial composition and functions after receiving green waste biochar while Steinbeiss et al. (2009) observed that a forest and an arable soil had similar microbial composition but different functional responses to yeast and glucose biochar. As such, one study indicated that dissimilar soil microbial communities may have similar preferences or abilities to utilize biochar while another suggested that similar soil microbial communities may have different preferences or capacities to utilize biochar. In the latter case, it is likely that a finer resolution study of the microbial community will be able to better resolve differences in the microbial diversity associated with the dissimilar functional outcomes.

In another study, Ameloot et al. (2014) observed effects on dehydrogenase enzyme activity even at 2 years after biochar input. They examined the microbial activity and PLFA composition in four sites that had received wood-based biochar between 7 months and 4 years and observed that microbial composition and PLFA biomass were generally unaffected in three out of four sites. However, dehydrogenase activity, an enzyme involved in oxidative breakdown of organic matter and an indicator of overall microbial activity, was still significantly lower in two of those three sites. The sorption property of biochar was proposed to be a possible cause of this outcome, whereby the substrate needed for enzyme activity was rendered unavailable when it was adsorbed and protected physically by biochar.

The sorption capacity of biochar brings attention to another interesting question: what happens when biochar adsorbs organic compounds such as signal molecules? Biochar sorption and later release of signal molecules intercepts plant–soil microbes and soil microbe–microbe communication (Warnock et al., 2007). Additionally, biochar addition can increase soil pH. The resultant soil pH can be either stimulatory or inhibitory to the production of signaling compounds from specific soil biota (Angelini et al., 2003). Such signaling interference may intensify, suppress, and, it is not inconceivable to suggest, cause mismatch in plant–soil microbes or soil microbes–microbes' response to signal molecules. As such the effect of biochar on communication in soil warrants further attention.

Finally, attempts have also been made to use fatty acids as biomarkers of diet, and thereby provide an insight into trophic interactions below ground. It is proposed that a consumer lipid profile reflects that of its food because of dietary routing (Ruess and Chamberlain, 2010). For example, Ruess et al. (2002) attempted to link the fatty acids of a nematode to its fungal diet. However, they found that the nematode lipid composition was more diverse than that of the consumed fungi and was only affected partially by their fungi diet. Notwithstanding this the authors suggest that this method may still be useful to determine general feeding habits (eg, between diet of mycorrhizal or saprophytic fungi) of nematodes. As far as we know, this approach has yet to be used to study subsurface trophic interactions in response to biochar.

MISUSE OF PLFA

Discussion of microbial PLFA responses to biochars would be incomplete without consideration of its potential misuses. Frostegard et al. (2011) provided a valuable discussion on the use and misuse of this method. Rather than providing a detailed discussion of the issues raised by Frostegard and colleagues, we restate a number of key points raised in their paper, and strongly encourage all users of the PLFA method to read the original paper.

1. **Elucidating the effects of treatments on specific groups of microorganisms using the PLFA method**. This approach makes the assumption that certain PLFA are markers for a particular group

of organisms, or at least indicative of changes in that group. The justification for assigning a specific marker to a group of organisms is typically based on information from culture-based studies. It relies on growth conditions to determine the lipid composition of a species, which can change under variable cultivation methods (Olsson et al., 1995; Spiegelman et al., 2005). As Frostegard et al. (2011) and Yao et al. (2015) (and references therein) pointed out, a number of proposed specific markers have been found in different organisms, suggesting that their specificity is questionable. Additionally, the use of ratios of specific markers as indicators of stress should be treated with caution. Microbial PLFA is affected by nutrient availability, growth stage, or temperature (Bååth, 2003; Vestal and White, 1989; White, 1993). As such, a change in the ratio may reflect a response to stress, but it may also reflect a change in community composition (Fischer et al., 2010; Frostegård et al., 2011).

2. **Rate of PLFA turnover.** As Frostegård et al. (2011) noted, there are relatively few studies investigating the rate of turnover of PLFA. The assumption is often made that when cells die the PLFA are turned over rapidly. If this is not the case, then changes in PLFA may not be a good indication of short-term changes in microbial communities.

3. **Calculating diversity indices using PLFA.** Frostegård et al. (2011) stated clearly that the "use of PLFA data to calculate diversity indices and then trying to interpret these are flawed approaches and should not be used." This approach assumes that each PLFA represents a single species, and this is not the case. For example, fungi contain relatively few PLFA and so a measure of diversity using this approach could grossly underestimate fungal diversity. Measures of diversity can also be affected by sample size, thus method of analysis can alter results.

CONCLUSIONS

This chapter has provided an overview of the potential to use PLFA to study microbial community responses to biochar addition to soil. In general, PLFA can provide insights into changes in microbial composition at a relatively rapid, quantitative, and standardized manner. With the current worldwide interest in soil biochar application, the response of soil biota and the corresponding functional dynamics must be determined.

As with any method there are advantages and disadvantages, which can be managed through an appropriate understanding of the methods and how they can be best (and safely) applied. Some examples of how the data arising from PLFA analyses can be statistically analyzed were provided to point out the possibilities that arise with bigger datasets. With these factors in mind, a number of studies have used this approach and have afforded valuable insights into microbe–biochar interactions. We envisage that this chapter will stimulate further work in this area.

Acknowledgments

TRC thanks the Australian Research Council for supporting his research (FT120100463).

References

Ameloot, N., Sleutel, S., Case, S.D.C., Alberti, G., McNamara, N.P., Zavalloni, C., Vervisch, B., delle Vedove, G., De Neve, S., 2014. C mineralization and microbial activity in four biochar field experiments several years after incorporation. Soil Biology and Biochemistry 78, 195–203. http://dx.doi.org/10.1016/j.soilbio.2014.08.004.

Angelini, J., Castro, S., Fabra, A., 2003. Alterations in root colonization and nodC gene induction in the peanut–rhizobia interaction under acidic conditions. Plant Physiology and Biochemistry 41 (3), 289–294. http://dx.doi.org/10.1016/S0981-9428(03)00021-4.

Ascough, P.L., Sturrock, C.J., Bird, M.I., 2010. Investigation of growth responses in saprophytic fungi to charred biomass. Isot. Environ. Health Stud. 46, 64–77. http://dx.doi.org/10.1080/10256010903388436.

Bååth, E., 2003. The use of neutral lipid fatty acids to indicate the physiological conditions of soil fungi. Microb. Ecol. 45, 373–383. http://dx.doi.org/10.1007/s00248-003-2002-y.

Balser, T.C., Firestone, M.K., 2005. Linking microbial community composition and soil processes in a California annual grassland and mixed-conifer forest. Biogeochemistry 73, 395–415.

Bardgett, R.D., Wardle, D.A., 2010. Aboveground-Belowground Linkages: Biotic Interactions, Ecosystem Processes and Global Change. Oxford University Press, Oxford, UK.

Biederman, L.A., Harpole, W.S., 2013. Biochar and its effects on plant productivity and nutrient cycling: a meta-analysis. GCB Bioenergy 5, 202–214.

Bird, J.A., Herman, D.J., Firestone, M.K., 2011. Rhizosphere priming of soil organic matter by bacterial groups in a grassland soil. Soil Biol. Biochem. 43, 718–725.

Böhme, L., Langer, U., Böhme, F., 2005. Microbial biomass, enzyme activities and microbial community structure in two European long-term field experiments. Agric. Ecosyst. Environ. 109, 141–152. http://dx.doi.org/10.1016/j.agee.2005.01.017.

Bossio, D.A., Scow, K.M., 1998. Impacts of carbon and flooding on soil microbial communities: phospholipid fatty acid profiles and substrate utilization patterns. Microb. Ecol. 35, 265–278.

Bowles, T.M., Acosta-Martínez, V., Calderón, F., Jackson, L.E., 2014. Soil enzyme activities, microbial communities, and carbon and nitrogen availability in organic agroecosystems across an intensively-managed agricultural landscape. Soil Biol. Biochem. 68, 252–262. http://dx.doi.org/10.1016/j.soilbio.2013.10.004.

Calderón, F.J., Jackson, L.E., Scow, K.M., Rolston, D.E., 2000. Microbial responses to simulated tillage in cultivated and uncultivated soils. Soil Biol. Biochem. 32, 1547–1559.

Cross, A., Sohi, S.P., 2011. The priming potential of biochar products in relation to labile carbon contents and soil organic matter status. Soil Biol. Biochem. 43, 2127–2134.

Dierksen, K., 2002. High resolution characterization of soil biological communities by nucleic acid and fatty acid analyses. Soil Biol. Biochem. 34, 1853–1860. http://dx.doi.org/10.1016/S0038-0717(02)00198-0.

Drenovsky, R.E., Feris, K.P., Batten, K.M., Hristova, K., 2008. New and current microbiological tools for ecosystem ecologists: towards a goal of linking structure and function. Am. Midl. Nat. 160, 140–159. http://dx.doi.org/10.1674/0003-0031(2008)160[140:NACMTF]2.0.CO;2.

Ennis, C.J., Evans, A.G., Islam, M., Ralebitso-Senior, T.K., Senior, E., 2012. Biochar: carbon sequestration, land remediation, and impacts on soil microbiology. Crit. Rev. Environ. Sci. Technol. 42, 2311–2364. http://dx.doi.org/10.1080/10643389.2011.574115.

Farrell, M., Kuhn, T.K., Macdonald, L.M., Maddern, T.M., Murphy, D.V., Hall, P.A., Singh, B.P., Baumann, K., Krull, E.S., Baldock, J.A., 2013. Microbial utilisation of biochar-derived carbon. Sci. Total Environ. 465, 288–297. http://dx.doi.org/10.1016/j.scitotenv.2013.03.090.

Fierer, N., Schimel, J.P., Holden, P.A., 2003. Variations in microbial community composition through two soil depth profiles. Soil Biol. Biochem. 35, 167–176.

Fierer, N., Strickland, M.S., Liptzin, D., Bradford, M.A., Cleveland, C.C., 2009. Global patterns in belowground communities. Ecol. Lett. 12, 1238–1249.

Fischer, J., Schauer, F., Heipieper, H.J., 2010. The trans/cis ratio of unsaturated fatty acids is not applicable as biomarker for environmental stress in case of long-term contaminated habitats. Appl. Microbiol. Biotechnol. 87, 365–371. http://dx.doi.org/10.1007/s00253-010-2544-0.

Frostegård, A., Bååth, E., 1996. The use of phospholipid fatty acid analysis to estimate bacterial and fungal biomass in soil. Biol. Fertil. Soils 22, 59–65.

Frostegård, Å., Tunlid, A., Bååth, E., 1991. Microbial biomass measured as total lipid phosphate in soils of different organic content. J. Microbiol. Methods 14, 151–163. http://dx.doi.org/10.1016/0167-7012(91)90018-L.

Frostegård, Å., Tunlid, A., Bååth, E., 2011. Use and misuse of PLFA measurements in soils. Soil Biol. Biochem. 43, 1621–1625.

Hirsch, P.R., Mauchline, T.H., Clark, I.M., 2010. Culture-independent molecular techniques for soil microbial ecology. Soil Biol. Biochem. 42, 878–887.

Jackson, L.E., Burger, M., Cavagnaro, T.R., 2008. Roots, nitrogen transformations, and ecosystem services. Annu. Rev. Plant Biol. 59, 341–363.

Kirkby, C.A., Kirkegaard, J.A., Richardson, A.E., Wade, L.J., Blanchard, C., Batten, G., 2011. Stable soil organic matter: a comparison of C:N:P:S ratios in Australian and other world soils. Geoderma 163, 197–208.

Kramer, C., Gleixner, G., 2008. Soil organic matter in soil depth profiles: distinct carbon preferences of microbial groups during carbon transformation. Soil Biol. Biochem. 40, 425–433. http://dx.doi.org/10.1016/j.soilbio.2007.09.016.

Kuzyakov, Y., Bogomolova, I., Glaser, B., 2014. Biochar stability in soil: decomposition during eight years and transformation as assessed by compound-specific ^{14}C analysis. Soil Biol. Biochem. 70, 229–236. http://dx.doi.org/10.1016/j.soilbio.2013.12.021.

Kuzyakov, Y., Subbotina, I., Chen, H., Bogomolova, I., Xu, X., 2009. Black carbon decomposition and incorporation into soil microbial biomass estimated by ^{14}C labeling. Soil Biol. Biochem. 41, 210–219.

Leckie, S.E., Prescott, C.E., Grayston, S.J., Neufeld, J.D., Mohn, W.W., 2004. Comparison of chloroform fumigation-extraction, phospholipid fatty acid, and DNA methods to determine microbial biomass in forest humus. Soil Biol. Biochem. 36, 529–532.

Legendre, P., Legendre, L., 2012. Numerical Ecology. Elsevier, Amsterdam.

Lehmann, J., 2007a. A handful of carbon. Nature 447, 143–144.

Lehmann, J., 2007b. Bio-energy in the black. Front. Ecol. Environ. 5, 381–387. http://dx.doi.org/10.1890/1540-9295(2007)5[381:BITB]2.0.CO;2.

Lehmann, J., Rillig, M.C., Thies, J., Masiello, C.A., Hockaday, W.C., Crowley, D., 2011. Biochar effects on soil biota - a review. Soil Biol. Biochem. 43, 1812–1836.

Lu, W., Ding, W., Zhang, J., Li, Y., Luo, J., Bolan, N., Xie, Z., 2014. Biochar suppressed the decomposition of organic carbon in a cultivated sandy loam soil: a negative priming effect. Soil Biol. Biochem. 76, 12–21. http://dx.doi.org/10.1016/j.soilbio.2014.04.029.

Marschner, B., Brodowski, S., Dreves, A., Gleixner, G., Gude, A., Grootes, P.M., Hamer, U., Heim, A., Jandl, G., Ji, R., Kaiser, K., Kalbitz, K., Kramer, C., Leinweber, P., Rethemeyer, J., Schäffer, A., Schmidt, M.W.I., Schwark, L., Wiesenberg, G.L.B., 2008. How relevant is recalcitrance for the stabilization of organic matter in soils? J. Plant Nutr. Soil Sci. 171, 91–110.

Muhammad, N., Dai, Z., Xiao, K., Meng, J., Brookes, P.C., Liu, X., Wang, H., Wu, J., Xu, J., 2014. Changes in microbial community structure due to biochars generated from different feedstocks and their relationships with soil chemical properties. Geoderma 226–227, 270–278. http://dx.doi.org/10.1016/j.geoderma.2014.01.023.

Ng, E.L., Patti, A.F., Rose, M.T., Schefe, C.R., Wilkinson, K., Cavagnaro, T.R., 2014a. Functional stoichiometry of soil microbial communities after amendment with stabilised organic matter. Soil Biol. Biochem. 76, 170–178. http://dx.doi.org/10.1016/j.soilbio.2014.05.016.

Ng, E.L., Patti, A.F., Rose, M.T., Schefe, C.R., Wilkinson, K., Smernik, R.J., Cavagnaro, T.R., 2014b. Does the chemical nature of soil carbon drive the structure and functioning of soil microbial communities? Soil Biol. Biochem. 70, 54–61. http://dx.doi.org/10.1016/j.soilbio.2013.12.004.

Nielsen, P., Petersen, S.O., 2000. Ester-linked polar lipid fatty acid profiles of soil microbial communities: a comparison of extraction methods and evaluation of interference from humic acids. Soil Biol. Biochem. 32, 1241–1249.

Noble, P.A., Almeida, J.S., Lovell, C.R., 2000. Application of neural computing methods for interpreting phospholipid fatty acid profiles of natural microbial communities. Appl. Environ. Microbiol. 66, 694–699. http://dx.doi.org/10.1128/AEM.66.2.694-699.2000.

Olsson, P.A., Baath, E., Jakobsen, I., Soderstrom, B., 1995. The use of phospholipid and neutral lipid fatty acids to estimate biomass of arbuscular mycorrhizal fungi in soil. Mycol. Res. 99, 623–629.

Papadopoulou, E.S., Karpouzas, D.G., Menkissoglu-Spiroudi, U., 2011. Extraction parameters significantly influence the quantity and the profile of PLFAs extracted from soils. Microb. Ecol. 62, 704–714.

Pietikäinen, J., Kiikkilä, O., Fritze, H., 2000. Charcoal as a habitat for microbes and its effect on the microbial community of the underlying humus. Oikos 89, 231–242.

Potthoff, M., Steenwerth, K.L., Jackson, L.E., Drenovsky, R.E., Scow, K.M., Joergensen, R.G., 2006. Soil microbial community composition as affected by restoration practices in California grassland. Soil Biol. Biochem. 38, 1851–1860. http://dx.doi.org/10.1016/j.soilbio.2005.12.009.

Ramette, A., 2007. Multivariate analyses in microbial ecology. FEMS Microbiology Ecology 62 (2), 142–160. http://dx.doi.org/10.1111/j.1574-6941.2007.00375.x.

Ruess, L., Chamberlain, P.M., 2010. The fat that matters: soil food web analysis using fatty acids and their carbon stable isotope signature. Soil Biol. Biochem. 42, 1898–1910.

Ruess, L., Häggblom, M.M., García Zapata, E.J., Dighton, J., 2002. Fatty acids of fungi and nematodes—possible biomarkers in the soil food chain? Soil Biol. Biochem. 34, 745–756. http://dx.doi.org/10.1016/S0038-0717(01)00231-0.

Santos, F., Torn, M.S., Bird, J.A., 2012. Biological degradation of pyrogenic organic matter in temperate forest soils. Soil Biol. Biochem. 51, 115–124. http://dx.doi.org/10.1016/j.soilbio.2012.04.005.

Schinner, F., Öhlinger, R., Kandeler, E., Margesin, R., 1996. Methods in Soil Biology.

Singh, B.P., Cowie, A.L., Smernik, R.J., 2012. Biochar carbon stability in a clayey soil as a function of feedstock and pyrolysis temperature. Environ. Sci. Technol. 46, 11770–11778.

Sinsabaugh, R.L., Hill, B.H., Follstad Shah, J.J., 2009. Ecoenzymatic stoichiometry of microbial organic nutrient acquisition in soil and sediment. Nature 462, 795–798.

Spiegelman, D., Whissell, G., Greer, C.W., 2005. A survey of the methods for the characterization of microbial consortia and communities. Can. J. Microbiol. 51, 355–386. http://dx.doi.org/10.1139/w05-003.

Steenwerth, K.L., Jackson, L.E., Calderón, F.J., Stromberg, M.R., Scow, K.M., 2002. Soil microbial community composition and land use history in cultivated and grassland ecosystems of coastal California. Soil Biol. Biochem. 34, 1599–1611. http://dx.doi.org/10.1016/S0038-0717(02)00144-X.

Steinbeiss, S., Gleixner, G., Antonietti, M., 2009. Effect of biochar amendment on soil carbon balance and soil microbial activity. Soil Biol. Biochem. 41, 1301–1310. http://dx.doi.org/10.1016/j.soilbio.2009.03.016.

Ventura, M., Alberti, G., Viger, M., Jenkins, J.R., Girardin, C., Baronti, S., Zaldeim, A., Taylor, G., Rumple, C., Tonon, G., 2014. Biochar mineralization and priming effect on SOM decomposition in two European short rotation coppices. GCB Bioenergy. http://dx.doi. org/10.1111/gcbb.12219.

Vestal, J.R., White, D.C., 1989. Lipid analysis in microbial ecology: quantitative approaches to the study of microbial communities. Bioscience 39, 535–541.

Wagg, C., Jansa, J., Schmid, B., van der Heijden, M.G.A., 2011. Belowground biodiversity effects of plant symbionts support aboveground productivity. Ecol. Lett. 14, 1001–1009.

Waldrop, M.P., Firestone, M.K., 2006. Response of microbial community composition and function to soil climate change. Microb. Ecol. 52, 716–724. http://dx.doi.org/10.1007/ s00248-006-9103-3.

Wardle, D.A., Bardgett, R.D., Klironomos, J.N., Setälä, H., Van Der Putten, W.H., Wall, D.H., 2004. Ecological linkages between aboveground and belowground biota. Science 304, 1629–1633.

Warnock, D.D., Lehmann, J., Kuyper, T.W., Rillig, M.C., 2007. Mycorrhizal responses to biochar in soil – concepts and mechanisms. Plant Soil 300, 9–20. http://dx.doi.org/10.1007/ s11104-007-9391-5.

Watzinger, A., 2015. Microbial phospholipid biomarkers and stable isotope methods help reveal soil functions. Soil Biol. Biochem. 86, 98–107. http://dx.doi.org/10.1016/j. soilbio.2015.03.019.

Watzinger, A., Feichtmair, S., Kitzler, B., Zehetner, F., Kloss, S., Wimmer, B., Zechmeister-Boltenstern, S., Soja, G., 2014. Soil microbial communities responded to biochar application in temperate soils and slowly metabolized 13 C-labelled biochar as revealed by 13 C PLFA analyses: results from a short-term incubation and pot experiment. Eur. J. Soil Sci. 65, 40–51. http://dx.doi.org/10.1111/ejss.12100.

White, D.C., 1993. In Situ Measurement of Microbial Biomass, Community Structure and Nutritional Status. Philos. Trans. R. Soc. Lond. A. http://dx.doi.org/10.1098/ rsta.1993.0075.

White, D.C., Davis, W.M., Nickels, J.S., King, J.D., Bobbie, R.J., 1979. Determination of the sedimentary microbial biomass by extractible lipid phosphate. Oecologia 40, 51–62. http://dx.doi.org/10.1007/BF00388810.

Williams, M., Rice, C., 2007. Seven years of enhanced water availability influences the physiological, structural, and functional attributes of a soil microbial community. Appl. Soil Ecol. 35, 535–545. http://dx.doi.org/10.1016/j.apsoil.2006.09.014.

Wu, Y., Ding, N., Wang, G., Xu, J., Wu, J., Brookes, P.C., 2009. Effects of different soil weights, storage times and extraction methods on soil phospholipid fatty acid analyses. Geoderma 150, 171–178.

Yao, H., Chapman, S.J., Thornton, B., Paterson, E., 2015. ^{13}C PLFAs: a key to open the soil microbial black box? Plant Soil 392, 3–15. http://dx.doi.org/10.1007/s11104-014-2300-9.

Zelles, L., 1997. Phospholipid fatty acid profiles in selected members of soil microbial communities. Chemosphere 35 (1–2), 275–294.

Zelles, L., 1999. Fatty acid patterns of phospholipids and lipopolysaccharides in the characterisation of microbial communities in soil: a review. Biol. Fertil. Soils 29, 111–129. http:// dx.doi.org/10.1007/s003740050533.

Zhang, H., Voroney, R.P., Price, G.W., 2015. Effects of temperature and processing conditions on biochar chemical properties and their influence on soil C and N transformations. Soil Biol. Biochem. 83, 19–28. http://dx.doi.org/10.1016/j.soilbio.2015.01.006.

Zimmerman, A.R., Gao, B., Ahn, M.Y., 2011. Positive and negative carbon mineralization priming effects among a variety of biochar-amended soils. Soil Biol. Biochem. 43, 1169–1179.

DGGE-Profiling of Culturable Biochar-Enriched Microbial Communities

T. Komang Ralebitso-Senior, C.J. Ennis, C.H. Orr, P. Barakoti, J. Pickering

Teesside University, Middlesbrough, United Kingdom

OUTLINE

Biochar Application
http://dx.doi.org/10.1016/B978-0-12-803433-0.00004-7

INTRODUCTION

Changes in the global environment by anthropogenic influences are known to affect animal and plant life although the severity of the impacts on microorganisms is largely unclear (eg, Ager et al., 2010). Nonetheless, the soil environment has a direct impact on numerous microbial species through parameters such as pH, temperature, nutrient concentration and availability, and water or moisture content (Guimarães et al., 2010; Lehmann et al., 2011).

In keeping with the models that are used to predict species decline in animals and plants, Ager et al. (2010) applied rank abundance to identify microbial species' decline caused by anthropogenic effects. The model was applied to water-filled tree holes because of the diversity of biota present. The results showed that general principles in ecology can also be applied to the microbial world and that the greater the impacts of environmental stresses, the higher the turnover of bacterial taxa. As a consequence their study afforded microbial ecologists additional data analyses tools to process and interpret profiling data, especially denaturing gradient gel electrophoresis (DGGE) derived, to then explore microbial community dynamics in greater detail.

As also discussed in Chapter 3, and similar to compost (Bougnom et al., 2010), it is possible to create "designer chars," which have specific characteristics that can be applied to certain soil types (Novak et al., 2009). The application of specific chars does not only change the chemical properties of the soil but also the biota, where both positive and adverse effects on specific species or functional communities can result. For example, Durenkamp et al. (2010) and Khodadad et al. (2011) both reported that microbial communities can be affected by different biochars while Khodadad et al. (2011) discussed further the possible "taxon-specific" shifts in diversity. Also, and as reported for other soil additions such as compost (Bougnom et al., 2010), it is possible to increase the abundance of certain species and/or microbial communities, including in degraded or contaminated ecosystems, by biochar supplementation.

A detailed review by Lehmann et al. (2011) explored "biochar effects on soil biota" by assembling a wide range of literature. As previously mentioned, biochar can have a negative effect upon microorganisms. The review discussed how this also applies to fungal species, in particular, and offered some possible explanations including: (1) species elimination because of the available nutrient content; (2) alteration in the chemical properties of the soil; and (3) increased amount of specific minerals.

Biochar has the potential to retain microorganisms because of adhesion (Jaafar et al., 2015; Lehmann et al., 2011) provided the pore sizes are correct. For example, both *Bacillus mucilaginosus* and *Acinetobacter* sp. require pore sizes of approximately 2–4 μm so if the sizes are larger than the occupying bacteria then no adhesion will result. Therefore, although the researchers indicated that there was limited evidence to validate this theory, they proposed that pore size allows microorganisms to shelter from competitors.

The overall effect of biochar on microbial community structure is not yet clear and further research is required particularly since contradictory trends have been reported in the literature (eg, Anderson et al., 2011; Maurathan et al., 2015). Thus particular species can proliferate because of an increase in nutrients, although some studies, including Khodadad et al. (2011), Kim et al. (2007), and Grossman et al. (2010), have reported declines in abundance. In situ biochar application studies are affected by multiple variables, which may confound a clearer understanding of the mechanisms involved. Therefore laboratory-based studies that would also be easily reproducible to mirror other investigations may be required initially before real soil environments.

As we discussed critically (Ennis et al., 2012), ongoing debates and achievements in culture-based analyses, to culture previously uncultivable and/or uncharacterized microbial strains and populations, have direct relevance for biochar-augmented ecosystems generally and soils in particular. An emerging example study within the biochar context includes that of Hammer et al. (2014) who used monoxenic cultures of the arbuscular mycorrhizal fungus *Rhizophagus irregularis* on carrot roots and reported that the strain used biochar as a physical medium for growth as well as a source of nutrients, particularly phosphorus. Kappler et al. (2014) grew pure strains in defined culture media and demonstrated how such culture-based experiments can afford additional insights of biochar mechanisms that underpin its effects on microbial species. In particular, the researchers used cultures of *Shewanella oneidensis* MR-1 in defined growth media and suggested potential insights of the electron-shuttling mechanisms of biochar to insoluble iron(oxy)hydroxides. Similar approaches would be relevant to elucidate the impacts of biochar on specific functional strains and communities including, for example, in biogeochemical cycling of both key (carbon, nitrogen) and trace (potassium, phosphorus, magnesium, etc.) elements.

In general, significant knowledge gaps could be addressed through media development initiatives to grow both known and hitherto unknown/uncultivable microbial strains and associations in studies designed specifically to explore how different biochars affect specific soil ecosystem functions. Hydroponic investigations (Graber et al., 2010) could, for example, be used to illustrate how biochar augmentation influences bulk and rhizosphere soil interactions between plants, bacteria, and mycorrhizal fungi relative to plant yield, phyto-/rhizoremediation, nutrient uptake and retention, biogeochemical cycling, and disease suppression. Furthermore, it is within some of these same contexts that biochar has also been identified as a bioaugmentation tool or microbial inoculum carrier (Douds et al., 2014; Hale et al., 2014, 2015), because of its porous structure, surface area, and capacity to adsorb microorganisms. Indeed the role of the "charosphere" (Quilliam et al., 2013b), where biochar impacts are pronounced in its immediate microgeography, would seem to justify the use of cultivable soil microbial species and associations, and relevant culture-based analyses, to elucidate fully the driving processes/ mechanisms. A study by Hale et al. (2014) illustrated this where a 300°C pinewood-derived biochar was applied to agricultural soil as a carrier for *Enterobacter cloacae*, a known plant growth-promoting rhizobacterium. Inoculum survival was determined by measuring the *Enterobacter* sp. root colonizing populations with colony-forming unit counts on Luria-Bertani agar and quantitative polymerase chain reaction (PCR). For fungi, Douds et al. (2014) tested the capacity of pelletized biochar as a carrier for the arbuscular mycorrhizal fungus *Rhizophagus intraradices* as a subsequent inoculum for vegetable and horticultural crops and recorded "promising" results. Thus colonization assays with bahiagrass (*Paspalum notatum* Flugge) as the host plant recorded increased *R. intraradices* propagules on the biochar pellets compared with light-expanded clay aggregates.

As discussed in Chapter 1, numerous microbial ecology techniques have been applied in biochar research with their respective strengths and disadvantages well documented in the literature. Of these, and although not strictly in response to biochar supplementation, DGGE is one of the more widely used platforms for rapid fingerprinting analysis of microbial population structure, diversity, and dynamics in complex microbial ecosystems (Green et al., 2009; Hill et al., 2000; Malik et al., 2008). Generally the technique does not require previous knowledge of the microbial associations under analysis as it can generate visual profiles and afford the monitoring of changes occurring, including in response to different treatments and/or soil modifications.

DGGE allows separation of same length amplified PCR products based on their sequence differences, which may vary as little as a single base pair (Dale and von Schantz, 2008; Muyzer, 1999; Muyzer et al., 1993). Principally, polyacrylamide is used as the matrix because of its ability to separate from 5 to 1000 base pairs (Reed et al., 2007). Thus DNA fragments are exposed to a linearly increasing gradient of denaturants at a constant temperature (60°C) as they migrate through the polyacrylamide gel. When the fragment encounters a specific denaturing threshold, the melting domain (T_m) with the lowest melting temperature begins to denature. As a result the migration of the specific DNA moiety in the gel halts while others continue until their respective denaturing concentration, relative to their T_m, is reached (Dale and Park, 2010; Muyzer et al., 1993). In addition, DNA sequences from different bacteria denature under different conditions to form eventually a pattern of multiple bands along the gel. Hence each band represents a different bacterial population or operational taxonomic unit (OTU) that was present in the original community. Additionally the incorporations of GC-rich sequences into one of the primers modify the melting behavior of the fragments of interest to the extent that close to 100% of all possible sequence variations can be detected (Muyzer et al., 1993). These provide a rough estimate of the richness and abundance of predominant microbial community members in any chosen environment (Dale and Park, 2010). It is for these reasons that the literature supports DGGE analysis of microbial communities generally (Jousset et al., 2010; Kirk et al., 2004) and within the biochar context (eg, Khodadad et al., 2011; Chapter 1) specifically.

The efficiency and efficacy of different nucleic acid (DNA/RNA) extractions from soil can cause biases while primer choice may affect further the outcomes of molecular techniques (Kirk et al., 2004; Maarit Niemi et al., 2001; Robe et al., 2003) and thus mandate closer investigation for biochar-augmented soils as illustrated by Hale and Crowley (2015) and Leite et al. (2014). While biochar pore architecture might be advantageous for the successful enhancement of specific pathways or environmental technologies, its potentially irreversible "trapping" of some microbial components could result in their exclusion from analyses. Similarly, biochar electron conductivity and ion exchange capacity could result in sorption of some of the extracted DNA with the respective source microbial species and communities then overlooked in studies to elucidate the responses of the total and functional soil microbial associations. For RNA-based characterization, which is often targeted to reflect the functional clades (eg, Manefield et al., 2002), additional steps would be required to mitigate for its typical susceptibility to degradation.

Despite their well-recognized limitations, many microbial community profiling techniques, including DGGE, are still applicable in different microecophysiology approaches (eg, Maarit Niemi et al., 2001; Ranjard et al., 2000), not least because they are relatively accessible with regards to ease of application and affordability. As a result they have been applied to analyze microbial community structure, composition, and diversity in different soils including in the Brazil Amazon, outwith and within the context of biochar (Cenciani et al., 2010; Navarrete et al., 2011; Taketani and Tsai, 2010). As also discussed in Chapter 1, DGGE analysis has been used to study bacterial, archaeal, and fungal community dynamics in diverse biochar-supplemented soil systems/environmental biotechnologies. For example, the tool has been adopted to explore the effects of biochar addition to: farmland (Sun et al., 2012); paddy-upland soils with and without chemical NPK fertilizer (He et al., 2014); crop growth and yield (Nzanza et al., 2011); saline soil (Lu et al., 2015); N-cycling communities in acidic or alkaline soils (Song et al., 2014; Wang et al., 2015); heavy metal- and polychlorinated biphenyl-contaminated systems (Liu et al., 2015); and greenhouse gas emission (Feng et al., 2012).

The use of biochar as a mode of inoculum delivery in agriculture and contaminant attenuation (Chen et al., 2012) will necessitate comprehensive physiological analyses of the culturable, functional strains and communities. For example, Chen et al. (2012) used biochar with high a sorptive capacity as a carrier for polycyclic aromatic hydrocarbon catabolic microbial communities and recorded increased pollutant biodegradation in soil.

Parallel to molecular analyses, the need for definitive studies of biochar impacts on microbial species/communities at the cellular level was exemplified by Masiello et al. (2013). Here the necessity for simple systems, ie, monocultures or model communities, for detailed investigations with controllable parameters and hence culture-based approaches was highlighted.

There are many established and potential attributes of biochar. It has the capacity to sequester carbon through storage, mitigate climate change, and modify soil properties. Although it is valuable, because of its ability to retain nutrients and water, and hence is useful in agriculture and bioremediation, its interactions with soil are extremely complex, both physicochemically and biologically.

Generally, complex exchanges occur in soils both between elements and electrons, creating electrically positive and negative areas. Also the relationship between pH and the point of zero net charge is pivotal in producing these positive and negative areas (Mukherjee et al., 2011) although biochar application may shift the charge. This can either benefit the soil or have additional implications because of nutrient movement through cation exchange (Mukherjee et al., 2011). Further research of biochar effects on soil nutrition will elucidate the chemistry and identify biochar type for specific applications. Confirmation may then be gained by studying the physiological properties of pertinent microbial strains and communities through culture-based analyses. As per previous deliberations (eg, Ennis et al., 2012; Handelsman, 2004), there is evidence and justification for the applicability of culture-based analyses outwith and, in particular, within the biochar context. This has been exemplified by numerous research initiatives to enrich, isolate, study, and apply microbial strains and communities with different functional capabilities in pristine and contaminated ecosystems (eg, Creamer et al., 2016; Luo et al., 2015). Therefore we present a case study where we investigated the influence of a wood-derived biochar on microbial community structure, composition, and diversity with different culture media.

THE CASE STUDY

Overview of Experimental Design and Analyses

Ten kilograms of fresh garden soil were collected from a domestic garden in Middlesbrough, UK (latitude 54°35′N and longitude 01°13′W),

TABLE 4.1   Physicochemical characteristics of the soil and mixed Broadleaf forestry-derived biochar

Parameter	Soil	Biochar
Al (g/kg^{-1})	28	7.6
Ca (g/kg^{-1})	18	26
Mg (g/kg^{-1})	9.2	2
K (g/kg^{-1})	5.3	1.8
Na (g/kg^{-1})	0.37	0.34
Electrical conductivity (μS/cm^{-1})	250	1400
Calorific value (MJ/kg^{-1})	1.2	16.6
Total organic carbon (%)	2.8	10
Total S (%)	0.03	0.03
P (mg/kg^{-1})	<0.10	57
Nitrate aqueous extract as NO_3 (mg/L^{-1})	1.5	26
pH	7.46	9.3

milled for homogenization and sieved (0.2 mm). The soil and locally produced biochar (mixed broadleaf forestry, pyrolysis—500°C < T < 600°C, pH 9.62; Yorkshire Charcoal Company) were characterized physicochemically (Derwentside Environmental Testing Services Ltd, County Durham, UK) (Table 4.1). The homogenized soil was then used in four different microcosms with (SB) or without (S) biochar, with the original soil moisture content of 8% (w/w) maintained with either sterile deionized water (H_2O) or basic mineral salts solution (MSS) as follows: (1) soil only + moisture content controlled with sterile deionized water (S + H_2O/SH); (2) soil with biochar + moisture content controlled with sterile deionized water (SB + H_2O/SBH); (3) soil only + moisture content controlled with basic MSS (S + MSS/SM); and (4) soil with biochar + moisture content controlled with basic MSS (SB + MSS/SBM). The microcosms were then maintained in the dark at room temperature (~25°C).

Reagents and Media Preparation

R2A agar (RA, 10% w/v; Fisher Scientific) was prepared according to the manufacturer's instructions but diluted to 10% (v/v) strength.

Congo red agar (CRA). The medium contained (g/L^{-1} sterile deionized water): DL-malic acid, 5; K_2HPO_4, 0.5; $MgSO_4 \cdot 7H_2O$, 0.2; KOH, 4.5; NaCl, 0.1; technical agar, 15; yeast extract, 0.5; $FeCl_3 \cdot 6H_2O$, 0.015; and Congo red (CR) solution (0.25%), 15 mL. The pH was adjusted to 7.0 with NaOH before autoclaving (120°C × 20 min × 15 psi) (Akbari et al., 2007).

Soil extract agar (SE). Study soil (1 kg) was mixed thoroughly with 2 L of 50 mM NaOH and left overnight at room temperature (~25°C). The mixture was coarse filtered (Whatman) and centrifuged (Eppendorf) at 4000 rpm × g for 1 h with the supernatant then series filtered (Nanopore) (1.6, 1.0, 0.45, and 0.2 μm) and mixed with 2× sterile molten technical agar (1:1 v/v) (Hamaki et al., 2005).

Mineral salts solution (MSS). The MSS contained (g/L^{-1} sterile deionized water): K_2HPO_4, 1.5; KH_2PO_4, 0.5; $(NH_4)_2SO_4$, 0.5, and $MgSO_4 \cdot 7H_2O$, 0.2 and was sterilized by autoclaving at 120°C and 15 psi for 20 min.

Soil pH and Moisture Determination

The soil pH was determined after mixing thoroughly with deionized water (1:2.5 w/v). The suspension was left to stand for 30 min before remixing for pH determination with a calibrated probe (Fisher Scientific) connected to a pH 213 microprocessor (Hanna Instruments, Bedfordshire, UK). As recommended by Hopkins et al. (2000), the probe was rotated gently and kept in the solution for 30 s for accuracy.

Soil moisture of 8% (w/w) was determined by heating for 8 h at 55°C. Moisture content maintenance was made subsequently at every sampling time by the addition of sterile deionized water or MSS.

Combined Culture-Based and Molecular Analyses

Soil/soil + biochar samples (0.1 g) were used in 10-fold serial dilutions (10^{-1}–10^{-7}) with sterile saline (0.9% (w/v) NaCl), with aliquots (100 μL) used subsequently for duplicate spread plates on the three solid media. After maintenance at room temperature (~25°C) for 72 h, viable cell counts and total colony plate scrapes (10^{-1} dilution plates) were made.

Total microcosm soil/soil + biochar and colony plate scrape DNA samples were isolated with the PowerSoil DNA Isolation kit (Mo Bio Laboratories, Inc., USA) as described by the manufacturer and stored at −20°C until required for PCR-DGGE. Red colonies on the 10^{-3} CRA dilution plates were removed with sterile toothpicks and resuspended in 200 μL molecular grade water, mixed by vortexing for 5 min to ensure uniform distribution and used to amplify the 16S rRNA gene for subsequent DGGE profiling.

According to Dale and von Schantz (2008) and Reed et al. (2007), nested-PCR uses two stages where a more specific amplification is achieved in the first stage with a second set of primers then used to amplify specific regions of the required sequences. This method can facilitate an improved analysis of microbial community diversity where low abundance sequences and/or OTUs are also amplified and included in the subsequent analyses. Thus Khodadad et al. (2011) concluded that the use of nested-PCR is beneficial for identifying microbial taxa in biochar-supplemented ecosystems. Similarly, we amplified the V3 region of the 16S rRNA gene of all DNA templates by the nested approach with the 11-F/1512-R (Felske et al., 1998) and GC356-F/519-R (Manefield et al., 2002) primer sets. These were followed by denaturing gradient (30–70%) gel electrophoresis and image analysis as described by Olakanye et al. (2014).

Results and Discussion

Some of the main challenges for soil ecologists are understanding and managing microbial community diversity, structure, and composition and their roles in ecosystem functioning (Waite et al., 2003). While various studies continue to consider biochar use and its role in crop yield increase, contaminated site remediation and carbon sequestration, typically by molecular analyses, this case study examined the impacts of biochar on cultivable and total soil microbial communities. A particular focus was to explore the potential role of growth media as a component of different approaches to address fundamental knowledge gaps in biochar-augmented soil ecosystems.

Microcosm pH and Moisture Content

The initial soil pH of 7.46 increased to 7.65 following biochar addition. Overall the pH values decreased on day 2 before rising again to reach maxima of 7.84 (water) and 7.67 (MSS) in the presence of biochar on day 8 in contrast to 7.55 and 7.47 of the equivalent controls. Throughout the study the higher pH values were recorded for the biochar microcosms (Fig. 4.1).

It is well accepted that soil pH has a direct impact on microbial community structure. Therefore the pH changes should have accounted partially for the recorded changes in microbial community structure, both in the presence and absence of biochar and relative to the moisture maintenance regime. Although not determined in the study, previous investigations, as reported in Chapters 1 and 2 (eg, Aciego Pietri and Brookes, 2009), have shown that these shifts can facilitate decreases and increases of Gram-positive and Gram-negative bacteria, respectively. In contrast, fungal communities can remain unchanged.

Soil moisture plays a key role in plant growth especially as it can affect soil microbial diversity with low moisture contents often leading to decreased diversity and activity (Barros et al., 1995; Chen et al., 2007). For this study the original soil moisture content was maintained with either sterile deionized water or basal MSS to explore the effects of nutrient availability or addition in a biochar-impacted soil. Because of the experimental

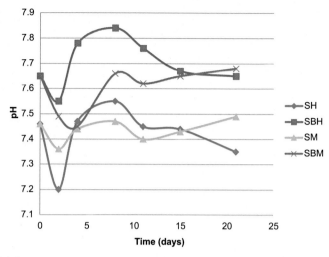

FIGURE 4.1   pH changes of the soil control and soil + biochar microcosms during 21 days' maintenance at room temperature (~25°C). *SH*, soil + H$_2$O; *SBH*, soil + biochar + H$_2$O; *SM*, soil+ mineral salts solution; *SMB*, soil + biochar + mineral salts solution.

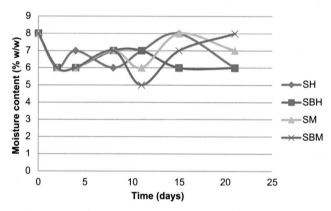

FIGURE 4.2   Moisture content (% w/w) changes of different soil and soil + biochar microcosms. *SH*, soil + H$_2$O; *SBH*, soil + biochar + H$_2$O; *SM*, soil+ mineral salts solution; *SMB*, soil + biochar + mineral salts solution.

design, fluctuations in moisture content were expected. With the exception of the highest decrease to 5% (w/w), which was recorded on day 11 in the SB + MSS microcosm (Fig. 4.2), general moisture content decreases by ~2% (w/w) were recorded for the microcosms with no obvious differences between the biochar microcosms and the controls. Although several researchers, such as Basso et al. (2013) and Laird et al. (2010), have suggested that biochar porosity can enhance moisture-holding capacity for (some) soils, this was not apparent in the case study. Nevertheless, moisture content assessment, including soil pore water or water-filled pore space determination, is a critical element of biochar research (Chapters 7 and 10) particularly in conjunction with microbial water activity assessment.

Viable Cell Counts on Different Culture Media

Conventionally, and largely because of expediency, viable cell counts, including on selective media, have been used to measure microbial community diversity. To address their accepted limitations where c. 1–5% of microbial strains are cultivable under laboratory conditions, molecular techniques are being applied increasingly. The efficacy and applicability of growth media to study complex microbial populations depends largely on medium composition and pH, and, to a lesser extent, on growth conditions such as incubation temperature, light regime, water activity, etc. Investigating specific groups that play key roles in the soil microbiome, such as nitrogen fixation, can identify how biochar affects particular properties. For example, Wessén and Hallin (2011) suggested that profiling ammonia-oxidizing bacteria acts as an improved method for soil monitoring compared to general community profiling. Therefore this study used three different media to explore biochar and minimal nutrient supplementation impacts on: general oligotrophic (R2A agar); potential nitrogen fixing (CRA); and indigenous site-specific (soil habitat agar) cultivable soil bacterial communities.

Originally developed by Reasoner, R2A medium has been used to identify heterotrophic bacteria in target samples. Since these species catabolize complex molecules such as oligosaccharides, R2A culturing could reveal oligotrophic bacteria (Massa et al., 1998). Typically, oligotrophs are slow growing and tend to be outcompeted by faster-growing strains on standard media. Thus R2A medium was used to select and identify a broader range of microbial species in the control and experimental microcosms. The general trends for culture-based analysis on R2A agar were similar to those recorded on CRA and SE with peaks on day 4. Specifically, R2A showed colony forming unit (CFU) counts that were between those detected on the selective and soil habitat media independent of biochar addition and moisture regime.

CRA was used to select for nitrogen-fixing bacteria (red colonies). According to Smalley et al. (1995) the dye binds to specific surface proteins while Daskaleros and Payne (1987) reported that it correlates with hemin binding, a property which has also been found in N_2-fixing legume nodule endosymbiotic rhizobia such as *Sinorhizobium meliloti* (Battistoni et al., 2002a,b). In the latter study this membrane characteristic facilitated Fe acquisition under Fe-limited conditions and thus increased nodule competitiveness for alfalfa plants.

CFU counts on CRA (Fig. 4.3) increased up to day 4 where they peaked before decreasing progressively. The counts on days 0, 2, and 4 were comparable for the two controls while differences for the biochar microcosms

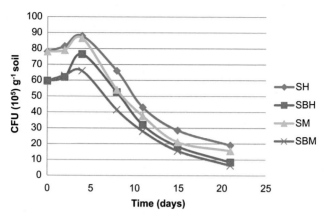

FIGURE 4.3 Changes in Congo red agar (CRA) colony forming unit (CFU) counts of soil and soil + biochar microcosms. *SH*, soil + H_2O; *SBH*, soil + biochar + H_2O; *SM*, soil+ mineral salts solution; *SMB*, soil + biochar + mineral salts solution.

were recorded on day 4. On day 21 the lowest viable cell counts characterized the MSS-maintained biochar microcosm.

In summary, these data provided preliminary evidence of biochar effects on an important functional community in relation to soil nitrogen dynamics. According to the emerging literature (eg, Orr and Ralebitso-Senior, 2014; Chapter 7), these effects can be both positive and negative relative to nitrogen moiety retention for, for example, fertilizer purposes and mitigation of greenhouse gas emissions *re* nitrous oxide.

Soil is a medium for a wide range of macro- and microorganisms where it provides nutrients such as organic matter together with carbon, nitrogen, vitamins, and minerals, all of which are essential for the growth of organisms that have specific niches. Therefore the third growth medium (SE) was prepared with microcosm soil since, as reported by other researchers (eg, Bergmann et al., 2015; Hamaki et al., 2005; Liebeke et al., 2009), it was deemed the most suitable habitat medium for the indigenous bacterial strains.

For all microcosms, particularly the controls, the maximum viable cell counts were recorded on day 4 (Fig. 4.4). Marked count decreases were then apparent up to day 11 with, subsequent, more gradual declines. On days 8, 11, and 15 the colony counts of the SB + H_2O and S + MSS microcosms were comparable while on day 21 the SB + H_2O and SB + MSS counts were approaching the detection limit of this ecological medium. As observed by other researchers (eg, Hamaki et al., 2005; Liebeke et al., 2009), colony size and morphology on SEs tend to afford little differentiation, which may be a key limitation of this particular culture-based approach for determining biochar impacts on culturable members of soil microbial communities.

Notwithstanding their disadvantage of colony differentiation relative to medium color, the CRA supported more colonies than the SE probably

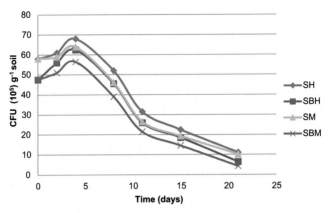

FIGURE 4.4   Changes in soil extract agar (SE) colony forming unit (CFU) counts of soil and soil+biochar microcosms. *SH*, soil + H_2O; *SBH*, soil + biochar + H_2O; *SM*, soil+ mineral salts solution; *SMB*, soil + biochar + mineral salts solution.

because of nutrient limitation of the latter. Here the hypothesis was that N_2-fixing bacteria would predominate the CRA while the SE would support the oligotrophic species.

DGGE-Based Analysis of Total and Culturable Communities

Total soil community 16S rRNA gene profiles for day 0 samples recorded low band numbers independent of biochar supplementation. Generally, band number varied temporally with the presence and absence of biochar, hence the highest was recorded on day 4 for the soil control with sterile deionized water. Similar band numbers resulted on day 11 for soil with the same moisture regime both in the presence and absence of biochar. Although the trends for the day 21 microcosm samples were similar to those recorded for day 0, increased dominances of some OTUs were apparent with some common to all microcosms.

The presence of both biochar and MSS increased the OTU numerical dominance of the total bacterial community. On day 4, taxa richness increased for the equivalent soil control. Subsequently, this decreased markedly from day 4 to day 8 then increased between days 11 and 15 before decreasing again before day 21. Comparable minimum band number for the biochar-supplemented microcosm was recorded on day 11.

For the cultivable community members the day 4 CFU count peaks were reflected in increased DGGE profile-based taxa richness. For the R2A cultures the low diversity profiles were characterized by high-GC bands but these varied in numerical dominance, particularly on days 8 and 15, for both the control and biochar microcosms.

As also recorded for the viable cell counts, and despite shifts in band abundance (Fig. 4.5A), the CRA generated the highest number of detectable

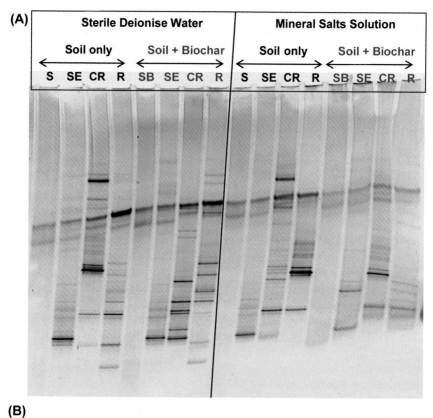

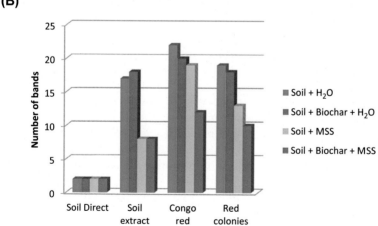

FIGURE 4.5 (A) Denaturing gradient gel electrophoresis (DGGE) profiles of soil control and biochar-supplemented microcosm initial communities cultivable on soil extract (SE) and Congo red (CR) agar and red colonies (R) from the latter. (B) DGGE bands of total and cultivable communities on day 0. *MSS*, Mineral salts solution; *S; SB*.

bands on day 0 with minimal differences between three (S + H_2O, SB + H_2O, S + MSS) of the four microcosms (Fig. 4.5B). The increased band numbers suggested that enrichment has resulted with this medium. For the red colonies, band number was marginally higher when the control moisture content was maintained with MSS. With sterile deionized water, similar taxa richness was observed both in the presence and absence of biochar.

The soil extract medium supported less diverse communities than the CRA (Fig. 4.6A). Although biochar augmentation increased some OTU abundances, the decreased diversity trend was more pronounced in the soil controls. Also the number of bands increased with the sterile deionized water moisture maintenance, with a slight increase caused by biochar addition. For the microcosm samples inoculated on soil extract medium, the lower band number resulted from the MSS-maintained moisture contents.

Day 21 recorded further shifts in OTU richness, numerical dominance, and cultivable community diversity (Fig. 4.6A). Band resolution from the SE-enriched fractions of the deionized water microcosms appeared close to the DGGE detection limit with very low abundances recorded. In contrast the MSS microcosms gave increased band abundances with comparable community compositions and structures. Unlike the soil habitat medium, CRA produced detectable bands for both moisture content maintenance strategies (Fig. 4.6A). Generally, community composition changed more in response to biochar addition than the moisture content maintenance approach for this medium.

For three of the microcosms, overall, taxa richness had decreased by day 21 (Fig. 4.6B) with minimal differences recorded for the microcosm samples cultured on CRA and their red colonies. Pronounced differences in richness increase were apparent following cultivation on SE for the S + MSS and SB + MSS microcosms and red colonies from the S + MSS microcosm.

In general, biochar application reduced band number but increased band intensity indicative of selective enrichment. For CRA plates, the bands from the DNA extracted from the 10^{-1} plate scrapes and the red colonies of the 10^{-3} dilution plate medium were comparable. The faint OTUs observed for the samples extracted directly from the plates were indicative of low DNA concentration although this can be increased to intensify the bands. It must be noted, however, that faint bands and fewer numbers of bands on different gels increase the chances of bias and may give misleading predications or conclusions.

According to Roelfsema and Peters (2005), DGGE has the capacity to analyze large numbers of unknown samples to detect single nucleotide mutations relatively rapidly, reliably, and reproducibly. The technique is applied to detect bacterial species type and number, and hence is semi-quantitative if a known concentration of (amplified) DNA is analyzed.

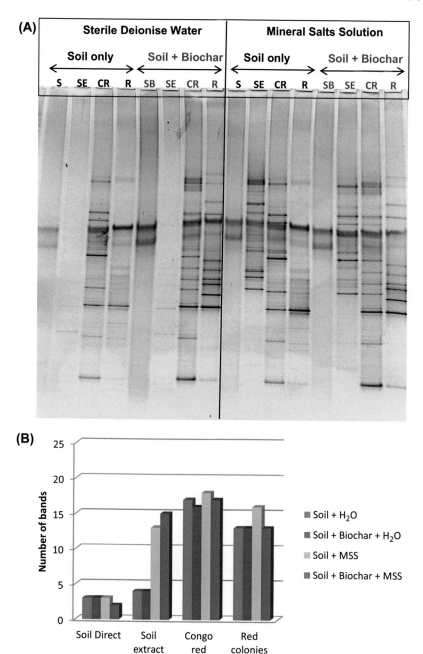

FIGURE 4.6   (A) Denaturing gradient gel electrophoresis (DGGE) profiles of cultivable communities on day 21. (B) Number of DGGE bands of total and cultivable communities on day 21. *MSS*, Mineral salts solution; *S*; *SB*.

Therefore the use of both culture-dependent and culture-independent techniques was important for this study as they provide specific information about bacterial community diversity where they potentially measure different fractions of the population. Consequently, and in light of considerable paucities in the literature, it was justifiable to adopt these complementary approaches toward a more complete analysis of community structure, composition, physiology, and phylogeny (Edenborn and Sexstone, 2007; Kirk et al., 2004) following biochar addition.

Study Conclusions and Knowledge Gaps

Compared to the soil controls the biochar-augmented microcosms recorded: increased pH; reduced viable cell counts on all media; but increased bacterial diversity, while the moisture content maintenance regime effected different trends. Also the CRA seemed the most effective culture medium since it afforded the highest colony recovery number and so facilitated improved subsequent analysis. Thus the number of CFUs supported on the different media was in a decreasing order of CR > RA > SE independent of the moisture regime. Notwithstanding this, colony DGGE profiling revealed that the soil habitat (SE) and selective CR media supported higher richness and overall diversity. Also the effects of biochar and nutrient (MSS) addition were more obvious than with commercial oligotrophic (RA) medium. Overall, richness decreases characterized the total soil DNA profiles compared with the colony-based analyses although biochar increased numerical dominance of some species in the culturable and total communities.

Nested PCR-DGGE profile analysis of the microcosm samples revealed temporal responses of the culturable microbial community structure in the presence of biochar with a numerical dominance increase of some species. Similarly, OTU abundance for the total community appeared to increase with the addition of biochar and MSS particularly on day 21. Therefore biochar effected shifts in microbial community diversity and specific OTU abundance possibly because of nutrient availability and alteration of both pH and water retention. Validation of these diversity changes would require comprehensive replication and analyses of functional populations such as nitrogen-fixing bacteria, particularly from the CRA-derived communities.

In summary, this preliminary study highlighted the necessity and applicability of simultaneous culture-dependent and culture-independent analyses to study comprehensively cultivable and total soil microbial community compositional and structural changes in response to biochar. Thus the complementary approach illustrated how the bias and/or limitations of a single technique can be addressed. For example, conventional viable cell counts and colony characterization facilitated measurement of the numerical occurrence of microbial strains adapted to and possibly unique to the specific soil/site (soil extract medium) and potential nitrogen-fixing bacteria (CRA). The effect of 4% (w/w) wood chip-based biochar addition on the composition and structure of these communities was then elucidated with PCR-DGGE profiling. Full cognizance is, nevertheless, made of the general limitations of both strategies where the culture-dependent techniques select for specific members of the community while the culture-independent profiling is now less favored than high-throughput microecophysiology platforms.

CHAPTER CONCLUSIONS AND FUTURE RECOMMENDATIONS

Measuring biochar impacts on microbial communities and their functional capacities is difficult in itself because of the number and range of variables involved. Also the assurance and efficacy of this undertaking is dependent heavily on the inherent strengths and limitations of the applied analytical method(s). As demonstrated by several studies (Chapters 1 and 2), and deliberated in some critical summations (eg, Anderson et al., 2011; Maurathan et al., 2015), it is difficult to make general conclusions on how biochar affects microbial diversity and abundance. This is because of the wide variety of feedstocks that can be used to create char, and the variable outcomes of different production conditions, which all effect unique properties to the resultant biochars. Therefore research should be made to identify if different feedstocks subsequently enhance microbial community diversity, abundance or both, relative to specific ecosystem functions.

As is well recognized, soil is characterized by a vast range of variables with the application of biochar affecting each to a different degree. A table of existing knowledge and priorities was presented by Lehmann et al. (2011) although research into the relationship between biochar and microorganisms was given the least priority. Instead the effects of char on fauna, pathogens, and the environment were deemed more important and placed at the top of the list. Notwithstanding this, it is well accepted that microbial communities can change in response to biochar addition, including in different soil horizons. Therefore research of biochar impacts in each zone should identify its effects on the complete microbiome across different soil transects. Also, because of soil lattice leaching, investigations of the effects on microbial communities below the primary biochar-augmented area should be made.

Additional analyses are necessary to determine the effects of different biochars on the soil microbiome specifically relative to the sizes/dimensions of the involved community members. Thus, together with biochar particle size and surface properties, pore architecture/size plays an important role in microbial community dynamics. While the dimensions of the microbial strains were not measured in our illustrative study, evidence from biochar and soil ecology literature justifies the need for specific investigations of surface interactions relative to the structure and composition of the microbial species/community that are selected preferentially by biochar depending on the size characteristics of both. Also the location of a molecule within the pores, relative to the size(s) of the specific catabolic strain(s) (du Plessis et al., 1998a,b; Chapter 7), warrant closer investigation. This could be achieved by, for example, permutations of sequential or simultaneous additions of the target molecule and degradative species. Naturally, a key third component of this scenario is the soil particles and their effects on the availability of pores for microbial attachment and colonization.

From its inception, biochar research has eluded to the hypothesis that micropores may serve as habitats for microorganisms where they are protected from grazers and/or desiccation (Steinbeiss et al., 2009). The importance of micropore architecture is dependent largely on biochar type and pore size distribution. For example, Quilliam et al. (2013a) reported that approximately 17% of pores of a wood-derived biochar with 1–2 cm fragments had diameters <1 μm and, as a result, were excluded to most microorganisms. For larger pores (>1 μm) low-density populations were recorded with external surface mineral aggregation presenting a barrier. Generally, labile organic carbon determined microbial colonization of the pores. The mechanisms and/or order of events in these tripartite configurations, especially for specific functional communities and strains, could be elucidated elegantly in simplified studies with known monocultures and/or model communities with culture-based analyses, singly or in combination with molecular tools.

The potential of biochar to support a wide range of microbial strains and communities in different ecosystems and environmental biotechnologies and therefore to underpin a range of ecosystem functions has been recognized and tested in agronomy, biogeochemical cycling, and bioremediation. Typically, its roles in these different processes and contexts are tested individually, despite obvious overlaps. This approach is probably justifiable considering the complexity of the material, the questions under investigation, and the myriad resolved and unresolved biochar-based mechanisms involved. Notwithstanding these, the feasibility and requirement to explore simultaneously biochar impacts on more than one process, within the same (experimental) system, are beginning to be recognized. Thus to study the link between the nitrogen biogeochemical cycle and polluted ecosystems, Stewart and Siciliano (2015) used a greenhouse trial to assess nitrogen fixation dynamics in tailings and mining-affected soils in northern Canada following growth of different herb species and the addition of biochar, biological soil crust slurries, and Rhizobia inocula.

Augmentation of the indigenous herbs with biochar and *Rhizobium* spp. showed the potential to enhance revegetation and reinstate the nitrogen biogeochemical cycle of the mine-affected site although the researchers recommended further investigations at field scale. Their approach could nonetheless be replicated for the amelioration of different mining-derived contaminants globally. Hence indigenous plants and nitrogen-cycle microbial species could be used in managed culture-based models initially to elucidate the fundamental underpinning processes/mechanisms. Thus, although not necessarily developed for remediation studies, experimental designs, such as those of Mia et al. (2014) and Oram et al. (2014), could be adopted and adapted to this end. In these studies the effects of different biochar additions on legume competitive ability and nitrogen fixation by: (1) red clover (*Trifolium pratense* L.) in mono- and mixed cultures with *Festuca rubra* L. and *Plantago lanceolate* L.; and (2) supplementation with N, P, K, and micronutrients were examined. In general, different nitrogen fixation rates, nutrient supplementation effects, plant biomass yields, and plant competitive advantage resulted depending on the biochar application and the biochar-mediated pH changes.

One of several emerging areas of research is the role of biochar in plant disease mitigation as exemplified in original studies and critical reviews by Bonanomi et al. (2015), Elad et al. (2010), Elmer and Pignatello (2011), Graber et al. (2014), Harel et al. (2012), Kolton et al. (2011), and Zwart and Kim (2012). Concomitant with several likely mechanisms by which biochar may exert disease control, these research teams identified key knowledge gaps and highlighted the need for further investigations. Hence, as also proposed for other contexts, models of plants + pathogenic microorganisms, complemented by DGGE profiling, as illustrated by Kowalchuk et al. (1997) with marram grass (*Ammophila arenaria*) root fungal infection, could be employed to increase an understanding of biochar impacts.

All of these established and emerging research areas justify the need for multiple and concerted efforts that engender the adoption of the full range of available microbial ecology techniques and protocols. Therefore improvements of culture-based methods necessitate particular investment in the biochar context, where they will facilitate analyses of the physiological effects of the material on the cultivable indigenous microbial populations, while also identifying its impacts on their specific diversities and abundances.

The illustrative study reported in this chapter applied a habitat medium (SE), two commercial media (R2A and CRA), and DGGE characterization. Theoretically, culture-independent PCR-DGGE visualizes more OTUs than culturing, which subsequently provides more information of microbial community richness, diversity, and evenness (Edenborn and Sexstone, 2007). Its results and analysis are, however, dependent on several key variables including: (1) primer choice/application and thermal cycling conditions; (2) polyacrylamide concentration and denaturing gradient; and (3) electrophoresis time and voltage. In particular, variations in the gradient and gel run time determine band resolution and hence differentiation. For example, short DGGE run times may result in band comigration or suboptimal separation within the analyzed amplicons (Miletto et al., 2007). Also DGGE profiling does not necessarily give the complete soil microbial profile in every case where band detection and number affect directly the subsequent fingerprint interpretation. As a consequence, alternative profiling methods may be considered and these include, for example, terminal restriction fragment length polymorphism, temperature gradient gel electrophoresis, single strand conformation polymorphism, and (automated) ribosomal RNA intergenic spacer analysis for total/occurring and cultivable microbial communities (Kirk et al., 2004; Reed et al., 2007), as have already been demonstrated by several researchers in biochar research (Chapter 1). Specifically, for example, Steinbeiss et al. (2009) analyzed phospholipid fatty acids and reported a decline in bacterial number when biochar was applied, while fungal communities increased. In contrast, Warnock et al. (2007) indicated that biochar supported fungal as well as bacterial and archaeal growth, probably because of nutrient availability.

Also several cutting-edge and high-throughput techniques have been adopted to elucidate the impacts of biochar on complex microbial communities and thus their roles in ecosystem function. Some of the tools targeting functional populations, such as the SIP-based RNA-DGGE, fluorescent in situ hybridization, and phospholipid fatty acid (PLFA) protocols, require parallel and/or complementary comprehensive microbial physiology analyses, even for proof-of-concept illustrations. For example, stable isotope-labeled nitrogen (^{15}N) biochar and/or substrates could be used to investigate N-dynamics in models inoculated with known nitrogen cycling microbial strains and communities. Thus, although not in culture-based experimental designs, de la Rosa and Knicker (2011) traced ^{15}N from a plant-derived biochar to establish its bioavailability as evidenced by ^{15}N presence in newly synthesized biomass 72 days after incubation in the biochar-augmented soils. Taghizadeh-Toosi et al. (2012) grew ryegrass (*Lolium* sp.) for 1 week in the presence of ^{15}N-NH$_3$-exposed Monterey pine (*Pinus radiata*)-derived biochar and recorded ^{15}N-labeled plant biomass, indicative of the bioavailability of the biochar-sorbed ammonia. Similarly, Hammer et al. (2014) traced ^{33}P uptake from biochar microsites by a carrot root-associated arbuscular mycorrhizal fungus, which facilitated plant nutrient uptake.

We propose that the knowledge gaps identified in these cutting-edge studies highlight that seminal investigations are still required to elucidate critically the multiple and complex mechanisms that underpin the positive and negative effects of biochar in different ecosystems, functions, and processes. These will be underpinned, not to an insignificant extent, by culture-based studies, which will also afford critical investigations of the effects of several additional elements such as cell-to-cell communication. Therefore simplified proof-of-concept experiments with known functional microbial strains or communities as inocula should be used. For plant-supported ecosystems these could entail common investigative plant species in initially soil-free systems such as hydroponic models, to understand fully the C/N dynamics and distributions, as also supported by mass balance determinations (see also Chapter 7).

Several studies, as exemplified by Jones et al. (2012), have highlighted that the results of laboratory investigations of the effects of biochar on microbial communities and plant yield are not always comparable to similar long-term field-scale studies. As a result the application and extrapolation of the laboratory findings from the culture-based models proposed earlier, including to monitor biochar-based water retention capacity, greenhouse gas emissions, biogeochemical cycling, and bioremediation, would require incrementally protracted testing through pot (eg, Kaštovská and Šantrůčková, 2011), greenhouse (eg, de la Rosa et al., 2015; Elmer and Pignatello, 2011; Stewart and Siciliano, 2015) and in situ pilot- and field-scale (eg, Elmer and Pignatello, 2011; Karhu et al., 2011) investigations.

It is accepted largely that biochar pore number, size, and distribution can provide habitat for a wide range of microbial species and functional communities. Consequently, the material is being considered as an ideal inoculum carrier (eg, Hale et al., 2015), a property that can be exploited to enhance multiple microbially driven pathways and processes such as analysis of nutrient (C, N, P, K. etc.) cycling in agronomic soils, enhanced crop output, and contaminant attenuation. As applied in other nonbiochar contexts, including extreme or unique habitats (eg, Sahni et al., 2011), the structure, composition, and diversity of functional microbial species and communities in ecosystems and environmental biotechnologies where biochar is added specifically as an inoculum carrier could be assessed by several microecophysiology tools including DGGE for bacteria, archaea (Robertson et al., 2005), and fungi (Anderson and Cairney, 2004; Marshall et al., 2003; Okubo and Sugiyama, 2009).

Although not yet published in the biochar literature, community-level physiological profiling (CLPP), with Biolog (Eco-Plates) and MicroResp systems, has recognized value in measuring microbial metabolic capacity and hence ecosystem function, as determined by changes in substrate utilization patterns. Thus Bossio et al. (2005) and Xue et al. (2008) used Eco-Plates, together with PLFA and DGGE profiling, to show CLPP applicability for comparative assessment of the effects of different land uses on soil microbial community structure and function. Also some independent research teams have developed microwell platforms parallel to the commercially produced systems to analyze simultaneously microbial community metabolic response to several parameter variations. The ultimate aim of all of these initiatives is increased throughput compared to traditional closed cultures. Independent of the platform, extensive literature on the applicability of CLPP or phenotypic screening, including critical analysis of applicability and limits for pristine and contaminated soil ecosystems, including across different biogeographical scales, is exemplified by numerous research papers and critical reviews (eg, Borglin et al., 2012; Creamer et al., 2016; Garland, 1997; Garland and Mills, 1991; Greetham, 2014; Hill et al., 2000; Stone et al., 2016). These deliberations, including those on the stringent requirements for data and statistical analyses to ensure relevant/appropriate extrapolation to the original ecosystem phenotypic capacities (eg, Wang et al., 2010), can be considered and applied subsequently as part of the microbial ecological studies and analyses that are essential for biochar-supplemented systems. Additional questions would probably require specific interrogation and management. For example, (semi-)quantitative Biolog-based community-level physiological profiling is typically reliant on optical density measurement, which would probably be hindered partly by the particulate nature and ash content of biochar.

Ash content also affects the successful recovery of high-quality, high-quantity microbial nucleic acids from biochar-augmented sites and soil microcosms/lysimeters, and so necessitates further optimization steps as observed by Herrera and Cockell (2007). The authors presented a critical discourse on the application of different DNA extraction methods for volcanic samples, which could afford knowledge transfer to biochar research in general, and that involving previously uncultivated/uncharacterized functional microbial strains and communities in particular. Some work is, however, emerging where researchers have investigated the efficacy of different extraction protocols in the biochar context (eg, Hale and Crowley, 2015)

Biochar has an as yet unexplored potential as a solid support medium for microbial biofilm formation in other environmental and/or waste management technologies such as malodorant gas biofiltration and wastewater treatment, including in upflow anaerobic sludge bed/blanket reactors. Preliminary studies on the applicability of this material in these contexts could be made with single molecule waste gases and model wastewaters, with catabolic species and/or model microbial communities. Thus studies such as Keyser et al. (2006), particularly where inert support matrices are used, afford opportunities for essential baseline determinations for subsequent knowledge transfer to the more complex biochar support.

Magnetic biochar, as an innovative extension to its nonmagnetized equivalent, and its applicability in physicochemical (adsorption/absorption) and biological remediation strategies, should be tested initially by fundamental studies with culture-based methods. These could entail monocultures, model populations, or catabolic microbial associations that have been enriched and isolated from contaminated ecosystems (see also Chapters 5, 9, and 10). Although the data are not yet published, we adopted both these strategies where:

1. Typical laboratory bacterial strains, *Alcaligenes faecalis, Bacillus subtilis, Bacillus cereus, Pseudomonas denitrificans, Staphylococcus epidermidis, Pseudomonas aeruginosa,* and *Chromobacterium violaceum* were used as a model bacterial community to inoculate aqueous R2A broth microcosms in the presence and absence of 450°C and 600°C willow (*Salix* sp.)-derived magnetic and nonmagnetic biochars (1% w/v). Generally, shifts in community richness, strain relative abundance, and Shannon–Wiener and Simpson community diversities were recorded with time and biochar addition. For example, the Shannon–Wiener index showed statistically significant increases in diversity, on day 4 of 7, because of biochar supplementation but nonstatistically significant differences between the magnetic and nonmagnetic moieties. Also, although day 2 showed decreased diversities for both temperature biochars, the magnetic type was not necessarily inhibitive to strain growth although it effected different responses to different strains.

2. Sediments from a closed gas works site with a protracted history of contamination were used to enrich and isolate a naphthalene-degrading microbial association. This was inoculated into aqueous Bushnell Haas broth-based microcosms with naphthalene in the presence and absence of the 450°C willow biochar (5% w/v). Subsequently, the microcosm communities were then inoculated onto Bushnell Haas agar plates augmented individually with $75.3\,mg/L^{-1}$ naphthalene, 2,3-dihydroxynaphthalene, 1,2-dihydroxynaphthalene, or salicylate. Overall, biochar supplementation effected shifts in the viable cell counts, diversity, and richness of the culturable bacterial community members that could tolerate naphthalene as the primary contaminant and its three metabolites.

In summary, CRA-, SE-, R2A agar-, naphthalene-, and naphthalene intermediates-based cultivation demonstrated the potential relevance of different media, including selective agars, in conjunction with DGGE profiling to characterize indigenous and potentially functional culturable communities. With the requisite improvements (eg, Tourlomousis et al., 2010), these approaches could ultimately produce other tools that are accessible/affordable to researchers and practitioners with most levels of expertise and in many laboratories worldwide to monitor ecosystem function following biochar supplementation. The realization of this potential will be dependent, however, on taking full cognizance of the need for comparison between commercial, quality controlled, and standardized references within and between studies/sites, and site-specific media. This could be achieved with the use of commercial nutrients and R2A broths/ agars parallel to specific ecological extract broths/agars.

Also, as a well-recognized approach in microbial ecology, where complementation of culture-based and molecular analyses maximizes the potentials and strengths of each, while using one to address the shortcomings of the other, it is directly relevant in the biochar context. It should therefore be applied and matched also with the (acknowledged) benefits for investment into culturing hitherto uncultured or uncharacterized strains.

Acknowledgments

Technical support from Karen Bradley, William Ling and Madelynne Hopper of the Life Sciences Laboratory, Teesside University, is acknowledged gratefully.

References

Aciego Pietri, J.C., Brookes, P.C., 2009. Substrate inputs and pH as factors controlling microbial biomass, activity and community structure in an arable soil. Soil Biol. Biochem. 41, 1396–1405.

Ager, D., Evans, S., Li, H., Lilley, A.K., van der Gast, C.J., 2010. Anthropogenic disturbance affects the structure of bacterial communities. Environ. Microbiol. 12, 670–678.

Akbari, A.G., Arab, M.S., Alikhani, H.A., Allahdadi, L., Arzanesh, M.H., 2007. Isolation and selection of indigenous *Azospirillum* spp. and the IAA of superior strains effects on wheat roots. World J. Agric. Sci. 3, 523–529.

Anderson, C.R., Condron, L.M., Clough, T.J., Fiers, M., Stewart, A., Hill, R.A., Sherlock, R.R., 2011. Biochar induced soil microbial community change: implications for biogeochemical cycling of carbon, nitrogen and phosphorus. Pedobiologia 54, 309–320.

Anderson, I.C., Cairney, J.W.G., 2004. Diversity and ecology of soil fungal communities: increased understanding through the application of molecular techniques. Environ. Microbiol. 6 (8), 769–779.

Barros, N., Gomez-Orellana, I., Feijóo, S., Balsa, R., 1995. The effect of soil moisture on soil microbial activity studied by microcalorimetry. Thermochim. Acta 249, 161–168.

Basso, A.S., Miguez, F.E., Laird, D.A., Horton, R., Westgate, M., 2013. Assessing potential of biochar for increasing water-holding capacity of sandy soils. Glob. Change Biol. Bioenergy 5, 132–143.

Battistoni, F., Platero, R., Duran, R., Cerveñansky, C., Battistoni, J., Arias, A., Fabiano, E., 2002a. Identification of an iron-regulated, hemin-binding outer membrane protein in *Sinorhizobium meliloti*. Appl. Environ. Microb. 68 (12), 5877–5881.

Battistoni, F., Platero, R., Noya, F., Arias, A., Fabiano, E., 2002b. Intracellular Fe content influences nodulation competitiveness of *Sinirhizobium meliloti* strains as inocula of alfalfa. Soil Biol. Biochem. 34 (5), 593–597.

Bergmann, R.C., Wright, D.A., Ralebitso-Senior, T.K., 2015. Ecological media reveal community structure shifts in a municipal wastewater treatment train. J. Bioremediat. Biodegr. 6, 3. http://dx.doi.org/10.4172/2155-6199.1000293.

Bonanomi, G., Ippolito, F., Scala, F., 2015. A "black" future for plant pathology? Biochar as a new soil amendment for controlling plant diseases. J. Plant Pathol. 28 (2), 223–234.

Borglin, S., Joyner, D., DeAngelis, K.M., Khudyakov, J., D'haedeleer, P., Joachimiak, M.P., Hazen, T., 2012. Application of phenotypic microarrays to environmental microbiology. Curr. Opin. Biotechnol. 23, 41–48.

Bossio, D.A., Girvan, M.S., Verchot, L., Bullimore, J., Borelli, T., Albrecht, A., Scow, K.M., Ball, A.S., Pretty, J.N., Osborn, A.M., 2005. Soil microbial community response to land use change in an agricultural landscape of Western Kenya. Microb. Ecol. 49, 50–62.

Bougnom, B.P., Knapp, B.A., Elhottová, D., Koubová, A., Etoa, F.X., Insam, H., 2010. Designer compost with biomass ashes for ameliorating acid tropical soils: effects on the soil microbiota. Appl. Soil Ecol. 45, 319–324.

Cenciani, K., Mazzetto, A.M., Lammel, D.R., Fracetto, F.J., Fracetto, G.G.M., Frazao, L., Cerri, C., Feigl, B., 2010. Genetic and functional diversities of microbial communities in Amazonian soils under different land uses and cultivation. In: Matovic, D. (Ed.), Biomass - Detection, Production and Usage. InTech Open. ISBN: 978-953-307-492-4.

Chen, B.L., Yuan, M.X., Qian, L.B., 2012. Enhanced bioremediation of PAH-contaminated soil by immobilized bacteria with plant residue and biochar as carriers. J. Soil Sediments 12, 1350–1359.

Chen, M., Zhu, Y., Su, Y., Chen, B., Fu, B., Marschner, P., 2007. Effects of soil moisture and plant interactions on the soil microbial community structure. Eur. J. Soil Biol. 43, 31–38.

Creamer, R.E., Hannula, S.E., Van Leeuwen, J.P., Stone, D., Rutgers, M., Schmelz, R.M., de Ruiter, P.C., Bohse Hendriksen, N., Bolger, T., Bouffaud, M.L., Buee, M., Carvalho, F., Costa, D., Dirilgen, T., Francisco, R., Griffiths, B.S., Griffiths, R., Martin, F., Martins da Silva, P., Mendes, S., Morais, P.V., Pereira, C., Philippot, L., Plassart, P., Redecker, D., Römbke, J., Sousa, J.P., Wouterse, M., Lemanceau, P., 2016. Ecological network analysis reveals the inter-connection between soil biodiversity and ecosystem function as affected by land use across Europe. Appl. Soil Ecol. 97, 112–124.

Dale, J.W., Park, S.F., 2010. Molecular Genetics of Bacteria, fifth ed. John Wiley and Sons Ltd, UK.

Dale, J.W., von Schantz, M., 2008. From Genes to Genomes: Concepts and Applications of DNA Technology, second ed. John Wiley and Sons Ltd, UK.

Daskaleros, P.A., Payne, S.M., 1987. Congo red binding phenotype is associated with hemin binding and increases infectivity of *Shigella flexneri* in the HeLa cell model. Infect. Immun. 55 (6), 1393–1398.

Douds Jr., D.D., Lee, J., Uknalis, J., Boateng, A.A., Ziegler-Ulsh, C., 2014. Pelletized biochar as a carrier for AM fungi in the on-farm system of inoculum production in compost and vermiculture mixtures. Compost Sci. Util. 22 (4), 253–262.

Durenkamp, M., Luo, Y., Brookes, P.C., 2010. Impact of black carbon addition to soil on the determination of soil microbial biomass by fumigation. Soil Biol. Biochem. 42, 2026–2029.

Edenborn, S.L., Sexstone, A.J., 2007. DGGE fingerprinting of culturable soil bacterial communities complements culture-independent analyses. Soil Biol. Biochem. 39, 1570–1579.

Elad, Y., David, D.R., Harel, Y.M., Borenshtein, M., Kalifa, H.B., Silber, A., Graber, E.R., 2010. Induction of systemic resistance in plants by biochar, a soil-applied carbon sequestration agent. Phytopathology 100, 913–921.

Elmer, W.H., Pignatello, J.J., 2011. Effect of biochar amendments on mycorrhizal associations and Fusarium crown and root rot of asparagus in replant soils. Plant Dis. 95, 960–966.

Ennis, C.J., Evans, G.A., Islam, M., Ralebitso-Senior, T.K., Senior, E., 2012. Biochar: carbon sequestration, land remediation and impacts on soil microbiology. Crit. Rev. Environ. Sci. Technol. 42 (22), 2311–2364.

Felske, A., Akkermans, A.D.L., De Vos, W.M., 1998. Quantification of 16S rRNAs in complex bacterial multiple competitive reverse transcription – PCR in temperature gradient gel electrophoresis fingerprints. Appl. Environ. Microbiol. 11, 4581–4587.

Feng, Y., Xu, Y., Yu, Y., Xie, Z., Lin, X., 2012. Mechanisms of biochar decreasing methane emission from Chinese paddy soils. Soil Biol. Biochem. 46, 80–88.

Garland, J.L., 1997. Analysis and interpretation of community-level physiological profiles in microbial ecology. FEMS Microbiol. Ecol. 24, 289–300.

Garland, J.L., Mills, A.L., 1991. Classification and characterization of heterotrophic microbial communities on the basis of patterns of community-level sole-carbon-source utilization. Appl. Environ. Microbiol. 57 (8), 2351–2359.

Graber, E.R., Frenkel, O., Jaiswal, A.K., Elad, Y., 2014. How may biochar influence the severity of diseases caused by soilborne pathogens? Carbon Manag. 5 (2), 169–183.

Graber, E.R., Harel, Y.M., Kolton, M., Cytryn, E., Silber, A., David, D.R., Tsechansky, L., Borenshtein, M., Elad, Y., 2010. Biochar impact on development and productivity of pepper and tomato grown in fertigated soilless media. Plant Soil 337, 481–496.

Green, S.J., Leigh, M.B., Neufeld, J.D., 2009. Denaturing gradient gel electrophoresis (DGGE) for microbial community analysis. In: Timmis, K.N. (Ed.), Microbiology of Hydrocarbons, Oils, Lipids, and Derived Compounds. Springer, pp. 4137–4158.

Greetham, D., 2014. Phenotype microarray technology and its application in industrial biotechnology. Biotechnol. Lett. 36 (6), 1153–1160.

Grossman, J.M., O'Neil, B.E., Tsai, A.M., Liang, B., Neves, E., Lehmann, J., Thies, J.E., 2010. Amazonian anthrosols support similar microbial communities that differ distinctly from those extant in adjacent, unmodified soils of the same mineralogy. Microb. Ecol. 60, 192–205.

Guimarães, B.C.M., Arends, J.B.A., Van Der Ha, D., Van De Wiele, T., 2010. Microbial services and their management: recent progress in soil bioremediation technology. Appl. Soil Ecol. 46, 157–167.

Hale, L., Crowley, D., 2015. DNA extraction methodology for biochar-amended sand and clay. Biol. Fertil. Soil 51, 733–738.

Hale, L., Luth, M., Crowley, D., 2015. Biochar characteristics relate to its utility as an alternative soil inoculum carrier to peat and vermiculite. Soil Biol. Biochem. 81, 228–235.

Hale, L., Luth, M., Kenny, R., Crowley, D., 2014. Evaluation of pinewood biochar as a carrier of bacterial strain *Enterocbacter cloacae* UW5 for soil inoculation. Appl. Soil Ecol. 84, 192–199.

Hamaki, T., Suzuki, M., Fudou, R., Jojima, Y., Kajiura, T., Tabuchi, A., Sen, K., Shibai, H., 2005. Isolation of novel bacteria and actinomycetes using soil-extract agar medium. J. Biosci. Bioeng. 99, 485–492.

Hammer, E.C., Balogh-Brunstad, Z., Jakobsen, I., Olsson, P.A., Stipp, S.L.S., Rillig, M.C., 2014. A mycorrhizal fungus grows on biochar and captures phosphorus from its surfaces. Soil Biol. Biochem. 77, 252–260.

Handelsman, J., 2004. Metagenomics: application of genomics to uncultured microorganisms. Microbiol. Mol. Biol. Rev. 4, 669–685.

Harel, Y.M., Elad, Y., Rav-David, D., Borenstein, M., Schulchani, R., Lew, B., Graber, E.R., 2012. Biochar mediates systemic response of strawberry to foliar fungal pathogens. Plant Soil 357 (1), 245–257.

He, L., Yang, H., Zhong, Z., Gong, P., Liu, Y., Lü, H., Yang, S., 2014. PCR-DGGE analysis of soil bacterium community diversity in farmland influenced by biochar. Shengtai Xuebao/Acta Ecol. Sin. 34 (15), 4288–4294.

Herrera, A., Cockell, C.S., 2007. Exploring microbial diversity in volcanic environments: a review of methods in DNA extraction. J. Microbiol. Methods 70, 1–12.

Hill, G.T., Mitkowski, N.A., Aldrich-Wolfe, L., Emele, L.R., Jurkonie, D.D., Ficke, A., Maldonado-Ramirez, S., Lynch, S.T., Nelson, E.B., 2000. Methods for assessing the composition and diversity of soil microbial communities. Appl. Soil Ecol. 15, 25–36.

Hopkins, D.W., Wiltshire, P.E.J., Turner, B.D., 2000. Microbial characteristics of soils from graves: an investigation at the interface of soil microbiology and forensic science. Appl. Soil Ecol. 14, 283–288.

Jaafar, N.M., Clode, P.L., Abbott, L.K., 2015. Soil microbial responses to biochars varying in particle size, surface and pore properties. Pedosphere 25, 770–780.

Jones, D.L., Rousk, J., Edwards-Jones, G., DeLuca, T.H., Murphy, D.V., 2012. Biochar-mediated changes in soil quality and plant growth in a three year field trial. Soil Biol. Biochem. 45, 113–124.

Jousset, A., Lara, E., Nikolausz, M., Harms, H., Chatzinotas, A., 2010. Application of the denaturing gradient gel electrophoresis (DGGE) technique as an efficient diagnostic tool for ciliate communities in soil. Sci. Total Environ. 408, 1221–1225.

Kaštovská, E., Šantrůčková, H., 2011. Comparison of uptake of different N forms by soil microorganisms and two wet-grassland plants: a pot study. Soil Biol. Biochem. 43, 1285–1291.

Kappler, A., Wuestner, M.L., Ruecker, A., Harter, J., Halama, M., Behrens, S., 2014. Biochar as an electron shuttle between bacteria and Fe(III) minerals. Environ. Sci. Technol. Lett. 1 (8), 339–344.

Karhu, K., Mattila, T., Bergström, I., Regina, K., 2011. Biochar addition to agricultural soil increased CH$_4$ uptake and water holding capacity – results from a short-term pilot field study. Agric. Ecosyst. Environ. 140, 309–313.

Keyser, M., Witthuhn, R.C., Lamprecht, C., Coetzee, M.P.A., Britz, T.J., 2006. PCR-based DGGE fingerprinting and identification of methanogens detected in three different types of UASB granules. Syst. Appl. Microbiol. 29, 77–84.

Khodadad, C.L.M., Zimmerman, A.R., Green, S.J., Uthandi, S., Foster, J.S., 2011. Taxa-specific changes in soil microbial community composition induced by pyrogenic carbon amendments. Soil Biol. Biochem. 43, 385–392.

Kim, J.S., Sparovek, G., Longo, R.M., De Melo, W.J., Crowley, D., 2007. Bacterial diversity of terra preta and pristine forest soil from the Western Amazon. Soil Biol. Biochem. 39, 684–690.

Kirk, J.L., Beaudette, L.A., Hart, M., Moutoglis, P., Klironomos, J.N., Lee, H., Trevors, J.T., 2004. Methods of studying soil microbial diversity. J. Microbiol. Met. 58, 169–188.

Kolton, M., Harel, Y.M., Pasternak, Z., Graber, E.R., Elad, Y., Cytryn, E., 2011. Impact of biochar application to soil on the root-associated bacterial community structure of fully developed greenhouse pepper plants. Appl. Environ. Microbiol. 14, 4924–4930.

Kowalchuk, G.A., Gerards, S., Woldendrop, J.W., 1997. Detection and characterization of fungal infection of *Ammophila arenaria* (marram grass) roots by denaturing gradient gel electrophoresis of specifically amplified 18S rDNA. Appl. Environ. Microbiol. 63 (10), 3858–3865.

Laird, D.A., Fleming, P., Davis, D.D., Horton, R., Wang, B., Karlen, D.L., 2010. Impact of biochar amendments on the quality of a typical Midwestern agricultural soil. Geoderma 158, 443–449.

Lehmann, J., Rillig, M., Thies, J., Masiello, C.A., Hockaday, W.C., Crowley, D., 2011. Biochar effects on soil biota - a review. Soil Biol. Biochem. 43, 1812–1836.

Leite, D.C.A., Balieiro, F.C., Pires, C.A., Madari, B.E., Rosado, A.S., Coutinho, H.L.C., Peixoto, R.S., 2014. Comparison of DNA extraction protocols for microbial communities from soils treated with biochar. Braz. J. Microbiol. 45 (1), 175–183.

Liebeke, M., Brözel, V.S., Hecker, M., Lalk, M., 2009. Chemical characterization of soil extract as growth media for the ecophysiological study of bacteria. Appl. Microbiol. Biotechnol. 83 (1), 161–173.

Liu, W., Wang, S., Lin, P., Sun, H., Hou, J., Zuo, Q., Huo, R., 2015. Response of CaCl$_2$-extractable heavy metals, polychlorinated biphenyls, and microbial communities to biochar amendment in naturally contaminated soils. J. Soils Sediment. http://dx.doi.org/10.1007/s11368-015-1218-z.

Lu, H., Lashari, M.S., Liu, X., Ji, H., Li, L., Zheng, J., Kibue, G.W., Joseph, S., Pan, G., 2015. Changes in soil microbial community structure and enzyme activity with amendment of biochar-manure compost and pyroligneous solution in a saline soil from Central China. Eur. J. Soil Biol. 70, 67–76.

Luo, F., Devine, C.E., Edwards, E.A., 2015. Cultivating microbial dark matter in benzene-degrading methanogenic consortia. Environ. Microbiol. http://dx.doi.org/10.1111/1462-2920.13121.

Maarit Niemi, R., Heiskanen, I., Wallenius, K., Lindström, K., 2001. Extraction and purification of DNA in rhizosphere soil samples for PCR-DGGE analysis of bacterial consortia. J. Microbiol. Met. 45, 155–165.

Malik, S., Beer, M., Megharaj, M., Naidu, R., 2008. The use of molecular techniques to characterize the microbial communities in contaminated soil and water. Environ. Int. 34, 265–276.

Manefield, M., Whiteley, A.S., Griffiths, R.I., Bailey, M.J., 2002. RNA stable isotope probing, a novel means of linking microbial community function to phylogeny. Appl. Environ. Microbiol. 11, 5367–5373.

Marshall, M.N., Cocolin, L., Mills, D.A., VanderGheynst, J.S., 2003. Evaluation of PCR primers for denaturing gradient gel electrophoresis analysis of fungal communities in compost. J. Appl. Microbiol. 95, 934–948.

Masiello, C.A., Chen, Y., Gao, X., Liu, S., Cheng, H.-Y., Bennett, M.R., Rudgers, J.A., Wagner, D.S., Zygourakis, K., Silberg, J.J., 2013. Biochar and microbial signalling: production conditions determine effects on microbial communication. Environ. Sci. Technol. 47, 11496–11503.

Massa, S., Caruso, M., Trovatelli, F., Tosques, M., 1998. Comparison of plate count agar and R2A medium for enumeration of heterotrophic bacteria in natural mineral water. World J. Microbiol. Biotechnol. 14 (5), 727–730.

Maurathan, N., Orr, C.H., Ralebitso-Senior, T.K., 2015. Biochar adsorption properties and the impact on naphthalene as a model environmental contaminant and microbial community dynamics – a triangular perspective. In: Borja, M.E.L. (Ed.), Soil Management: Technical Systems, Practices and Ecological Implications. Nova Publishers, New York, pp. 83–122.

Mia, S., van Groenigen, J.W., van de Voorde, T.F.J., Oram, N.J., Bezemer, T.M., Mommer, L., Jeffery, S., 2014. Biochar application rate affects biological nitrogen fixation in red clover conditional on potassium availability. Agric. Ecosyst. Environ. 191, 83–91.

Miletto, M., Bodelier, P.L.E., Laanbroek, H.J., 2007. Improved PCR-DGGE for high resolution diversity screening of complex sulfate-reducing prokaryotic communities in soils and sediments. J. Microbiol. Met. 70, 103–111.

Mukherjee, A., Zimmerman, A.R., Harris, W., 2011. Surface chemistry variations among a series of laboratory-produced biochars. Geoderma 163, 247–255.

Muyzer, G., 1999. DGGE/TGGE a method for identifying genes from natural ecosystems. Curr. Opin. Microbiol. 2, 317–322.

Muyzer, G., De Waal, E., Uiterlinden, A.G., 1993. Profiling of complex microbial populations by denaturing gradient gel electrophoresis analysis of polymerase chain reaction-amplified genes coding for 16S rRNA. Appl. Environ. Microbiol. 3, 695–700.

Navarrete, A.A., Taketani, R.G., Mendes, L.W., de Souza Cannavan, F., de Souza Moreira, F.M., Tsai, S.M., 2011. Land-use systems affect archaeal community structure and functional diversity in western Amazon soils. R. Bras. Ci. Solo 35, 1527–1540.

Novak, J.M., Lima, I., Xing, B., Gaskin, J.W., Steiner, C., Das, K.C., Ahmedna, M., Rehrah, D., Watts, D.W., Busscher, W.J., Schomberg, H., 2009. Characterization of designer biochar produced at different temperatures and their effects on a loamy sand. Ann. Environ. Sci. 3, 195–206.

Nzanza, B., Marais, D., Soundy, P., 2011. Effect of arbuscular mycorrhizal fungal inoculation and biochar amendment on growth and yield of tomato. Int. J. Agric. Biol. 14 (6), 965–969.

Okubo, A., Sugiyama, S., 2009. Comparison of molecular fingerprinting methods for analysis of soil microbial community structure. Ecol. Res. 24, 1399–1405.

Olakanye, A.O., Thompson, T.J.U., Ralebitso-Senior, T.K., 2014. Changes to soil bacterial profiles as a result of Sus scrofa domesticus decomposition. Forensic Sci. Intern. 245, 101–106.

Oram, N.J., van de Voorde, T.F.J., Ouwehand, G.-J., Bezemer, T.M., Mommer, L., Jeffery, S., Groenigen, J.W.V., 2014. Soil amendment with biochar increases the competitive ability of legumes via increased potassium availability. Agric. Ecosyst. Environ. 191, 92–98.

Orr, C.H., Ralebitso-Senior, T.K., January 18, 2014. Tracking N-cycling genes in biochar-supplemented ecosystems: a perspective. OA Microbiol. 2 (1), 1.

du Plessis, C.A., Senior, E., Hughes, J.C., 1998a. Growth kinetics of microbial colonisation of porous media. S. Afr. J. Sci. 94, 33–38.

du Plessis, C.A., Senior, E., Hughes, J.C., 1998b. Effects of microbial colonization of porous media on resistance to potentially inhibitory chemical challenges. S. Afr. J. Sci. 94, 487–492.

Quilliam, R.S., Deluca, T.H., Jones, D.L., 2013a. Biochar application reduces nodulation but increases nitrogenase activity in clover. Plant Soil 366 (1–2), 83–92.

Quilliam, R.S., Glanville, H.C., Wade, S.C., Jones, D.L., 2013b. Life in the 'charosphere' – does biochar in agricultural soil provide a significant habitat for microorganisms? Soil Biol. Biochem. 65, 287–293.

de la Rosa, J.M., Knicker, H., 2011. Bioavailability of N released from N-rich pyrogenic organic matter: an incubation study. Soil Biol. Biochem. 43 (12), 2368–2373.

de la Rosa, J.M., Paneque, M., Hilber, I., Blum, F., Knicker, H., Bucheli, T., 2015. Assessment of polycyclic aromatic hydrocarbons in biochar and biochar-amended agricultural soil from Southern Spain. J. Soil Sediment. 1–9. http://dx.doi.org/10.1007/s11368-015-1250-z.

Ranjard, L., Poly, F., Nazaret, S., 2000. Monitoring complex bacterial communities using culture-independent molecular techniques: application to soil environment. Res. Microbiol. 151 (3), 167–177.

Reed, R., Holmes, D., Weyers, J., Jones, A., 2007. Practical Skills in Biomolecular Science. Pearson Education Limited, UK.

Robe, P., Nalin, R., Capellano, C., Vogel, T.M., Simonet, P., 2003. Extraction of DNA from soil. Eur. J. Soil Biol. 39, 183–190.

Robertson, C.E., Harris, J.K., Spear, J.K., Pace, N.R., 2005. Phylogenetic diversity and ecology of environmental archaea. Curr. Opin. Microbiol. 8 (6), 638–642.

Roelfsema, J.H., Peters, D.J.M., 2005. Denaturing gradient gel electrophoresis (DGGE). In: Medical Biomethods Handbook, sixth ed. Humana Press. Chapter 8.

Sahni, S.K., Jaiswal, P.K., Kaushik, P., Thakur, I.S., 2011. Characterization of alkalotolerant bacterial community by 16S rDNA based denaturing gradient gel electrophoresis method for degradation of dibenzofuran in soil. Int. Biodeter. Biodegr. 65 (7), 1073–1080.

Smalley, J.W., Birss, A.J., McKee, A.S., Marsh, P.D., 1995. Congo red binding by *Porphyromonas gingivalis* is mediated by a 66 kDa outer-membrane protein. Microbiology 141 (Pt 1), 205–211.

Song, Y.-J., Zhang, X.-L., Gong, J., 2014. Effects of biochar amendment on the abundance and community structure of nitrogen-fixing microbes in a coastal alkaline soil. Chin. J. Ecol. 33 (8), 2168–2175.

Steinbeiss, S., Gleixner, G., Antonietti, M., 2009. Effect of biochar amendment on soil carbon balance and soil microbial activity. Soil Biol. Biochem. 41, 1301–1310.

Stewart, K.J., Siciliano, S.D., 2015. Potential contribution of native herbs and biological soil crusts to restoration of the biogeochemical nitrogen cycle in mining impacted sites in Northern Canada. Ecol. Restor. 33 (1), 30–42.

Stone, D., Blomkvist, P., Hendriksen, N.B., Bonkowski, M., Jørgensen, H.B., Carvalho, F., Dunbar, M.B., Gardi, C., Geisen, S., Griffiths, R., Hug, A.S., Jensen, J., Laudon, H., Mendes, S., Morais, P.V., Orgiazzi, A., Plassart, P., Römbke, J., Rutgers, M., Schmelz, R.M., Sousa, J.P., Steenbergen, E., Suhadolc, M., Winding, A., Zupan, M., Lemanceau, P., Creamer, R.E., 2016. A method of establishing a transect for biodiversity and ecosystem function monitoring across Europe. Appl. Soil Ecol. 97, 3–11.

Sun, D., Jun, M., Zhang, W., Guan, X., Huang, Y., Lan, Y., Gao, J., Chen, W., 2012. Implication of temporal dynamics of microbial abundance and nutrients to soil fertility under biochar application - field experiments conducted in a brown soil cultivated with soybean, north China. Adv. Mat. Res. 518–523, 384–394.

Taghizadeh-Toosi, A., Clough, T., Sherlock, R., Condron, L., 2012. Biochar adsorbed ammonia is bioavailable. Plant Soil 350 (1–2), 57–69.

Taketani, R.G., Tsai, S.M., 2010. The influence of different land uses on the structure of archaea communities in Amazonian anthrosols based on 16S rRNA and *amoA* genes. Microb. Ecol. 59, 734–743.

Tourlomousis, P., Kemsley, E.K., Ridgway, K.P., Toscano, M.J., Humphrey, T.J., Narbad, A., 2010. PCR-denaturing gradient gel electrophoresis of complex microbial communities: a two-step approach to address the effect of gel-to-gel variation and allow valid comparisons across a large dataset. Microb. Ecol. 59, 776–786.

Waite, I.S., O'Donnell, A.G., Harrison, A., Davies, J.T., Colvan, S.R., Ekschmitt, K., Dogan, H., Wolters, V., Bongers, T., Bongers, M., Bakonyi, G., Nagy, P., Papatheodorou, E.M., Stamou, G.P., Boström, S., 2003. Design and evaluation of nematode 18S rDNA primers for PCR and denaturing gradient gel electrophoresis (DGGE) of soil community DNA. Soil Biol. Biochem. 35, 1165–1173.

Wang, Q., Dai, J., Wu, D., Yu, Y., Shen, T., Wang, R., 2010. Statistical analysis of data from BIOLOG method in the study of microbial ecology. Shengtai Xuebao/Acta Ecol. Sin. 30 (3), 817–823.

Wang, Z., Zong, H., Zheng, H., Liu, G., Chen, L., Xing, B., 2015. Reduced nitrification and abundance of ammonia-oxidizing bacteria in acidic soil amended with biochar. Chemosphere 138, 576–583.

Warnock, D.D., Lehmann, J., Kuyper, T.W., Rillig, M.C., 2007. Mycorrhizal response to biochar in soil – concepts and mechanisms. Plant Soil 300, 9–20.

Wessén, E., Hallin, S., 2011. Abundance of archaeal and bacterial ammonia oxidizers – possible bioindicators for soil monitoring. Ecol. Indic. 11, 1696–1698.

Xue, D., Yao, H., Ge, D., Huang, C., 2008. Soil microbial community structure in diverse land use systems: a comparative study using Biolog, DGGE, and PLFA analyses. Pedosphere 18, 653–663.

Zwart, D.C., Kim, S.-H., 2012. Biochar amendment increases resistance to stem lesions caused by *Phytophthora* spp. in tree seedlings. HortScience 47 (12), 1736–1740.

CHAPTER

5

Next-Generation Sequencing to Elucidate Biochar-Effected Microbial Community Dynamics

F.S. Cannavan[1], F.M. Nakamura[1], M.G. Germano[1,2], L.F. de Souza[1], S.M. Tsai[1]

[1]University of São Paulo, Piracicaba, São Paulo, Brazil; [2]Brazilian Agricultural Research Corporation, Embrapa Soybean, Londrina, Paraná, Brazil

OUTLINE

Biochar Application
http://dx.doi.org/10.1016/B978-0-12-803433-0.00005-9

109

INTRODUCTION

Biochar is the result of the pyrolysis of recalcitrant carbon by means of natural or anthropogenic combustion, which brings some benefits to soil. Biochar fragments are the most recalcitrant structures in soil organic matter (SOM) as they are resistant to thermal, chemical, and photooxidation (Skjemstad et al., 1996), resulting in highly aromatic humic acids. Also biochar can absorb soluble organic compounds, retain water, and act as host for soil microorganisms (Benites et al., 2005), being an important factor for soil fertility. The fragments of biochar differ in size, morphology, density, and surface area to be oxidized (Poirier et al., 2000). In addition, biochar can be affected by the environment, which can determine its size and drive the dynamics of its fragments in soil.

Soil microorganisms are fundamental to fertility as observed by O'Neill et al. (2009), and are essential to ecosystems function, being directly involved in the maintenance of biogeochemical cycles (see also Chapter 7). Studies of soil microorganisms are frequently based on molecular methods (Fig. 5.1). These tools allow the study of composition, structure, and function of microbial communities, as well as how they respond to environmental change (Rosado et al., 1997; Tiedje et al., 1999; Handelsman and Smalla, 2003; Kirk et al., 2004; Schütte et al., 2008). Such studies require manipulation of genetic material using methods based on DNA amplification and/or sequencing. Otherwise, isolation methods can provide an assessment of the organism's phenotype and nutritional requirements. Nevertheless, such selective methods do not reveal the extent of the natural diversity, with cultivated microorganisms constituting about 1–10% of typical diversity (Amann et al., 1995).

Because of their versatile metabolism, bacterial species present phenotypic variation depending on the carbon source used in the culture media. Consequently they may present a diverse morphological status, which can confound identification when solely based on their phenotypical characteristics. Thus molecular analyses based on conserved genes are needed to complement the data. Nowadays, both methods are needed to reach a satisfactory approach (see also Chapter 4).

Ribosomal genes (rRNA) are universally distributed within different groups of living organisms, such as prokaryotes and eukaryotes, being the molecules that represent the major degree of evolutionary variation proportional to species classification (conserved genes).

Prokaryote ribosomes consist of ribosomal RNA and proteins. Prokaryote rRNA genes (Bacteria and Archaea domains) are divided into subunits 40S (divided further into 23S and 5S genes with 31 proteins) and 30S, where the 16S rRNA gene is located. This fragment is approximately 1500 base pairs and is normally used as a target to

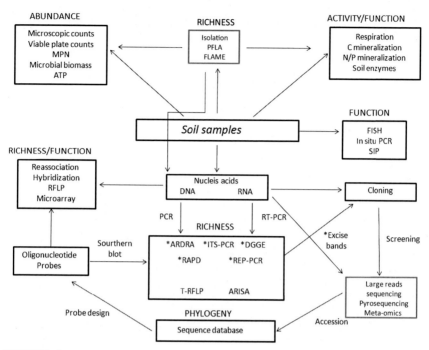

FIGURE 5.1 Techniques used on microbial ecology. *PLFA*, Phospholipid fatty acids; *FAME*, fatty acid methyl ester; *FISH*, fluorescence in situ hybridization; *SIP*, stable isotope probing; *RFLP*, restriction fragment length polymorphism; *PCR*, polymerase chain reaction; *RT-PCR*, reverse transcriptase polymerase chain reaction; *ARDRA*, amplified ribosomal DNA restriction analysis; *ITS-PCR*, internal transcribed spacer polymerase chain reaction; *DGGE*, denaturing gradient gel electrophoresis; *RAPD*, random amplified polymorphic DNA; *T-RFLP*, terminal restriction fragment length polymorphism; *REP-PCR*, repetitive element palindromic polymerase chain reaction. *Adapted from Theron J., Cloete T.E., 2000. Molecular techniques for determining microbial diversity and community structure in natural environments. Crit. Rev. Microbiol., London, 26(1), 37–57 and Navarrete A.A., Cannavan F.S., Taketani R.G., Tsai S.M., 2010. A molecular survey of the diversity of microbial communities in different Amazonian agricultural model systems. Diversity, Basel, 2, 787–809.*

measure phylogenetic diversity (Amann et al., 1995; Weisburg et al., 1991) and qualitative and quantitative detection of specific populations (Head et al., 1998).

Pyrosequencing constitutes one of the molecular approaches that are commonly employed to evaluate microbial diversity (Ronaghi et al., 1996) (Fig. 5.2). In this approach, four enzymes are utilized in the synthesis of a complementary strand. The technique is based on the detection of a pyrophosphate molecule (PPi) produced during polymerization of DNA, as this PPi is converted into ATP and leads, consequently, to emission of light, which can be detected (Ronaghi, 2001) when nucleotides are added

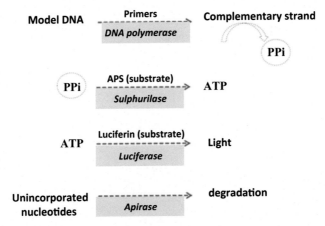

FIGURE 5.2   Representative scheme of enzymatic reactions on pyrosequencing process. *ATP*, Adenosine triphosphate; *PPi*, pyrophosphate molecule. *Adapted from Germano M.G., Cannavan F.S., Mendes L.W., Lima, A.B., Teixeira W.G., Pellizari V.H., Tsai S.M., 2012. Functional diversity of bacterial genes associated with aromatic hydrocarbon degradation in anthropogenic dark earth of Amazonia. Pesqui. Agropecuária Bras., Brasília, DF, 47, 654–664.*

to the sequence. The method was denominated "sequencing by synthesis," as the target sequence is determined while it is being amplified; the pyrosequencing methods utilize primers that match a barcode-type sequence on the gene target, allowing the analysis of many samples simultaneously, following processing with bioinformatics tools (Parameswaran et al., 2007).

ANTHROPOGENIC BIOCHAR

The term black carbon relates to all residues with a substantial amount of carbon derived from combustion. Likewise, vegetal char is defined as the result of burn and charcoal is the char used for firewood. Thus the biochar commonly referred to in this chapter is the char produced by partial pyrolysis and related to environmental well-being.

Pyrolysis is the process that results in carbon residues like biochar. This is in contrast to burning where the whole organic carbon is consumed, evaporating CO_2 and dioxins, and resulting in ash composed mostly of inorganic compounds (Peacocke and Joseph, undated). Therefore demineralization takes place along with the withdrawal of H and O atoms, generating compounds with condensed chains made of 70–80% carbon. Partial pyrolysis of lignocellulose from organic matter is common in biochar produced at temperatures of around 180°C (Meira, 2002; Rohde, 2007).

The ash content in biochar relates to residues of oxidized minerals after combustion, while fixed carbon is related to the carbon that remains after combustion and represents the highest content in biochar when compared to original organic matter (Meira, 2002). Woods with high contents of aromatic matter-like extracts and lignin results in dense biochars (Oparina et al., 1971, apud Brito and Barrichello, 1977), presenting more recalcitrance in terms of physical and chemical properties (Satanoka, 1963).

BIOCHAR AND SOIL PROPERTIES

Each soil has its own pedogenesis, with different rock matrices, abiotic and biotic factors, and anthropogenic inputs shaping the environment (Schulz et al., 2013). This is also true for biochar where diverse organic materials or feedstocks can undergo differing pyrolysis conditions and produce biochars of different sizes, morphologies, densities (Poirier et al., 2000), and surface areas to be oxidized.

Biochar in soil can be formed naturally or by anthropogenic fires, as happens in Brazilian soils from Cerrado (Roscoe et al., 2001), in highlands cave environments (Benites et al., 2005), and one of the most fertile soils in the world, the Amazonian Dark Earths (ADE), locally known as *Terra Preta de Índio*. Fires on forest and agricultural systems with consequent biochar production may have a particular influence on soil fertility of these environments (Wardle et al., 1998; Kleinman et al., 1995).

Incorporation of biochar fragments into soil has an important role in carbon sequestration (Schmidt and Noack, 2000; Glaser et al., 2001; Masiello, 2004). The presence of functional groups produced by negative charges on its surface can be the reason for the high cation exchange capacity (Novotny et al., 2009) and consequently for the low pH of ADE (Glaser et al., 2003) as well as the high total acidity in SOM. These features can promote the interaction of SOM, soil aggregates, and water percolation, which may help to reduce nutrient leaching.

Liang et al. (2006) used X-ray absorption spectroscopy and observed oxidized surface structures on biochar particles, which they believed to be absorbed organic matter. In addition, water retention by the porous structure consisted of 70–80% of its volume. Thus this evidence suggested that the presence of biochar fragments could shape not only the structure, humidity, and density of soils, but also affects the soil aggregates by chemical interactions (see also Chapters 2, 6, and 11).

The oxidized surface of biochar particles associated with adsorbed organic molecules may suggest that biochar can provide a stable habitat for microorganisms (Pietikainen et al., 2000) and this feature can be beneficial to agricultural practices (Zanetti et al., 2003). Hence biochar and humus, and their interactions with clay particles, are aspects that may have positive influences on agriculture, by promoting eutrophic characteristics to soil (Brady and Weil, 2008).

As a practical example, biochar particles of 2–5 mm moistened in extract of wood were used as an organic fertilizer in Japan, with excellent results on the yields of rice (Tsuzuki et al., 1989), sugar cane (Uddin et al., 1995), sweet potato (Du et al., 1998), and melon (Du et al., 1997). The application of organic material to soil takes place by the processing of carbon (material on the soil surface, gas, and erosion and leaching) by the soil macro-, meso-, and microfauna that decompose the carbon sources and provide material (small fractions, coprolites) to the microorganisms' activity. To process and incorporate the carbon into soil, microorganisms require nitrogen for the enzymatic oxidizing or methanogenic fermentation of archaea individuals. About 60–80% of carbon generated is lost as CO_2 gas, while 15–20% of the total carbon is converted to humic and nonhumic substances. The soil microbiota then metabolize the nutrients into stable forms, providing nutrients to plants (Teixeira et al., 2009), whereas 3–8% of the carbon remains immobilized by microbial biomass.

Decomposition of organic matter by microorganisms (Fig. 5.3) depends on the nature of the material, such as sugars, hemicellulose, cellulose, lignin, fat acids, polyphenols, and biochar, which can be labile or recalcitrant. In this context, biochar constitutes the most recalcitrant structure of SOM. In the case of lignin, besides the biochemical

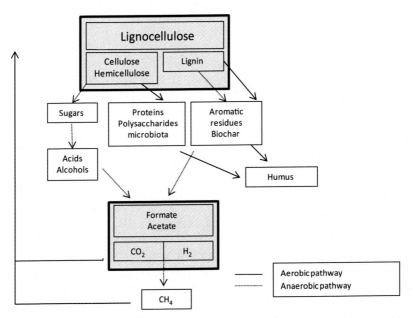

FIGURE 5.3   Carbon cycle in tropical soil of Amazonian Dark Earths (ADE). *Adapted from Tuomela et al. (2000).*

nature, the fibrous molecular structure grants an increased resistance to degradation.

THE MICROBIAL DIVERSITY IN BIOCHAR

Microorganisms are fundamental to soil fertility as observed by O'Neill et al. (2009), being essential to ecosystem function because of their decomposing activities, nutrient cycling, toxin removal, and pathogen suppressive activity (Moreira and Siqueira, 2002). Microorganisms within soil environments exhibit notable metabolic diversity and genetic adaptability, making them an important source of genetic resources (Kurtboke et al., 2004). Also individuals within the soil microbiota interact through synergism, antagonism, mutualism, frequent parasitism, and saprophytic processes. Some microbial species are widely distributed in almost all soil types, while others are restricted to specific environments. A common example is the variation in fungal communities and filamentous bacteria (Actinobacteria) whose occurrence differs with pH.

Soil microorganisms have assumed great importance for the sustainability of the tropical environment, once it is largely known that they are responsible for many biological processes, such as nutrient cycling, plant nutrition, and aggregates formation, as well as biochar fragments, which have a role in water purification and also maintenance of soil structure (Filip, 2002).

Because of its physicochemical properties, biochar can absorb soluble organic compounds, stabilize SOM (Glaser et al., 2000), increase soil pH levels and carbon stocks (Lehmann, 2007), and potentially contribute to high cation exchange capacity (Liang et al., 2006). Altogether these features may sustain soil fertility and hence the agricultural productivity and sustainability (Steiner et al., 2008). Biochar fragments may also serve as a shelter for some soil microorganisms (Pietikainen et al., 2000).

BIOCHAR AND THE AMAZONIAN DARK EARTH: A CASE STUDY

ADE is a soil of a third pedogenesis, aged from 300 BC to the 11th century, which is found in the Amazon basin as a result of organic matter accumulation and burning. It is currently classified as *Latossolo* or *Argissolo Antrópico Coeso* (EMBRAPA, 2013). The typical dark color found in ADE is caused by the high concentration of pyrogenic carbon (Glaser, 2007), with 50% of total carbon present in the more stable, not oxidized state. As a consequence, ADE soils consist of humic fractions that are chemically stable because of aromatic groups and higher polymerization rates and are therefore less susceptible to microbial attack (Teixeira et al., 2009). The structure of humic acids from ADE soils is composed mostly of biochar particles (around 50%) affected by burning, which may have transformed its structure because of continuous, long-term chemical reactions in soil. Therefore humic acids from ADE are more resistant to temperature variations, but also have more chemical reactivity compared with humic acids derived from adjacent (ADJ) soils with no anthropogenic history. Studies have demonstrated that there is a positive correlation between humic acids from ADE and some soil standards correlated to fertility and high cation exchange capacity (Novotny et al., 2009). The humic acids arising from ADE soils are more recalcitrant to decomposition and leaching, given the formation of organometallic complexes, which can promote the accumulation of organic C into ADE soil systems (Novotny et al., 2007).

On the other hand, the non-ADE soils, which consistently lack both SOM and biochar, are frequently related to losses in agriculture production in the Amazon region (Teixeira et al., 2009). Moreover, fire inputs may alter the soil structures promoting particle aggregation and leading to a decrease in nutrient leaching (Teixeira and Martins, 2003). In addition, polycyclic aromatic structures are a common result of pyrolysis (Kramer et al., 2004), and the content of biochar fragments is 70 times higher in ADE in comparison to their ADJ infertile soils.

The particular structure of biochar remains protected inside the organo–mineral complexes and it can promote aggregate formations and water retention (Golchin et al., 1997). Likewise, biochar structure can work as a stable microhabitat to microorganisms (Steiner et al., 2008; Grossman et al., 2010; Navarrete et al., 2010; Cannavan, 2011; Germano et al., 2012; Nakamura, 2014).

Residual hydrocarbons and materials adhering to the particle surface, such as bio-oils produced by pyrolysis (Maggi and Delmon, 1994; Bridgewater and Boocok, 1997; Guan, 2004), may support microbes directly (Germano et al., 2012). These hydrocarbons were classified by Schnitzer et al. (2007) into six chemical groups, including butene that acts as a germination promoter of native species (Dixon, 1998, apud Lehmann and Joseph, 2009), sesquiterpenes, which can hasten microbial growth (Akiyama and Hayashi, 2006), and smoke vinegar, known to have biocidal properties (Guan, 2004, apud Lehmann and Joseph, 2009).

Studies performed on ADE soils have shown higher diversity within the Fungi, Bacteria, and Archaea domains, in comparison to ADJ soils (Kim et al., 2007; Otsuka et al., 2008; O'Neill et al., 2009; Graber et al., 2010; Grossmann et al., 2010; Jin, 2010; Taketani and Tsai, 2010; Khodadad et al., 2011). Bacterial diversity was 25% greater in ADE soils in culture-independent (Kim et al., 2007) and culture-dependent (O'Neill et al., 2009) studies. Still, according to Khodadad et al. (2011), soils amended with oak- and grass-derived biochars showed lower diversity in comparison to ADE soils, showing that differences in relation to the type of soil itself, to the source of organic matter transformed into biochar, and to the incorporation rates of the biochar fragments into soil may directly influence microbial diversity.

Interactions of microorganisms with biochar are not well understood, and the microbial communities involved in the biochar degradation are still largely unknown. On the other hand, the high biological activity in the presence of biochar indicates that its carbon can be used by soil microbial communities as a nutrient source, creating an exchange platform for the microorganisms (Grossman et al., 2006). Even so, these interactions need to be studied further (Germano, 2011) to: elucidate the role of microbial groups directly involved in the maintenance of biogeochemical cycles; promote possible uses of vegetal residues as a soil conditioner generally, and in Brazil as an example context (Steiner et al., 2008); and determine the capacity or mechanisms for the suppression of plant diseases (Mendes et al., 2011). Therefore we present here three seminal studies developed with ADE soils in Brazil in order to illustrate how the assessment of soil microbial community dynamics by molecular techniques can contribute to the evaluation of the overall biodiversity of Amazon tropical agroecosystems.

Study I: Biochar Bacterial Diversity from Amazonian Anthrosols Revealed by T-RFLP and Pyrosequencing (Bacteria)

It is commonly known that the local environment in which soil organisms dwell is never constant. For instance, conditions such as temperature, water, and nutrient availability often fluctuate over time (Schloter et al., 2003). The extent to which external drivers affect process rates will certainly be dependent on the type of process and associated microbial players. Specifically, such fluctuations may affect the dynamics and activities of soil organisms and the interactions between them and consequently the functioning of soil ecosystems (Bascompte, 2009). This natural variation can be depicted as a sequential occurrence of maxima and minima in relevant parameters that define soil process rates. Taken together, such ups and downs determine the "natural" limits of variation in soil functioning, on the basis of which a normal operating range (NOR) can be defined. Depending on the nature and intensity of the external drivers, higher or lower limits of variation in soil processes may be expected. This description, when used over time, will allow an assessment of the dynamics in the soil status, providing a background against which out-of-range situations are compared. Consequently, Navarrete et al. (2010) applied genomic approaches to study the bacterial community of the ADE anthrosols and biochar by comparing them with neighboring nonanthropogenic ADJ soils in order to assess their NOR.

Thus the particles of biochar were carefully and manually separated under sterile conditions at 4°C, using magnification to facilitate their screening from ADE soil samples. Total microbial community DNA from biochar fragments and ADE and ADJ soils was extracted and amplified by polymerase chain reaction (PCR) using the primers for the 16S rRNA region 27F (Edwards et al., 1989) and 1492R (Woese et al., 1990). Following purification, the PCR products were digested with the enzymes *Msp*I and *Hha*I, and their restriction fragments separated by capillary electrophoresis (ABI Prism 3100 Genetic Analyser) for terminal restriction fragment length polymorphism (T-RFLP) analysis. Analyses of the biochar and soil DNA were performed using the Genome Sequencer FLX System (454 Life Sciences, Bradford, Connecticut, USA) at the Microbial Ecology Center, Michigan State University, Michigan, USA. The sequence target was the V4 region of the 16S rRNA gene (http://pyro.cme.msu.edu/).

Based specifically on the ordination of the terminal restriction fragment profiles obtained with the enzyme *Msp*I, the results by Navarrete et al. (2010) suggested that bacterial communities differed in structure between the biochar, ADE, and nonanthropogenic samples, and that these differences were linked especially to phosphorus (Fig. 5.4). When the ordination obtained using the enzyme *Hha*I was compared, the bacterial structure of the ADE soil and biochar could not be separated, while the ADJ soil formed its own group. These results demonstrated that soil characteristics like phosphorus and organic matter (from the enzyme *Hha*I

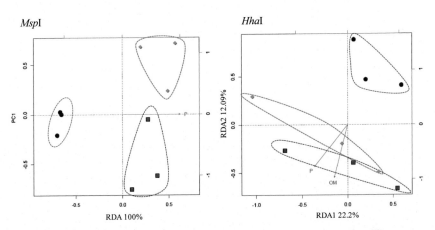

FIGURE 5.4   Redundancy analysis of Amazonian Dark Earth (ADE) soils [■]; adjacent (ADJ) soils [●] and biochar (BC) [◆] based on the structure of soil bacterial community as determined by terminal restriction fragment length polymorphism (T-RFLP) and soil attributes. The ordinations were performed using data obtained with the enzymes *Msp*I and *Hha*I (Navarrete et al., 2010).

data) and phosphorus (from the enzyme *Msp*I data) concentrations were related strongly to differences in the soil bacterial community structures.

Anderson et al. (2011) also used T-RFLP to evaluate the influence of black carbon (biochar) applications in agricultural soils on the bacterial community structure. The results indicated that biochar can promote bacterial phosphate solubilization, as well as altering the structure of the soil microbial community, most notably with the reduction of plant pathogenic bacteria. To date, studies on the application of carbon in soils have used biochar from biomass samples, such as *Pinus radiata* (Anderson et al., 2011) and citrus trees (Kolton et al., 2011) produced at different temperatures. However, the biochar from ADE soils is formed from the burning of various plant materials, and its distribution in the Amazon soils occurred hundreds of years ago. Therefore caution should be taken when comparing biochar from recent slash-and-burn systems to the char from ADE.

The effect of biochar on sorghum plants in highly weathered xanthic ferralsol from the Amazon region was evaluated by Steiner et al. (2008) who used stable isotope ^{15}N as tracer to study plant resource acquisition. The authors noted that the plots that received biochar showed an increase of 40% in plant growth and 50% in productivity when compared to the portion that received fertilizers only. These results highlighted the role of biochar in the retention of nutrients, probably because of its sorption capacity and consequently the observed positive effect on crop yields. The authors also recorded that biochar addition enhanced microbial activity along with the decrease in nutrient losses by leaching.

By using 16S rRNA gene pyrosequencing, Navarrete et al. (2010) aimed to identify the bacterial community associated with the ADE soil and its biochar. In general, Acidobacteria, Proteobacteria, Verrucomicrobia, and Actinobacteria represented the major phyla in ADE soil and, more specifically, within the biochar fragments. Although these groups are usually dominant in most soil types (Janssen, 2006), the results, which were secured from a larger number of sequences (10,857 reads), corroborated the data obtained by clone libraries from bacterial communities in soils with anthropogenic horizon (Kim et al., 2007; Tsai et al., 2008).

Furthermore, phylogenetic analysis of the data (Fig. 5.5) highlighted the differences in community composition linked to ADE soil and biochar with the distinct differences found in the Acidobacteria, Proteobacteria, Actinobacteria, and Firmicutes. Bacteria of the phylum Acidobacteria can represent approximately 20% of the bacterial communities in soil with increased abundance in tropical environments and acid soils (Kim et al., 2007; Navarrete et al., 2010; Grossman et al., 2010; Taketani et al., 2013). Although a further finding was that biochar hosted species of bacteria in numbers not much lower than the ADE soil, Navarrete et al. (2010) further described 512 operational taxonomic units (OTUs) found exclusively in ADE and 343 only from biochar samples. The OTUs that these samples

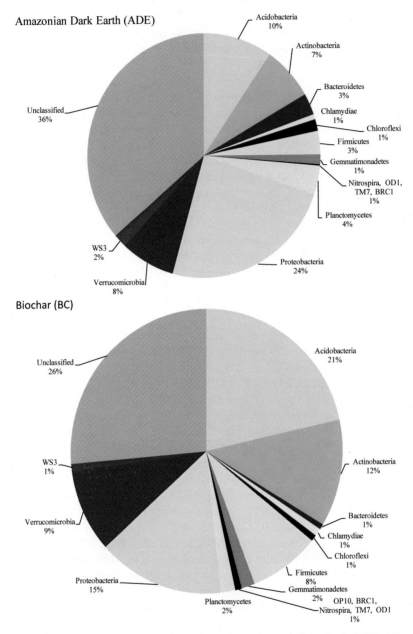

FIGURE 5.5   Relative abundance of phyla of the Amazonian Dark Earth (ADE) soil and biochar (BC) library, in which 16S sequences were classified according to the nearest neighbor in the Ribosomal Database Project pipeline (Navarrete et al., 2010).

have in common (492) suggested that the biochar environment may be responsible for specific processes.

The Actinobacteria is a main phylum within the Bacteria domain, composed from a single class divided into three common subclasses in soils: Actinobacteridae, Acidimicrobidae, and Rubrobacteridae (Jansen, 2006). Members belonging to this phylum can have enhanced antibiotic production capabilities (Silveira et al., 2006), and perform a key role in SOM decomposition and humus formation (Navarrete et al., 2010; Taketani et al., 2013). According to Navarrete et al. (2010), this phylum was often more abundant in biochar samples. When comparing the effect of biochar on bacterial diversity in Florida soils, Khodadad et al. (2011) found an increase in the relative abundance of the phylum Actinobacteria in soils treated with biochar. The authors showed that these results proved a marked effect of biochar on the soil bacterial community. Thus, although biochar is a relatively inert material in soil, given that the stability depends on the status of formation and transformations over the years, the fragments have the ability to improve physical (Teixeira and Martins, 2003), chemical (Lehmann et al., 2003), and therefore the biological properties of soil. Also biochars in ADEs can provide a buffer-like effect, protecting the system against environmental changes. However, this statement needs to be tested further with alternative methods (Lima et al., 2015, accepted for publication).

Study II: Functional Diversity of Bacterial Genes Associated with Aromatic Hydrocarbon Degradation in Anthropogenic Dark Earth of Amazonia

Germano (2011) also studied the functional microbiota of biochar and the ADE soil from which it was derived and recorded an abundance of Actinobacteria, Proteobacteria, and Firmicutes phylum in the biochar fraction. These phyla are largely described as bacterial aromatic hydrocarbon degraders (Nunes, 2006).Therefore, because of high humic acid content and biochar fragments in ADEs, Germano et al. (2012) found a high diversity of aromatic ring-hydroxylating dioxygenases (ARHD), the bacterial enzymes responsible for aromatic hydrocarbon degradation. The authors used PCR with α-ARHD-specific primers $ARHD_2F$ (5'-TTYRYITGYAIITAY-CAYGGITGGG-3') and $ARHD_2R$ (5'-AAITKYTCIGCIGSIRMYTTCCA-3') (Bellicanta and Pellizari, 2004), with subsequent gene library construction. The α-ARHD genes libraries for biochar and soil from the ADE samples revealed mixed groups of toluene/biphenyl dioxygenase sequences, together with enzymes for aromatic hydrocarbon degradation, specifically naphthalene and phenanthrene. These reflected the extensive diversity of functional genes according to the Kweon dioxygenases group (Kweon et al., 2008) (Fig. 5.6). In addition, the α-ARHD gene-expressing strains were classified into 20 groups of many yet uncultured bacteria, with

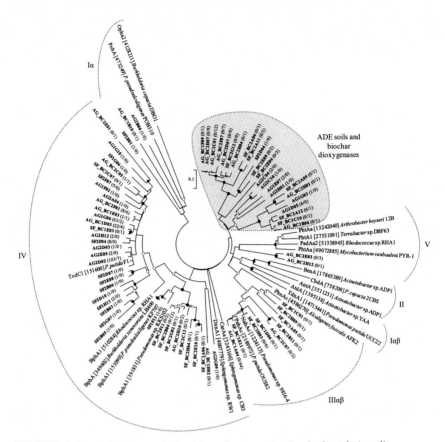

FIGURE 5.6 Phylogenetic relationships of α-aromatic ring-hydroxylating dioxygenase (ARHD) genes and their functionality under Kweon dioxygenase groups Iα, IV, IIIαβ, Iαβ, II, and V (Kweon et al., 2008). *From Germano M.G., Cannavan F.S., Mendes L.W., Lima, A.B., Teixeira W.G., Pellizari V.H., Tsai S.M., 2012. Functional diversity of bacterial genes associated with aromatic hydrocarbon degradation in anthropogenic dark earth of Amazonia. Pesqui. Agropecuária Bras., Brasília, DF, 47, 654–664.*

Proteobacteria, Actinobacteria, and Firmicutes as the predominant phyla (Fig. 5.7).

Additionally, Germano et al. (2012) observed major operational protein families that were shared between biochar and disturbed ADE (cultivated or damaged areas), with higher diversity on biochar samples than preserved and disturbed ADE (Fig. 5.8). Also there was higher diversity in ADE from secondary forests than highly disturbed ADE areas. The data suggested that ADE microbial communities are highly heterogeneous and responsible for dynamic changes in the environment, mostly to facilitate their adaptation to changing environmental parameters. Furthermore, these results highlighted the role of biochar as the main

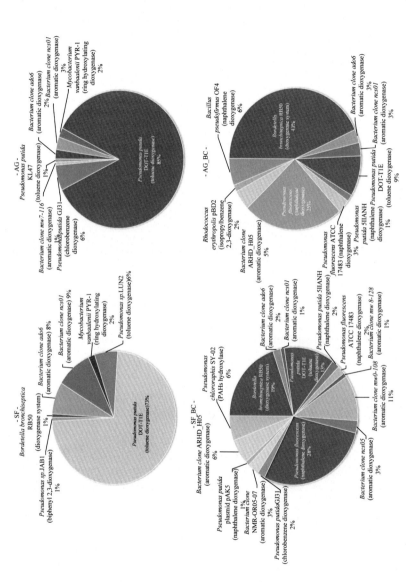

FIGURE 5.7 Classification of aromatic ring-hydroxylating dioxygenase (ARHD) gene bacterial owners (Germano et al., 2012).

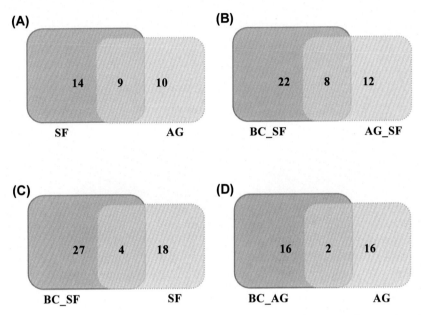

FIGURE 5.8 Venn diagrams of clones showing operational protein clusters shared between Amazonian Dark Earth (ADE); biochar (BC); secondary forest (SF); and agriculture area (AG).

factor for the maintenance of microbial diversity in ADE soils, which is associated with central metabolic processes such as nutrient cycling and SOM dynamics.

Study III: Bioprospection of Bacterial Aromatic Hydrocarbon Degraders Isolated from Biochar of Amazonian Dark Earths from Central Amazon

To study the functional capacity of bacteria isolated from biochar fragments, Nakamura et al. (2014) utilized five culture media, microaerophilic incubation, and 16S rRNA gene sequencing and identified 24 bacterial genera including *Burkholderia, Ralstonia, Pseudomonas, Achromobacter, Pantoea, Leclercia, Acinetobacter, Cupriavidus, Enterobacter, Variovorax, Leifsonia, Parapusiliimonas, Bordetella, Bacillus, Paenibacillus, Lysinibacillus, Brevibacillus, Rummeliibacillus, Arthrobacter, Streptomyces, Rhodoccocus,* and *Fibrobacterium*. In addition, many yet uncultured bacteria were found related to the Proteobacteria, Firmicutes, Actinobacteria, and Bacteroidetes phyla.

Furthermore, a screening bioassay with phenanthrene, biphenyl, and diesel oil as substrates was applied to assess the functional capacities of the aromatic hydrocarbon degraders. Although some of the isolates were capable of degrading all or most substrates, others had the ability

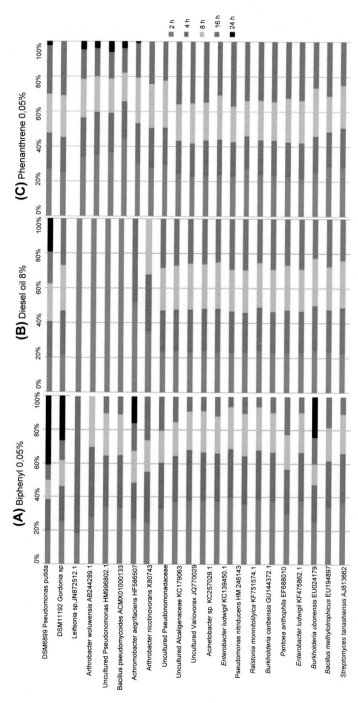

FIGURE 5.9 Degradation rates of different substrates by bacterial strains isolated from biochar fragments of Amazonia Dark Earth for 2, 4, 8, 16, and 24 h as visualized spectrophotometrically by 2,6-dichlorophenol-indophenol absorption at 620 nm: (A) degradation of biphenyl at 0.05%; (B) degradation of diesel oil at 8%; (C) degradation of phenanthrene at 0.05%. Isolates are identified by species and National Center for Biotechnology Information accession number. Percentages length bars indicate the quantity of the substrate that was degraded, and bars colors indicate the time of degradation for the specific substrate concentration.

to degrade just one aromatic molecule, but at a more rapid rate (Fig. 5.9). For example, two isolates were capable of degrading 100% of substrates, in relation to the bacterial standards, and they were *Pseudomonas putida* DSM6899, which degraded only phenanthrene in 16h, and *Gordonia* sp. DSM11192, which degraded both diesel oil and phenanthrene in 16h. The remainder had the following degradative capacities:

1. Three substrates in 16h: uncultured *Alcaligenaceae* KC179063.1, uncultured *Variovorax* JQ770029.1, *Acinetobacter* sp. KC257028.1, *Enterobacter ludwigii* KC139450.1, *Pseudomonas nitrireducens* HM246143, *Ralstonia mannitolilytica* KF751574.1, *Burkholderia caribensis* GU144372.1, *Pantoea anthophila* EF688010, *E. ludwigii* KF475862.1, *Bacillus methylotrophicus* EU194897, *Streptomyces tanashiensis* AJ781362;
2. Biphenyl and phenanthrene in 16h and diesel oil in 8h: *Arthrobacter nicotinovorans* X80743;
3. Diesel oil in 8h: *Achromobacter* aegrifaciens HF586507;
4. Diesel oil in 2h and biphenyl in 16h: uncultured *Pseudomonas* HM996802.1 and *Bacillus pseudomycoides* ACMX01000133;
5. Diesel oil in 2h and biphenyl in 4 and 8h, respectively: *Leifsonia* sp. JN872512.1 and *Arthrobacter woluwensis* AB244289.1.

Moreover, this study demonstrates that ADE biochar may serve as a microhabitat for bacterial communities with specific metabolic functions. Thus the generated data highlighted the potential metabolic activity of microorganisms associated with biochar application, which may influence the resilience, functional capacity, and fertility of anthropogenic soils in tropical regions.

FUTURE RESEARCH

The studies presented here highlight the importance of biochar as a niche for bacterial survival and maintenance with a direct role in biogeochemical cycles, as well as on improving the physical and chemical properties of soil. Although metagenomics provided more information, further analyses are being made to understand the biodiversity of biochar fragments from ADE in comparison to char in Amazon croplands and pastures where burning is a current aspect of soil management. Also enrichments with stable isotopes in consumption bioassays, such as stable isotope probing (Dumont and Murrell, 2005), are complementing metagenomics to trace metabolic pathways for complex compounds in soil, with particular focus on evaluating microbial enzyme activity. Furthermore, metatranscriptomics should expand existing knowledge by providing information on the identity and function of the occurring microorganisms in a punctuated time.

Finally, biochar molecular characterization by methods like pyrolysis and gas chromatography with mass spectrometry could identify the bio-oils and other functional groups associated with biochar surfaces in order to evaluate the environment that microbiota is in contact with.

Acknowledgments

The group would like to sincerely express their thanks and gratitude to CNPq—Conselho Nacional de Pesquisa e Desenvolvimento Tecnológico (Project No 474455/2007-6), FAPESP—Fundação de Amparo à Pesquisa do Estado de São Paulo (Project No 2011/50914-3). We also would like to thank the fellowships given by FAPESP and CNPq. The group also is in debt to James Tiedje (Center of Microbial Ecology—Michigan State University, Michigan, USA) for the DNA pyrosequencing support.

References

Akiyama, K., Hayashi, H., 2006. Strigolactones: chemical signals for fungal symbionts and parasitic weeds in plant roots. Ann. Bot. 97, 925–931.

Amann, R.I., Ludwig, W., Schleifer, K.H., 1995. Phylogenetic identification and *in situ* detection of individual microbial cells without cultivation. Microbiol. Rev. 59, 143–216.

Anderson, C.R., Condron, T.M., Clough, T.J., Friers, M., Stewart, A., Hill, R.A., Sherlock, R.R., 2011. Biochar induced soil microbial community change: implications for biogeochemical cycling of carbon, nitrogen and phosphorus. Pedobiologia, Jena 54, 309–320.

Bascompte, J., 2009. Disentangling the web of life. Science 325 (5939), 416–419.

Bellicanta, G.S., Pellizari, V.H., 2004. Development of degenerated primers for detection of distinct aromatic ring-hydroxy lating dioxygenases. In: International Symposium on Microbial Ecology (ISME), 10, 2004. Cancun. Proceedings Cancun: Asociación Mexicana de Microbiología, 1 CD-ROM.

Benites, V.M., Madari, B.E., Machado, P.L.O., 2005. A. Matéria Orgânica do Solo. In: Wadt, P.G.S. (Ed.), Manejo do Solo e Recomendação de Adubação para o Estado do Acre. Embrapa Acre, Rio Branco, pp. 93–120.

Brady, N.C., Weil, R.R., 2008. An Introduction to the Nature and Properties of Soils, 14th ed. Prentice Hall, Upper Saddle River. 881 p.

Bridgewater, A.V., Boocock, D.G.B. (Eds.), 1997. Developments in Thermochemical Biomass Conversion. Blackie Academic and Professional, London, UK.

Brito, J.O., Barrichelo, L.E.G., 1977. Correlações entre características físicas e químicas da madeira e a produção de biocarvão vegetal: I. Densidade e teor de lignina da madeira de eucalipto. n. 14. IPEF, Piracicaba, pp. 9–20.

Cannavan, F.S., 2011. Structure and Composition of Bacterial Communities (*Bacteria* and *Archaea*) in Fragments of Pyrogenic Charcoal from Amazonian Dark Earth in the Central Amazon. 2011. Centro de Energia Nuclear na Agricultura, Universidade de São Paulo, Piracicaba-SP, Brazil (Ph.D. thesis). http://www.teses.usp.br/teses/disponiveis/64/64133/tde-20092012-102334/pt-br.php.

Dixon, K., 1998. Smoke Germination of Australian Plants. RIRDC report (98/108, KPW–1A). Kings Park and Botanic Garden Plant Science Division, Perth, Australia.

Du, H.G., Mori, E., Terao, H., TsuzukiI, E., 1998. Effect of the mixture of charcoal with pyroligneous acid on shoot and root growth of sweet potato [*Ipomoea*]. Tokyo Jpn. J. Crop Sci. 67 (2), 149–152.

Du, H.G., Ogawas, M., Ando, S., Tsuzuki, E., Murayama, S., 1997. Effect of mixture of charcoal with pyroligneous acid on sucrose content in netted melon (*Cucumismelo* L. var. *reticulatus* Naud.) fruit. Tokyo Jpn. J. Crop Sci. 66 (3), 369–373.

Dumont, M.G., Murrell, J.C., 2005. Stable isotope probing – linking microbial identity to function. Nat. Rev. Microbiol. 3, 499–504.

Edwards, U., Rogall, T., Blocker, H., Emde, M., Bottger, E.C., 1989. Isolation and direct complete nucleotide determination of entire genes. Characterization of a gene coding for 16S ribosomal RNA. London Nucleic Acids Res. 17, 7843–7853.

EMBRAPA, 2013. Sistema brasileiro de classificação de solos, third ed. Brasília, DF, 353p.

Filip, Z., 2002. International approach to assessing soil quality by ecologically-related biological parameters. Agric. Ecosyst. Environ., Amsterdam 88, 169–174.

Germano, M.G., 2011. Functional Diversity in Amazonian Dark Earth Soils and Black Carbon. Centro de Energia Nuclear na Agricultura, University of São Paulo, Piracicaba-SP, Brazil (Ph.D. Thesis). http://www.teses.usp.br/teses/disponiveis/64/64133/tde-30092011-105127/pt-br.php.

Germano, M.G., Cannavan, F.S., Mendes, L.W., Lima, A.B., Teixeira, W.G., Pellizari, V.H., Tsai, S.M., 2012. Functional diversity of bacterial genes associated with aromatic hydrocarbon degradation in anthropogenic dark earth of Amazonia. Pesqui. Agropecuária Bras., Brasília, DF 47, 654–664.

Glaser, B., 2007. Prehistorically modified soils of central Amazonia: a model for sustainable agriculture in the twenty-first century. Phil. Trans. R. Soc., London 362, 187–196.

Glaser, B., Guggenbberger, G., Haumaier, L., Zech, W., 2000. Persistence of soil organic matter in archaeological soils (Terra Preta) of the Brazilian Amazon region. In: Rees, R.M., Ball, B.C., Campbell, C.D., Watson, C.A. (Eds.), Sustainable Management of Soil Organic Matter. CAB International, Wallingford, pp. 190–194.

Glaser, B., Guggenberger, G., Zech, W., Ruivo, M.L., 2003. Soil organic matter stability in Amazonian Dark Earths. In: Lehmann, J., Kern, D.C., Glaser, B., Woods, W.I. (Eds.), Amazonian Dark Earths: Origin, Properties, Management. Kluwer Academic Publishers, Dordrecht, pp. 141–158.

Glaser, B., Haumaier, L., Guggenberger, G., Zech, W., 2001. The "Terra Preta" phenomenon: a model for sustainable agriculture in the humic tropics. Berlin Die Naturwiss. Nat. Sci. 88 (2), 37–41.

Golchin, A., Clarke, P., Baldock, J.A., Hogashi, T., Skejemtad, J.O., Oades, J.M., 1997. The effects of vegetation and burning on the chemical composition of soil organic matter in a volcanic ash soil as shown by 13C NMR spectroscopy: I whole soil and humic acid fraction. Amsterdam Geoderma 76.

Graber, E.R., Harel, Y.M., Kolton, M., Cytryn, E., Silber, A., David, D.R., Tsechansky, L., Borenshtein, M., Elad, Y., 2010. Biochar impact on development and productivity of pepper and tomato grown in fertigated soilless media. Plant Soil 337, 481–496.

Grossman, J.M., Sheaffer, C., Wyse, D., Bucciarelli, B., Vance, C., Grahman, P.H., 2006. An assessment of nodulation and nitrogen fixation in inoculated *Inga oerstediana*, a nitrogen-fixing tree shading organic coffee in Chiapas, Mexico. Oxford Soil Biol. Biochem. 38 (4), 769–784.

Grossman, J.M., O'Neill, B.E., Tsai, S.M., Liang, B., Neves, E., Lehmann, J., Thies, J.E., 2010. Amazonian anthrosols support similar microbial communities that differ distinctly from those extant in adjacent, unmodified soils of the same mineralogy. Microb. Ecol., New York 60, 192–205.

Guan, M., 2004. Manual for Bamboo Charcoal Production and Utilization. Bamboo Engineering Research Center, Nanjing Forestry University, China.

Handelsman, J., Smalla, K., 2003. Conversations with the silent majority. Curr. Opin. Microbiol., London 6, 271–273.

Head, I.M., Saunders, J.R., Pickup, J., 1998. Microbial evolution, diversity and ecology: a decade of ribosomal rRNA analysis of uncultivated organisms. Microb. Ecol., New York 35, 1–21.

Janssen, P.H., 2006. Identifying the dominant soil bacterial taxa in libraries of 16S rRNA and 16S rRNA genes. Appl. Environ. Microbiol., Washington 72, 1719.

Jin, H., 2010. Characterization of Microbial Life Colonizing Biochar and Biochar-amended Soils. Cornell University, Ithaca, NY (Ph.D. Dissertation).

Khodadad, L.M., Zimmerman, A.R., Green, S.J., Uthandi, S., Foster, J.S., 2011. Taxa-specific changes in soil microbial community composition induced by pyrogenic carbon amendments. Soil Biol. Biochem., Oxford 43, 385–392.

Kim, J.-S., Sparovek, S., Longo, R.M., De Melo, W.J., Crowley, D., 2007. Bacterial diversity of terra preta and pristine forest soil from the Western Amazon. Soil Biol. Biochem., Oxford 39, 648–690.

Kirk, J.E., Beaudette, L.A., Hart, M., Moutglis, P., Klironomos, J.N., Lee, H., Trevors, J.T., 2004. Methods of studying soil microbial diversity. J. Microbiol. Methods, Amsterdam 58, 169–188.

Kleinman, P.J.A., Pimentel, D., Bryant, R.B., 1995. The ecological sustainability of slash- and-burn agriculture. Agric. Ecosyst. Environ.. Amsterdam 52, 235–249.

Kolton, M., Harel, Y.M., Pasternak, Z., Graber, E.R., Elad, Y., Cytrin, E., 2011. Impact of biochar application to soil on the root-associated bacterial community structure of fully developed greenhouse pepper plants. Appl. Environ. Microbiol., Baltimore 77, 4924–4930.

Kramer, R.W., Kujawnski, E.B., Hatcher, P.G., 2004. Identification of black carbon derived structures in a volcanic ash soil humic acid by fourier transform ion cyclotron resonance mass spectrometry. Environ. Sci. Technol., Washington DC 38, 3387–3395.

Kurtboke, D.I., Swings, J., Storms, V., 2004. Microbial Genetic Resources and Biodiscovery. WFCC Publishing, Oxford, UK. 400 p.

Kweon, O., Kim, S.J., Baek, S., Chae, J.C., Adjei, M.D., Baek, D.H., Kim, Y.C., Cerniglia, C.E., 2008. A new classification system for bacterial Rieske non-heme iron aromatic ring-hydroxylating oxygenases. BMC Biochem., London 9, 11. http://dx.doi.org/10.1186/1471-2091-9-11.

Lehmann, J., Joseph, S., 2009. Biochar for Environmental Management: Science and Technology. Earthscan, London. 449 p.

Lehmann, J., Silva, J.P., Steiner, C., Nehls, T., Zech, W., Glaser, B., 2003. Nutrient availability and leaching in an archaeological Anthrosol and a Ferralsol of the Central Amazon basin: fertilizer, manure and charcoal amendments. Plant Soil, Dordrecht 249, 343–357.

Lehmann, J., 2007. Bio-energy in the black. Front. Ecol. Environ., Washington DC 5, 381–387.

Liang, B., Lehmann, J., Solomon, D., Kinyangi, J., Grossman, J., O'Neil, B., Skejemstad, J.O., Thies, J., Luizão, F.J., Petersen, J., Neves, E.G., 2006. Black carbon increases cation exchange capacity in soils. Soil Sci. Soc. Am. J., Madison 70, 1719–1730.

Lima, A.B., Cannavan, F.S., Germano, M.G., Dini-Andreote, F., Paula, A.M., Franchini, J.C., Teixeira, W.G., Tsai, S.M., 2015. Effects of vegetation and seasonality on bacterial communities in Amazonian Dark Earth and adjacent soils. Afr. J. Microbiol. Res. 9 (40), 2119–2134.

Maggi, R., Delmon, B., 1994. Comparison between "slow" and "flash" pyrolysis oils from biomass. Fuel, London 73 (5), 671–677.

Masiello, C.A., 2004. New directions in black carbon organic geochemistry. Mar. Chem., Amsterdam 92, 201–213.

Meira, A.M., 2002. Diagnóstico sócio-ambiental e tecnológico da produção de biocarvão vegetal no município de Pedra Bela Estado de São Paulo. 2002. 105 f. Dissertação (Mestrado em Recursos Florestais) - Escola Superior de Agricultura "Luiz de Queiroz", da Universidade de São Paulo, Piracicaba, Brazil.

Mendes, R., Kruijt, M., de Bruijn, I., Dekkers, E., van der Voort, M., Schneider, J.H.M., Piceno, Y.M., DeSantis, T.Z., Andersen, G.L., Bakker, P.A.H.M., Raaijmakers, J.M., 2011. Deciphering the rhizosphere microbiome for disease-suppressive bacteria. Science, Washington DC 332 (6033), 1097–1100.

Moreira, F.M.S., Siqueira, J.O., 2002. Microbiologia e bioquímica do solo. UFLA, Lavras. 625 p.

Nakamura, F.M., 2014. Bioprospecting the Bacterial Degraders of Aromatic Hydrocarbon from Biochars from Anthropogenic Dark Earth of the Central Amazon (M.Sc. Dissertation). Centro de Energia Nuclear na Agricultura, Universidade de São Paulo, Piracicaba.

Nakamura, F.M., Germano, M.G., Tsai, S.M., 2014. Capacity of aromatic compound degradation by bacteria from Amazon Dark Earth. Diversity, Basel 6 (2), 339–353.

Navarrete, A.A., Cannavan, F.S., Taketani, R.G., Tsai, S.M., 2010. A molecular survey of the diversity of microbial communities in different Amazonian agricultural model systems. Diversity, Basel 2, 787–809.

Novotny, E.H., Hayes, M.H.B., Madari, B.E., Bonagamba, T.J., Azevedo, E.R., Souza, A.A., Song, G., Nogueira, C.M., Mangrich, A.S., 2009. Lessons from the Terra Preta de Índios of the Amazon region for the utilization of charcoal for soil amendment. J. Braz. Chem. Soc., São Paulo 20 (6), 1003–1010.

Novotny, E.H., De Azevedo, E.R., Bonagamba, T.J., Cunha, J.F.T., Madari, E.B., Benites, V.M., Hayes, M.H.B., 2007. Studies of the compositions of humic acids from Amazonian dark earth soils. Environ. Sci. Technol. Lett., Washington DC 41, 400–405.

Nunes, G.L., 2006. Diversidade e estrutura de comunidades de Bacteria e Archaea em solo de mangue contaminado com hidrocarbonetos de petróleo. 2006. 84 p. Dissertação (Mestrado em Microbiologia Agrícola) – Escola Superior de Agricultura "Luiz de Queiroz", Universidade de São Paulo, Piracicaba.

O'Neill, B., Grossman, J., Tsai, M.T., Gomes, J.E., Lehmann, J., Peterson, J., Neves, E., Thies, J.E., 2009. Bacterial community composition in Brazilian Anthrosols and adjacent soils characterized using culturing and molecular identification. Microb. Ecol., New York 58, 23–35.

Oparina, L.V., et al., 1971. Sb. Tr. Vses. Nauch.-Issled. Inst. Gidroliza Rast. Mater. 19, 176–186.

Otsuka, S., Sudiana, I., Komori, A., Isobe, K., Deguchi, S., Nishiyama, M., Shimizu, H., Senoo, K., 2008. Community structure of soil bacteria ina tropical rainforest several years after fire. Microbes Environ., Sapporo 23, 49–56.

Parameswaran, P., Jalili, R., Tao, R., Shokralla, S., Gharizadeh, B., Ronaghi, M., Fire, A.Z., 2007. A pyrosequencing-tailored nucleotide barcode design unveils opportunities for large-scale sample multiplexing. Nucleic Acids Res., London 35 (19), e.130.

Peacocke, C., Joseph, S., undated. Notes on terminology and technology in thermal conversion. Int. Biochar Initiat.: www.biochar-international.org; Biochar-international.org/images/Terminology_and_Technology_final_vCP_sj (accessed 30.10.15.). http://www.biochar-international.org/images/Terminology.doc

Pietikainen, J., Kiikkila, O., Fritze, H., 2000. Charcoal as a habitat for microbes and its effect on the microbial community of the underlying humus. Oikos, Copenhagen 89, 231–242.

Poirier, N., Derenne, S., Rouzaud, J.-N., Largeau, C., Mariotti, A., Balesdent, J., Maquet, J., 2000. Chemical structure and sources of the macromolecular, resistant, organic fraction isolated from a forest soil (Lacadée, south-west France). Org. Geochem., Oxford 31 (9), 813–827.

Rohde, G.M., 2007. O Mito da combustão espontânea do biocarvão. Revistada Madeira, Florianópolis. n. 106, jul. 2007. Disponível m: http://www.remade.com.br/br/revistadamadeira_materia.php?num=1112&subject=E%20mais&title=O%20mito%20da%20combust%E3o%20espont%E2nea%20do%20carv%E3o%20vegetal.

Ronaghi, M., Karamohamed, S., Pettersson, B., Uhlen, M., Nyren, P., 1996. Real-time DNA sequencing using detection of pyrophosphate release. Anal. Biochem., New York 242, 84–89.

Ronaghi, M., 2001. Pyrosequencing sheds light on DNA sequencing. Genome Res., Woodbury 11, 3–11.

Rosado, A.S., Duarte, G.F., Seldin, L., Van Elsas, J.D., 1977. Molecular microbial ecology: a mini review. Rev. Microbiol., Rio de Janeiro 28, 135–147.

Roscoe, R., Buurman, P., Velthorst, E.J., Vasconcellos, C.A., 2001. Soil organic matter dynamics in density and particle size fractions as revealed by the 13C/12C isotopic ratio in a Cerrado'soxisol. Geoderma, Amsterdam 104, 185–202.

Satanoka, S., 1963. Enshurin Hokuku. Hokkaido Daigaku 22 (2), 609–814.

Schloter, M., Dilly, O., Munch, J.C., 2003. Indicators for evaluating soil quality. Agric. Ecosyst. Environ. 98, 255–262.

Schmidt, M.W., Noack, A.G., 2000. Black carbon in soils and sediments: analysis, distribution, implications, and current challenges. Glob. Biochem. Cycles, Washington, DC 14, 777–793.

Schnitzer, M.I., Monreal, C.M., Facey, G.A., Fransham, P.B., 2007. The conversion of chicken manure to biooil by fast pyrolysis I. Analyses of chicken manure, biooils and char by C-13 and H-1 NMR and FTIR spectrophotometry. J. Environ. Sci. Health B, Oxford 42, 71–77.

Schulz, S., Brankatschk, R., Dümig, A., Kogel-Knabner, I., Schloter, M., Zeyer, J., 2013. The role of microorganisms at different stages of ecosystem development for soil formation. Biogeosciences, Katlenberg-Lindau 10, 3983–3996.

Schütte, U.M.E., Abdo, Z., Bent, S.J., Shyu, C., Williams, C.J., Pierson, J.D., Forney, L.J., 2008. Advances in the use of terminal restriction fragment length polymorphism (T-RFLP) analysis of 16S rRNA genes to characterize microbial communities. Appl. Microbiol. Biotechnol., Berlin 80, 365–380.

Silveira, E.L., Pereira, R.M., Scaquitto, D.C., Pedrinho, E.A., Val-Moraes, S.P., Wickert, E., Carareto-Alves, Lemos, E.G.M., 2006. Bacterial diversity of soil under eucalyptus assessed by 16S rDNA sequencing analysis. PesquisaAgropecuáriaBrasileira, Brasília, DF 41, 1507–1516.

Skjemstad, J.O., Clarke, P., Taylor, J.A., Oades, J.M., McClure, S.G., 1996. The chemistry and nature of protected carbon in soil. Aust. J. Soil Res., Melbourne 34, 251–271.

Steiner, C., Das, K.C., Garcia, M., Forster, B., Zech, W., 2008. Charcoal and smoke extract stimulate the soil microbial community in a highly weathered xanthic ferralsol. Pedobiologia, Amsterdam 51, 359–366.

Taketani, R.G., Lima, A.B., Jesus, E.C., Teixeira, W.G., Tiedje, J.M., Tsai, S.M., 2013. Bacterial community composition of anthropogenic biochar and Amazonian anthrosols assessed by 16S rRNA gene 454 pyrosequencing. Antonie van Leeuwenhoek, Wageningen 104, 233–242.

Taketani, R.G., Tsai, S.M., 2010. The influence of different land uses on the structure of archaeal communities in Amazonian Anthrosols based on 16S rRNA and amoA genes. Microb. Ecol., New York 59, 734–743.

Teixeira, W.G., Kern, D.C., Madari, B.E., Lima, H.N., Woods, W., 2009. As terras pretas de índio da Amazônia: sua caracterização e usodeste conhecimento na criação de novas áreas. Embrapa Amazônia Ocidental, Manaus. 420 p. 1 CD-ROM.

Teixeira, W.G., Martins, G.C., 2003. Soil physical characterization. In: Lehmann, J., Kern, D.C., Glaser, B., Woods, W.I. (Eds.), Amazonian Dark Earths: Origin, Properties, Management. Kluwer Academic Publishers, Dordrecht, pp. 272–286.

Theron, J., Cloete, T.E., 2000. Molecular techniques for determining microbial diversity and community structure in natural environments. Crit. Rev. Microbiol., London 26 (1), 37–57.

Tiedje, J.M., Asuming- Brempong, S., Nüsslein, K., Marsh, T.L., Flynn, S.J., 1999. Opening the black box of soil microbial diversity. Appl. Soil Ecol., Amsterdam 13, 109–112.

Tsai, S.M., O'Neill, B., Cannavan, F.S., Saito, D., Falcão, N.P.S., Kern, D., Grossman, J., Thies, J., 2008. Microbial world of terra preta. In: Woods, W.I. (Ed.), Amazonian Dark Earths: WimSombroek Vision, first ed. Springer, New York, pp. 299–308. Lehmann, J., 2007. Bioenergy in the black. Front. Ecol. Environ., Washington DC, 5, 381–387.

Tsuzuki, E., Wakiyama, Y., Eto, H., Handa, H., 1989. Effect of pyroligneous acid and mixture of charcoal with pyroligneous acid on the growth and yield of rice plant. Jpn. J. Crop Sci., Tokyo 58 (4), 592–597.

Tuomela, M., Vikman, M., Hatakka, A., Itavaara, M., 2000. Biodegradation of lignin in a compost environment: a review. Bioresource Technology, Essex, 72, 169–183.

Uddin, S.M.M., Murayama, S., Ishimine, Y., Tsuzuki, E., Harada, J., 1995. Effect of the mixture of charcoal with pyroligneous acid on dry mather production and root growth of summer planted sugarcane (*Saccharumofficinarum* L.). Jpn. J. Crop Sci., Tokyo 64 (4), 747–753.

Wardle, D.A., Zackrisson, O., Nilsson, M.C., 1998. The charcoal effect in Boreal forests: mechanisms and ecological consequences. Oecologia, Berlin 115, 419–426.

Weisburg, W.G., Barns, S.M., Pelletier, D.A., Lane, D.J., 1991. 16S Ribosomal DNA amplification for phylogenetic study. J. Bacteriol., Baltimore 173 (2), 697–703.

Woese, C.R., Kandler, O., Wheelis, L., 1990. Towards a natural system of organisms: proposal for the domains *Archaea, Bacteria* and *Eucarya*. Proc. Natl. Acad. Sci. U. S. A., Washington, DC 87, 4576–4579.

Zanetti, M., Cazetta, J.O., Júnior, D.M., Carvalho, S.A.C., 2003. Uso de subprodutos de carvão vegetal na formação do porta-enxerto limoeiro 'cravo' em ambiente protegido. Rev. Bras. Frutic., Jaboticabal - SP 25 (3), 508–512.

Examining Biochar Impacts on Soil Abiotic and Biotic Processes and Exploring the Potential for Pyrosequencing Analysis

R. Chintala[1], S. Subramanian[2], A.-M. Fortuna[3], T.E. Schumacher[2]

[1]Nutrient Management & Stewardship, Innovation Center for U.S. Dairy, Rosemont, IL, United States; [2]South Dakota State University, Brookings, SD, United States; [3]North Dakota State University, Fargo, ND, United States

Biochar Application
http://dx.doi.org/10.1016/B978-0-12-803433-0.00006-0

133

INTRODUCTION

Soil microbial indices that include functional gene copy numbers, community structure, and enzyme activities are sensitive to management changes, which include but are not limited to biochar additions, where they detect shifts in soil processes more readily than many other commonly used measures of soil health (Dose et al., 2015; Lehmann et al., 2011; Pankhurst et al., 1995). The health of a soil is evaluated using inherent and dynamic soil properties that serve as indicators of soil function. Inherent soil properties are not affected by management and form over geological time. Dynamic, or management dependent, soil properties are affected by land use and natural disturbances over a growing season, decade, or century (Doran and Parkin, 1994). Micro organisms and their activity are affected by and integrate with chemical, biological, and physical properties reflecting short- and long-term management. Developments in high-throughput metagenomics analysis of fungal and bacterial communities, coupled with measures of microbial activity via reverse transcriptase polymerase chain reaction (RT-PCR) and enzyme assays, offer effective means of assessing the short- and long-term impacts of biochar additions on soil health (Chen et al., 2013; Dose et al., 2015; Rincon-Florez et al., 2013).

The production of second-generation biofuels from lignocellulosic biomass and nonedible oil seed feedstocks using pyrolytic processes generates secondary products including syngas and biochar. Biochars are complex carbonaceous materials composed of mineral and organic chemical species and their structure and activity within soils depend on the choice of feedstock and pyrolytic conditions (Ennis et al., 2012). When biochars are incorporated into soils they interact with the mineral and organic fractions forming secondary products including organomineral–biochar complexes. The amphiphilic carbon framework of biochar exerts unique electrochemical properties in soil. Long- and short-range chemical reactions occurring at the complex interface include protonation, deprotonation, cation and anion exchange, sorption, π-electron bond formation, hydrogen bonding, and complexation (Chintala et al., 2013; Xu et al., 2015). Chemical reactions at the solid–solution interface also promote self-assembly of organic and inorganic fractions of soil ultimately influencing the structural geometry of soil particles and their physical functionality (Masiello et al., 2015). The physical properties affected include bulk density, porosity, water-holding capacity, aggregate stability, water infiltration, and hydraulic conductivity (Ennis et al., 2012; Thies et al., 2015). Biochar-induced abiotic processes (physical and chemical pathways) can exert influences on soil microbial community function and diversity (Lehmann et al., 2011).

As discussed previously, the properties of an individual biochar will determine how the material will interact with microorganisms, organic molecules (signal molecules, pesticides, soil organic matter [SOM], etc.), and soil minerals. Biochars have many properties that influence microbial communities by providing exchangeable cation exchange capacity (CEC), micronutrients, an energy source, organic carbon (C), electron acceptors/ donors, physical pore space, and surface area for microbes to colonize. Biochars contain a majority of mineral C and variable amounts of organic C, the proportion of which is determined partially by the temperature and conditions under which they were manufactured (Kleber et al., 2015). In general, as pyrolysis temperature is elevated (350, 400, 450, 500, 550, and 600°C), aromatic structures on biochar surface areas increase because of decreases in oxygen to carbon and hydrogen to carbon ratios (Chintala et al., 2014c; Kleber et al., 2015). These conditions cause the biochar surfaces to become less negatively charged and more hydrophobic, as well as increase the volume of micropores (pores >1 nm in size; Chia et al., 2015; Suliman et al., 2016a).

Biochars synthesized at different temperatures have also been shown to have different capacities to serve as electron donors and acceptors (Klüpfel et al., 2014). High-temperature pyrolysis was shown to have the greatest potential because of the production of more quinones and aromatics that possibly served as electron acceptors. In contrast, intermediate and lower temperatures produced quinones and phenolics, respectively, with the least potential to accept electrons. Oxidation of biochars at low temperature (350°C) resulted in greater CEC (Gul et al., 2015; Suliman et al., 2016a). In summary, production temperature affects surface area, pore volume, pH, volatile matter, oxygen content, and hydrophobicity. Also properties of the feedstocks control ash content, total carbon, fixed carbon, and mineral concentrations of the produced biochar (Xu et al., 2015; Suliman et al., 2016b).

Microbial communities are critical for soil ecosystem functioning including soil health, fertility, carbon and nutrient cycling, and plant productivity (Devare et al., 2007; Thies and Rillig, 2009). In order to meaningfully predict the effects of biochar addition on microbial communities in soil we must define a range of biochar properties and relate these to the myriad chemical, physical, and biological properties that constitute the soil environment. Therefore the examples provided will be discussed within the context of biochar type, soil texture, SOM content, and management. Understanding the influence of biochar on microbial community diversity and function can provide the necessary knowledge base for the development of biochar-based technologies to improve degraded soils and maintain soil health.

Traditional culture-dependent microbiological methods (eg, plate count method) have significant limitations to capture the large microbial genetic diversity of soil. Most soil microorganisms are not yet cultivable in the laboratory. In order to overcome these limitations, novel culture-independent metagenomic approaches have been developed. These methods can not only determine the phylogenetic structure of microbial communities, but also predict functional components of the microbiome through direct sequencing of microbial DNA and RNA extracted from biochar-supplemented soils.

These high-throughput technologies have the potential to enhance significantly our understanding of the impact of biochar on microbial communities in the bulk soil and within the rhizosphere microbiome.

ABIOTIC PROCESSES

Soil Physical and Hydrological Properties

Abiotic processes involving water, temperature, and aeration are critical in the development and maintenance of soil microbial communities (Thies et al., 2015). For example, the abundance and distribution of soil water is a critical factor controlling microbial community structure. Thus an examination of microbial community structure using phospholipid fatty acid analysis (PLFA) found that soil moisture content was the main physical factor controlling the structure and enzyme activities of soil microbial communities across a regional climate gradient in western Canada (Brockett et al., 2012). On a field scale, changes in microbial communities were found to be associated with changes in hydraulic conductivity caused by soil degradation induced by varying intensities of tillage events (Chaer et al., 2009). Microbial communities were characterized using three different methods including PLFA, terminal restriction fragment length polymorphism (T-RFLP), and community level physiological profiles. On a smaller scale, anaerobic microsites associated with differences in water-filled pore space, because of localized differences in pore size distribution, resulted in different microbial communities at the scale of millimeters (Hansel et al., 2008).

Generally the degree to which physical factors control soil microbial communities is site specific, hence abiotic chemical factors such as pH, nutrient content, and soil organic carbon may take precedence depending on soil, climate, and past history (Grayston et al., 2004; He et al., 2012). Additionally, biotic and abiotic factors interact to determine soil microbial community structure.

Microbial colonization can also induce changes in water movement in soils. Using bioluminescence to monitor microbial growth and a bromophenol dye solution to highlight hydraulic flow paths within sand, microbial colonization was found to change the localized flow paths and spatial distribution of water (Yarwood et al., 2006). A dynamic interaction between microbial activity and water flow appeared to determine the distribution of the microbial colony. Biofilm formation in soils can also alter soil hydraulic properties. Biofilms can effectively change spatial patterns of pore size distribution altering soil flow paths and conductivity (Rosenweig et al., 2014).

Biochars have been shown to modify a number of static soil physical properties including surface area, bulk density, porosity, and tensile strength, which affect soil biota (Chan et al., 2007; Lehmann et al., 2011). The effects of biochar on these relatively static properties impact several dynamic soil properties such as moisture content, hydraulic conductivity, and water retention. Although biochar can potentially affect microbial community structure through changes in the soil physical environment, it is also possible for biochar-induced changes in microbial growth to alter soil physical properties. Thies et al. (2015) provide a detailed review of six potential mechanisms by which biochar could influence soil microbial communities with changes in abiotic factors. However, the authors pointed out that these mechanisms are not mutually exclusive and are more likely to act concurrently. Therefore positive and negative feedback between mechanisms would mean that biochar effects on the soil physical environment could alter microbial communities which, in turn, may modify soil physical properties, and vice versa.

Pore size distribution and geometry determine the temporal and spatial distribution of soil physical and hydrological properties and this is likely to elicit changes in the occurring microbial communities. Biochar alters pore size distribution and geometry of the supplemented soil through the addition of pore structures and surface area, contained within the biochar material, and through interaction with other soil components to form secondary pore structures external to the material (Lehmann et al., 2011). Also the effects of biochar on soil properties depend not only on its characteristics but also on the soil type. For example, biochar tends to increase saturated flow of water in fine textured soils, and is decreased in coarse textured soils (Barnes et al., 2014; Lim et al., 2015).

Most of the biological activity in soil takes place within aggregates that contain primary particles (sand, silt, clay), water, and air-filled pore space, roots, soil fauna including microorganisms, as well as inorganic (hydroxides) and organic (polysaccharides) binding agents. Aggregates are comprised of microaggregates formed mainly by chemical processes associated with aluminosilicates, iron oxides and calcium carbonates, and macroaggregates that are bound together largely through biological activity (Whalen and Sampedro, 2010). The component most similar to biochar that comprises soil aggregates is SOM. Biochars, in contrast to SOM, have a highly ordered mineral C structure whereas humic fractions isolated from SOM are amorphous. Because of pyrolysis, most biochars also contain more recalcitrant carbon materials relative to SOM. Biochars are defined separately from soil or sediment humus (ie, SOM) as "a series of high molecular weight, yellow or black colored substances formed by secondary synthesis reactions...that are dissimilar to the biopolymers of plants and microorganisms" (Stevenson, 1994). Yet, methods used to measure functional groups and surface areas, chemical equilibria, organomineral complexes, cation bridging, and partition of organic molecules such as pesticides into SOM fractions can be adapted to study biochars and the microbial communities associated with these materials.

Biochar structure is highly dependent on the original feedstock as well as the pyrolysis process (Chia et al., 2015), thus biochars derived from plant materials often retain the cell structure of the feedstock (Fig. 6.1). Additionally, biomass feedstocks undergo shrinkage, fusion, melting, and cracking during pyrolysis, which increases the porosity of the resultant biochars. However, rapid heating rates often degrade feedstock structure resulting in lower porosities (Chia et al., 2015). The surface morphology of pyrogenic biochar materials is highly heterogeneous with honeycomb-like porosity (micro- and mesopores) with faces, edges, and ordered sheets (lattice structure) formed during the thermal conversion. Furthermore, biochar materials contain pore sizes ranging from a few nanometers to hundreds of micrometers (Chia et al., 2015). As a result the incorporation of porous biochar into soils provides additional surface area affecting water and nutrient retention as well as creating protected environments for microorganisms (Lehmann et al., 2011; Thies et al., 2015).

Biochar additions contribute to soil aggregate formation, the basis of soil structure. According to Fortuna (2012), "A soil aggregate or ped is a naturally formed assemblage of sand, silt, clay, organic matter, root hairs, microorganisms and their secretions, and resulting pores." These

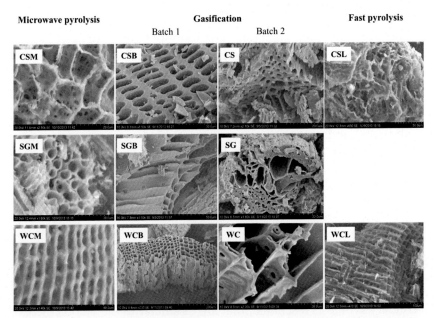

FIGURE 6.1   Scanning electron microscope (SEM) images of biochar materials at various magnifications produced from corn stover, switchgrass, and Ponderosa pine wood residue (Chintala et al., 2014c). Microwave pyrolysis—corn stover biochar (CSM), switchgrass biochar (SGM), and Ponderosa pine wood residue biochar (WCM). Gasification—batch 1: corn stover biochar (CSB), switchgrass biochar (SGB), and Ponderosa pine wood residue biochar (WCB). Batch 2: corn stover biochar (CS), switchgrass biochar (SG), and Ponderosa pine wood residue biochar (WC). Fast pyrolysis using electricity: corn stover biochar (CSL) and Ponderosa pine wood residue biochar (WCL).

components allow for the formation of microaggregates that contribute to "soil structure the manner in which primary particles, organic matter and pore space are arranged within three-dimensional space, a fundamental characteristic of soil" (Whalen and Sampedro, 2010). Organic material in soil includes: litter (macroorganic matter); light fraction (undecayed plant and animal tissue and their partial decomposition products); soil biomass (organic matter in live microbial tissue); and SOM or humus (Stevenson, 1994). Humus represents up to 80% of the organic material contained in soils (Whalen and Sampedro, 2010).

Biochars interact with soils to change pore size distributions within the soil matrix by affecting inter- and intra -aggregate pore size, shape, and connectivity (Masiello et al., 2015). Also the addition of biochar to coarse textured soils has been shown to alter soil structure and hydrological properties in the short-term (Hardie et al., 2014; Mollinedo et al., 2015; Sun et al., 2013). Biochar's effects on macropores were illustrated in a study that examined the application of five different biochars to a sandy loam soil (Maddock series) at an application rate of $40\,g\,kg^{-1}$ soil (4%w/w). Four of the added biochars resulted in an increase in water content at water tensions associated with pore diameters at the upper end of the water retention curve indicating an increase in macropores in this soil (Mollinedo et al., 2015) (Fig. 6.2).

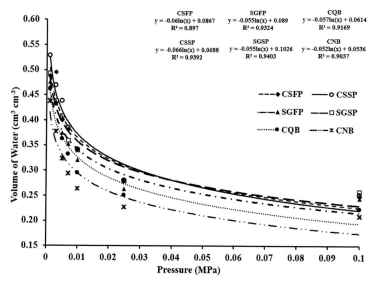

FIGURE 6.2 Water retention curves from 0.001 to 0.1 MPa for soil–biochar treatments for Maddock* soil series (Mollinedo et al., 2015). *CSFP*, Corn stover fast pyrolysis biochar; *CSSP*, corn stover slow pyrolysis biochar; *SGFP*, switchgrass fast pyrolysis biochar; *SGSP*, switchgrass slow pyrolysis biochar; *DB*, Dynamotive CQuest biochar; *CNB*, control no biochar. *Sandy, mixed, frigid Entic Hapludolls; $n = 4$; fast pyrolysis produced under microwave technology; slow pyrolysis produced using two-container nested retort; biochar application rate of $40\,g\,kg^{-1}$ soil (4%w/w).

Several of the biochars also increased plant available water-holding capacity (PAWC) at a tension range of 0.01–1.5 MPa in the Maddock soil, which was equal to additional 80,000–120,000 L ha^{-1} of water available for transpiration relative to the no-biochar control (assuming a transpiration rate of 5 mm day^{-1}) (Table 6.1). Other studies assessing different types of biochar have not recorded effects on water retention within the plant available water tension range even though impacts on water tensions associated with macropores were observed (Hardie et al., 2014; Ojeda et al., 2015). The effects of the different biochars on soil water retention reinforced the likelihood that soil microbial communities may differ depending on biochar type.

A comprehensive mechanistic model of biochar action on microbial communities is lacking and will require significant sophistication in experimental design and advanced techniques in microbiological and molecular analysis. A detailed review of microbiological and molecular methods of analysis for evaluating biochar effects on soil microbial

TABLE 6.1 ΔPAWC (plant available water attributable to biochar addition) for biochar treatments with PAWC found to be significantly different than the control. ΔPAWC is also presented on a hectare basis assuming a biochar application rate of 104 Mg ha^{-1} incorporated to a depth of 0.2 m with a final bulk density of 1300 kg ha^{-1} (Mollinedo et al., 2015)

Soil	Feedstock–pyrolysis processing	ΔPAWC (0.01–1.5 MPa)	
		cm^3 cm^{-3}	L ha^{-1}
Maddock			
	Corn stover fast pyrolysis	0.06	120,000
	Switchgrass fast pyrolysis	0.05	100,000
	Corn stover slow pyrolysis	0.05	100,000
	Switchgrass slow pyrolysis	0.04	80,000
	CQuest biochar	–	–
	Control no-biochar	–	–

Normal PAWC range used for coarse textured soils (Maddock).
Sandy, mixed, frigid Entic Hapludolls.
Fine-loamy, mixed, superactive, frigid Typic Eutrudepts. Soils collected Eastern South Dakota.
$n=4$.
Fast pyrolysis produced under microwave technology.
Slow pyrolysis produced using two-container nested retort.

populations is given in Ennis et al. (2012). A critical need is to be able to relate microbial community structure to function within the soil system. The metagenomics approach to evaluate function within representative microbial communities relative to biochar type, soil type, and other edaphic factors is promising as it can help to establish the role of the soil physical environment in biochar–soil–microbial–plant relationships.

Soil Solution Chemistry and Chemical Interactions at the Biochar Complex Interface

Biochars can exhibit highly complex and heterogeneous lattice structures composed of mineral (oxides, cations, anions, free radicals, etc.) and organic fractions (labile and recalcitrant organic molecules). The chemical properties of these biochars, including pH, electrical conductivity, calcium carbonate content, electrical charge density, volatile and nonvolatile organic molecules, and elemental composition, depend on the biomass feedstock type and extent of its condensation during the pyrolytic process. Generally, aromatic C (eg, syringyl units produced from lignin fractions) is dominant in biochar materials compared to aliphatic C (acetyl and methyl C groups), carboxyl, and carbonyl C (formed from cellulose transformations), irrespective of feedstock and pyrolytic process. The aromatic carbon (nuclear magnetic resonance spectrum at 108–165 ppm) provides lower polarity with hydrophobic fractions and is expected to interact (eg, formation of π-electron bonds, van der Waals forces, and hydrogen bonding) preferentially with nonpolar or weakly polar organic species. Carboxyl and carbonyl C of biochar materials are associated with surface functional groups (eg, oxygenated ligands/radicals) and their electrochemical properties respond to the soil solution chemistry (Fig. 6.3).

Functional groups ($-OH$, $-COOH$, $C\equiv N$, $C-O$, $C-H$, $-CH_3$, etc.) on biochar surfaces can exhibit acid–base properties and develop a variable surface charge in response to the chemical reactivity (soil solution pH, charge potential of free radicals at interface, etc.) of the soil solution. Cation and anion exchange capacities (CEC and AEC) of biochars change like any other organic molecule to soil solution pH (Fig. 6.4).

The pH values of these biochar materials are higher than point of zero net charge, which indicates the existence of overall negative charge potential on their surface. This variable charge develops on biochar surfaces in response to pH and ionic strength of soil solution because of several chemical reactions including protonation, deprotonation, and ligand exchange by amphoteric functional groups. Also the biochars' surface functional groups and dominant trace metals, including Fe and Mn oxides (mineral

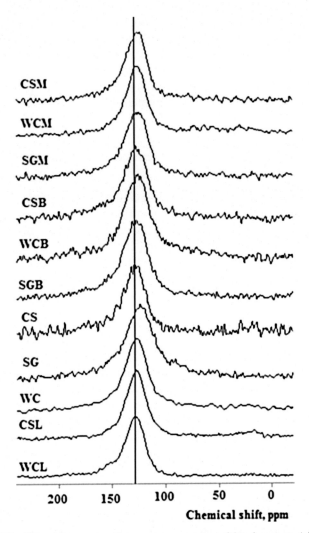

FIGURE 6.3 ^{13}C nuclear magnetic resonance spectra of biochar materials produced from corn stover, switchgrass, and Ponderosa pine wood residue (Chintala et al., 2014c). Microwave pyrolysis—corn stover biochar (CSM), switchgrass biochar (SGM), and Ponderosa pine wood residue biochar (WCM). Gasification—batch 1: corn stover biochar (CSB), switchgrass biochar (SGB), and Ponderosa pine wood residue biochar (WCB). Batch 2: corn stover biochar (CS), switchgrass biochar (SG), and Ponderosa pine wood residue biochar (WC). Fast pyrolysis using electricity: corn stover biochar (CSL) and Ponderosa pine wood residue biochar (WCL).

fraction), can undergo redox transformations to facilitate electron transport between oxidant and reductant species (redox couples). There are two tentatively identified redox-active fractions of biochars, which include quinone–hydroquinone moieties and condensed aromatic structures with

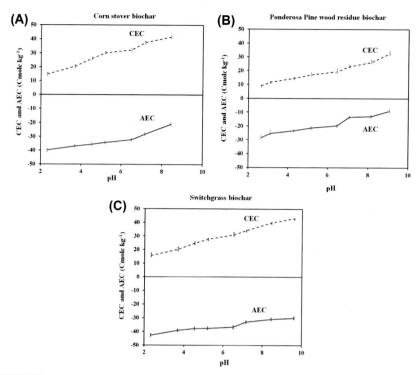

FIGURE 6.4 pH-dependent cation and anion exchange capacities (CEC and AEC) of biochars within the pH investigated (2.0≤pH≤9.0). Symbols represent data ±SE. Each data point is the average of four replications (Chintala et al., 2013).

conjugated π-electron systems (Klupfel et al., 2014; Xu et al., 2015). The contribution of the conjugated π-electron systems is increased in high-temperature biochars (>600°C) because of their higher aromaticity (Xu et al., 2015). The decrease of nitrous oxide emissions in biochar-amended soil systems was attributed to biochar's role as electron donor (reductant) for heterotrophic microbial populations (Briones, 2012; Cayuela et al., 2013; Chintala et al., 2014b). The application of biochar materials reduced the concentration of Cr (VI) from soil solution and also decreased chromate toxicity to sunflower plants (Choppala et al., 2012) (see also Chapter 9). This remediation of heavy metal by biochar was attributed to its reducing capability. Similarly the phenolic compounds of biochars have been shown to reduce the oxides of Mn and Fe to more soluble Mn and Fe forms (Graber et al., 2014). Biochars have also been found to mediate catalytic pathways to dissipate (reductive transformation) organic contaminants including dinitro herbicides and nitro explosives in soil (Oh et al., 2013). Overall, the incorporation of biochars with high concentrations of redox-active moieties (quinone–hydroquinone moieties) and conjugated π-electron systems of condensed aromatic carbon has the potential to facilitate electron transfer reactions, which could alter the microbial community structure and its functionality in the rhizosphere.

Additions of biochars, especially those produced from high-temperature pyrolytic process, exhibit liming effects by reducing exchangeable acidity (H^+ and Al^{3+}) and increasing soil pH (Fig. 6.5). The liming potential is caused by the release of base cations (Ca^{2+}, K^+) from biochar materials into soil, which replaces exchangeable Al^{3+} and H^+ from soil surfaces. These can then be precipitated from the soil solution as insoluble hydroxyl Al species as soil pH increases. Redox-active biochars can induce changes to key soil variables including pH and redox potential, which influence the microbial phylogenetic structure and their biosynthetic pathways of nutrient cycling, catalysis of organic matter and contaminants (mineralization/dissipation, reductive transformations), and greenhouse gas emissions (eg, via methanogenesis, denitrification, etc.) (see also Chapter 7).

Biochar materials and their associated complexes in soil have great potential to interact with organic and inorganic chemical species at the interface. Organic chemical species such as pesticides, organic acids,

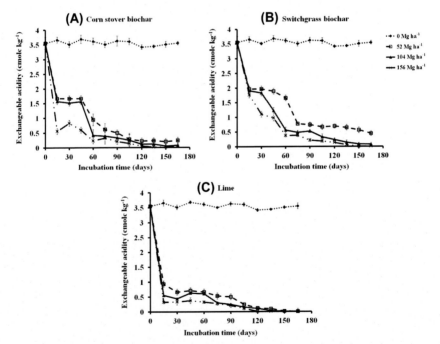

FIGURE 6.5   Effect of (A) corn stover biochar, (B) switchgrass biochar, and (C) lime on exchangeable acidity of an acidic soil depending on amount and incubation time. Each data point represents the mean of three replications with standard error. Biochars were produced using microwave pyrolysis at 650°C with a residence time of 18 min (Chintala et al., 2014a).

and organic pollutants exhibit strongly the partitioning into biochars by physical and chemical interactions (eg, hydrogen bond, π–π electron interaction). The partitioning of organic chemical species can be a physical process through deposition into micropores of the biochar lattice matrix (intraparticular occlusion). Hydrogen bonds could form because of strong valence forces between O-containing hydrophilic polar moieties of biochar and H atoms of organic chemical species. The aromatic rings of both these biochar properties can interact with organic chemical species in soil by developing π electron donor–π electron acceptor bonds.

Inorganic chemical species (cations and anions) can demonstrate long- and short-range interactions with the biochar surface in soil. Long-range interactions (physical forces) include the van der Waals and electrostatic outersphere complexes. Short-range interactions (chemical forces) can be performed by the formation of inner-sphere complexes caused by ligand exchange, covalent bonds, and hydrogen bonds. The liming effect of biochars can increase pH as well as the negative charge of soil particles by decreasing zeta potential, which causes an enhanced partitioning of cations into biochar complexes. Cations (base cations and trace metals) can demonstrate both long- and short-range interactions based on their charge density with biochar surface functional groups (carboxylic, alcohol, hydroxyls, etc.). Similarly, the partitioning of anions into biochar complexes depends on soil solution pH, ionic strength, charge density of ions, and the presence of competitive ions at the interface. In a batch experiment a decrease in partitioning of nitrate ions was observed because of deprotonation of biochar surface functional groups as the solution pH increased above the point of zero net charge, which resulted in a net negative charge on the biochar surface and caused electrostatic repulsion for negatively charged nitration ions. Moreover, the competition for sorption sites from OH^- ions could be enhanced in high-pH solution systems. The presence of competitive ions, including phosphate and sulfate with high charge density at the interface, can also hinder nitrate partitioning because of preferential sorption by biochar surface sites (Fig. 6.6).

Acidic soil amended with 4% (w/w) biochars increased the lability of sorbed phosphorus (0.5M $NaHCO_3$ extractable/labile P) by decreasing the partitioning coefficients and increasing the equilibrium P concentration. The improvement of lability of P could be caused by lowering the point of net zero charge of acidic soil and increasing the negative charge on the surface. The proton consumption reactions including ligand exchange between functional groups of biochar surface with aluminol/ferrol surfaces could potentially release oxy anions of phosphate into solution and, as a result, precipitate free radicals of Fe and Al as hydroxides with reduced high energy sorption sites. Moreover, the incorporation of biochars would increase the ionic strength of solution because of the release of

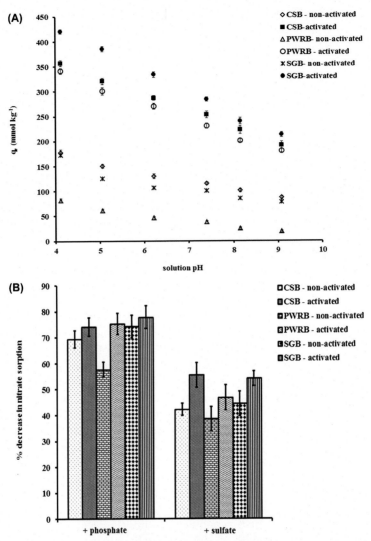

FIGURE 6.6 Effect of solution pH and competitive ion concentration on nitrate sorption (q-sorbed nitrate concentration) in nonactivated and activated biochars produced from corn stover (CSB), Ponderosa pine wood residue (PWRB), and switchgrass (SGB). The initial nitrate concentration of solution was $1.29\,mmol\,L^{-1}$ at pH that ranged from 4.0 ± 0.1 to 9.0 ± 0.1. Values are the mean of four samples. Error bars indicate standard error of the mean. Biochars were produced using microwave pyrolysis at 650°C with a residence time of 18 min (Chintala et al., 2013).

soluble salts, which can decrease the positive electric potential of acidic soil surface because of a screening effect (Fig. 6.7).

These kinds of biochar-induced abiotic processes (physical and chemical pathways) can exert influence on soil microbial community functions and diversity. Microbial communities are critical for soil ecosystem

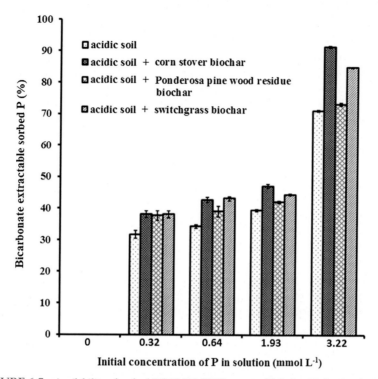

FIGURE 6.7 Availability of sorbed P (0.5M NaHCO₃ extractable) after 30-day incubation of acidic soil (Grummit series) with corn stover, Ponderosa pine wood residue, and switchgrass biochars at 40 g kg⁻¹ soil (4% w/w). Each bar represents the mean of four replications, with standard error bars. Biochars were produced using microwave pyrolysis at 650°C with a residence time of 18 min (Chintala et al., 2014d).

functioning that determine its health, fertility, carbon and nutrient cycling, etc. Moreover, the mechanistic understanding of biochar-induced physicochemical conditions and their intriguing influence on microbial community phylogeny and functioning will help to tease out the precise causation and predictability for biochar's potential as a climate change mitigation tool in enhancing soil functional resilience.

BIOTIC PROCESSES

Biochar Influence on Soil Microbial Community Structure

Biochar-induced physicochemical changes of soil can impact the microbial community abundance, diversity, and structure, which ultimately influence the soil–plant–microbe interactions (Quilliam et al., 2013). The highly porous structure of biochar can facilitate a protective microhabitat for soil microbial populations and provide necessary organic metabolites and minerals for microbial growth (Anderson et al., 2011; Saito and Muramoto, 2002; Warnock et al., 2007). The shifts in pH and oxidation potential because of the presence of biochar in soil can influence the microbial activity and community structure (Chen et al., 2013; McCormick et al., 2013; Seo and DeLaune, 2010). There is a strong need to investigate the relationship between biochar and soil microbial communities to identify the mechanisms of ecological services provided by the biochar as a soil amendment.

Most organic substrates in soil are made of complex polymers that are broken into monomeric compounds by extracellular enzymes that are used to maintain metabolic processes of soil fauna, mostly microorganisms (Brockett et al., 2012). Much of the organic material in biochar is recalcitrant mineral C bound in neatly structured aromatic rings unavailable for use in microbial processes (Chintala et al., 2014c). As a result, extracellular enzymes may bind (physically or chemically) to biochar, mineral surfaces, and organomineral complexes that may include biochar components. The bound extracellular enzymes can become inactive. Moreover, the biochar materials can combine with humic materials and microbial metabolites that undergo polymerization and condensation reactions to produce hydrogen-bonded hydrophobic polyphenol conglomerates. These conglomerates are highly resistant to decomposition, a process known as biochemical stabilization.

Biochemically stabilized organic materials form organomineral complexes. The organomineral interactions take place in three zones of the organomineral complex, which include contact, hydrophobic, and kinetic (Kleber et al., 2007; Whalen and Sampedro, 2010). How quickly and to what extent biochars and other organic materials are chemically stabilized is determined by their biochemical constituents and the type and amount of clay minerals present in a soil. Components of biochar are most strongly stabilized when they are chemically bound to mineral surfaces, typically irreversibly, leading to little or no decomposition, which generally these interactions occur in contact zone. In contrast, organic materials associated with the hydrophobic zone are exchangeable but strongly bound via hydrophobic reactions. Finally, organic materials in the kinetic zone are exchangeable but held via interactions such as hydrogen bonding and cation bridging and can therefore move freely in and out of soil solution (Kleber et al., 2007; Whalen and Sampedro, 2010).

Polyvalent cations play an important role in chemical stabilization of organomineral complexes and, in turn, the formation of soil aggregates (Zimmerman et al., 2011). In soils such as Mollisols where organic material and pH are high, cations such as Ca^{2+} and Mg^{2+} play a critical role in the chemical stabilization of organic matter by acting as bridging or binding agents between organic and mineral soil components (Zimmerman et al., 2011). Iron oxides that are largely amorphous serve a similar purpose in highly weathered soils low in pH and organic matter. The oxides are variably charged and can bind via electrostatic charge, acting as ion bridges to clay minerals, or bind with and adsorb to mineral and organic materials to form organomineral complexes (Maestrini et al., 2014; Masiello et al., 2015).

The characteristics of organic materials added to soil, including biochar, affect decomposition and interactions with other soil components. As an example, materials with high surface area and variable charge facilitate microbial colonization, sorption of nutrient ions, and form organomineral complexes (Chintala et al., 2013; Gomez et al., 2014; Mollinedo et al., 2015). How quickly organic materials interact with and are incorporated into aggregates, and the long-term stability of the formed aggregates, is dependent upon the chemical composition, decomposition, and incorporation of the organic material. This concept is known as the hierarchical model of aggregate formation (Monnier, 1965; Whalen and Sampedro, 2010). The more recalcitrant a biochar is, such as those formed at higher pyrolysis temperature with higher mineral C, the longer the material will take to be incorporated into soil aggregates and the greater the retention time for the aggregates (Zimmerman et al., 2011).

Previous studies showed the positive impact of biochars on microbial activity and populations in a wide range of soil textural classes (Domene et al., 2014; Smith et al., 2010; Steiner et al., 2008; Wang et al., 2014). Thus bacteria population density (*Enterobacter cloacae* UW5strain) increased by 16% after 4 weeks of incubation in a sandy loam soil with the incorporation of pine wood biochar produced at 600°C (Hale et al., 2015) because of its liming effect. Similarly, there was a 29% and 62% increase of microbial biomass carbon in sandy loam (Ameloot et al., 2013) and clay loam soils (Luo et al., 2013) with the application of high-temperature biochars (produced at 700°C) including willow wood biochar and *Miscanthus giganteus* residue-derived biochars, respectively. The positive priming effect of biochar incorporation on microbial activity and population is attributed to its ability to provide easily degradable organic metabolites and nutrients for microbial growth (Maestrini et al., 2014; Smith et al., 2010; Watzinger et al., 2014). The perfect pore size and geometry of the biochar lattice can protect soil microbes by excluding the wide size range of competitors and predators (such as collembola and protozoans) (Thies and Rillig, 2009). Water retention properties of soil can also be improved with the addition of highly porous biochar materials with high internal surface area, which may ultimately protect the soil microbes from desiccation. Highly interactive biochar surfaces can either sorb microbiocidal compounds (eg, catechol, cresols, xylenols, formaldehyde, acrolein, etc.) in soils and reduce their bioavailability to soil microbes (Painter, 2001) or sorb the microbial communities and avoid their leaching losses. The latter may ultimately be responsible for enhanced microbial abundance and activity (Chen et al., 2013). Several studies have tentatively identified these mechanisms for positive priming effects on microbial abundance and their activity especially with the incorporation of low pyrolytic temperature biochars.

Some studies showed either a negative priming effect or no effect on microbial abundance and activity. For instance, Chintala et al. (2014c) quantified fluorescein diacetate activity (FDA) and observed a negative priming effect on microbial activity (20% reduction compared to control) in a sandy loam with the addition (10 and 50 g kg^{-1}) of three different high-temperature biochars. These were derived from corn stover, switchgrass, and Ponderosa pine wood residue, all produced using a carbon-optimized gasification process at reactor temperatures ranging from 150 to 850°C with a residence time of 4 h and 4 min. In contrast the addition of biomass (corn stover, Ponderosa pine wood residue, and switchgrass) increased FDA activity in soil (Fig. 6.8). In this study the inhibitory effect on microbial activity (FDA activity) was observed because of the inaccessibility of readily metabolizable carbon and nutrient substrates following biochar amendments with high sorbing potential.

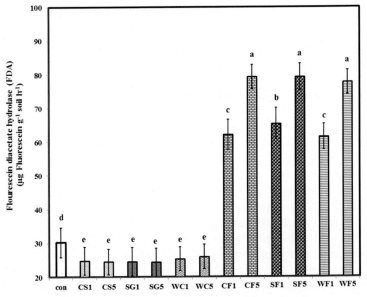

FIGURE 6.8 Effect of biomass feedstocks and their corresponding biochars on microbial activity in a sandy loam soil (Chintala et al., 2014c). Biochars: corn stover biochar (CS), switchgrass biochar (SG), and Ponderosa pine wood residue biochar (WC). Biomass feedstocks: corn stover (CF), switchgrass (SF), and Ponderosa pine wood residue (WF). Application rate: 1–10 g kg^{-1}; 5–50 g kg^{-1}. Sandy loam soil: Maddock soil series (sandy, mixed, frigid Entic Hapludolls) at crest position of landscape. Different letters represent significant differences by Tukeys test α = 0.05.

Soil microbial community structure is dynamic because of continuous shifts in bacterial and fungal composition and always correlates with the changes in soil properties. The addition of biochars can affect the adsorption dynamics of readily mineralizable carbon and nutrient substrates, which can influence the bacterial and fungal community structure and activities. Muhammed et al. (2014) found that total dissolved organic carbon (total N and C:N ratios of sandy loam soil incubated with slow pyrolyzed biochars [produced at 500°C using swine manure, fruit peals, *Brassica rapa* residue, and reed grass] for 90 days) positively influenced the fungal:bacterial ratio. A similar correlation was observed between the fungal:bacterial ratio and C:N ratio of four different textured soils incubated with fast pyrolysis biochar for 12 months (Gomez et al., 2014). PLFA biomarkers showed the stimulation of Gram-negative bacteria and actinomycetes populations in soils amended with ^{13}C-depleted wheat straw biochar (Watzinger et al., 2014). Amplified ribosomal intergenic spacer analysis was used to observe an increase in the actinomycetes population in a biochar-amended soil (Khodadad et al., 2011). A quantitative polymerase chain reaction (qPCR) study showed that gene copies of bacterial 16S rRNA genes were increased and copy numbers of fungal 18S rRNA were decreased in paddy soil with application of wheat straw biochar. But the overall microbial biomass C and N were not significantly different with control treatment despite shifts in bacterial and fungal community structure with the application of biochar (Chen et al., 2013). In a plate count study, the fungi and actinomycetes populations were increased and fungal population decreased in heavy metal (Cd, Pb)-contaminated field soil incorporated with wheat straw biochar. Generally the shifts in soil microbial community structure depends on several soil, plant, and environmental factors including characteristics of soil, amendment (biochar), and crop, climate, moisture, and temperature regimes of soil. A range of molecular approaches including PCR fingerprinting (T-RFLP and denaturing gradient gel electrophoresis [DGGE]), PCR-based clone libraries, and PLFA profiling are used widely in current studies to capture the shift in different microbial populations and their activities.

Biochar Effect on Microbial-Mediated Pathways in Soil

Biochars may act as a quorum quencher in soil by adsorbing signaling compounds used for intercellular communication and gene expression among soil microbial communities. Soil microbial populations communicate via the production of extracellular acyl homoserine lactones and other signal molecules allowing them to sense their population density and "make decisions," including when to produce many extracellular enzymes (Atkinson and Williams, 2009; Masiello et al., 2013; Miller and Bassler, 2001). Biochar can bind acyl homoserine lactones and/or other quorum-sensing molecules and consequently disrupt the signaling/communication pathways of soil microorganism. Specifically, these signaling molecules can bind with biochar materials using either polar or nonpolar interactions. Masiello et al. (2013) found that wood-derived biochar (700°C) inhibited N-(3-oxododecanoyl)-L-homoserine lactone-mediated communication between Gram-negative soil bacteria. This inhibitory effect was observed to increase with the addition of high-temperature biochars (>600°C) especially in soils with low organic matter contents. Recently, there has been focus on biochar's synergetic impact on signal exchanges, such as flavonoids, sesquiterpenes, and strigolactones, between plants roots and arbuscular mycorrhizae (AM), which facilitate spore germination and fungal growth (Akiyama et al., 2005). The adsorption of allelopathic toxins such as coumaric, caffeic, and ferulic acids (Zhu and Pignatello, 2005), parallel to the slow release of positive signaling molecules by biochars, are believed to be helpful for AM fungi colonization in the rhizosphere. Real-time qPCR studies showed that biochars can induce systemic plant resistance to necrotrophic *Botrytis cinerea* (gray mold) by mediating the salicylic acid-induced and jasmonic acid/ethylene-induced signaling pathways in strawberry (Meller-Harel et al., 2012) and tomato plants (Mehari et al., 2013). Salicylic acid has been implicated in modulating root microbiome community structure in *Arabidopsis thaliana* rhizospheres by affecting bacterial taxa in different ways (eg, growth signal or carbon source) (Lebeis et al., 2015). Biochar has been shown to adsorb salicylic acid (Essandoh et al., 2015). Thus there is a possibility that biochars could alter salicylic acid signaling resulting in changes in the root microbiome. A metagenomics approach combined with well-designed experimental approaches would be useful in evaluating the various mechanisms that have been proposed for biochar activity within the rhizosphere.

METAGENOMIC SEQUENCING TECHNOLOGIES

High-throughput sequencing coupled with bioinformatics enables researchers to determine how the community structure of the soil biota is affected by biochar applications. Soil biota consist of microorganisms (bacteria, fungi, archaea, and algae), soil animals (protozoa, nematodes, mites, springtails, spiders, insects, and earthworms), and plants (Doran and Parkin, 1994) living all or part of their lives in or on the soil or pedosphere (Fortuna, 2012). The soil zone closest to the plant roots termed the "rhizosphere" has the most density and diversity of microbes (Clark, 1949). Rhizosphere microbial communities have become a major focal point of research in recent years, especially regarding how they affect plants, and vice versa (Philippot et al., 2013). The microbiome present in the rhizosphere can be considered in part to be an active extension of the plant root system (reviewed by Bais et al., 2006; Philippot et al., 2013; Shi et al., 2011; Turner et al., 2013). Although influenced largely by the plant, other factors, such as the soil type and environmental conditions, can affect the composition and activity of the rhizosphere microbiome (Lundberg et al., 2012). Culture-based methods were used typically to isolate and identify the diversity of microbiomes (see also Chapter 4). However, because of specific and distinct nutrient and growth condition needs of various microorganisms, culture-based techniques often missed the vast majority of microbial diversity.

Culture-independent methods, where microbial diversity is identified using molecular techniques, are able to analyze both the diversity as well as activity of microbiomes. The majority of these methods depend on the use of molecular fingerprints of conserved gene sequences, which are unique among different microbial species. Diversity is determined by sequencing analysis that until recently required Sanger sequencing or hybridization of each unique fragment. Current next-generation sequencing techniques allow for sequencing of entire microbial communities. Further to this, estimating the number of copies of a given sequence of DNA or RNA can be accomplished via qPCR and RT-PCR, respectively. The accuracy of these estimates is dependent upon the number of copies of a given sequence in the targeted organism or organisms' genome(s) (see also Chapter 5).

The most commonly used conserved marker for fingerprinting of bacterial communities is the16S rRNA sequence (Woese and Fox, 1977), which has been adapted for next-generation sequencing. Nine different variable regions of this gene have been identified as useful in distinguishing bacterial species. Markers commonly used to differentiate between various groups of fungi include: the 18S rRNA gene; internal transcribed spacer (ITS) region (ITS1 and ITS2) or ITS2; and a portion of the 28S nuclear ribosomal large subunit rRNA gene (LSU), all of which have been successfully adapted to next-generation sequencing platforms (Schoch et al., 2012; Lim et al., 2010; Mello et al., 2011). Variable regions have been amplified with bacterial and fungal primers and amplicons with sequence variations then distinguished using techniques such as DGGE (see also Chapter 8) and thermal gradient gel electrophoresis. Subsequent identification of sequences using hybridization or sequencing methods has led to a much better understanding of composition and diversity of microbiomes (Roncon-Flores et al., 2013). While these methods helped obtain a much deeper understanding of rhizosphere microbial ecology, the major bottleneck was their low-throughput nature.

There are a number of next-generation sequencing platforms available with the basic concepts and workflow differing between platforms (Lindahl et al., 2013; Thomas et al., 2012). In pyrosequencing, which is the focus of our discourse, the target DNA or RNA is extracted, amplified, and the sequences then "tagged" with an additional molecular marker often called a barcode—a set of sequences that can be used to identify a treatment or set of conditions that will be amplified along with the target material. These tagged samples will then be pooled to create a library for sequencing on a given platform. Bioinformatic software is used to sort the samples by their barcodes. The sequences are then removed or sorted based on their similarity to one another and a set of known sequences. Similarity is determined based on the alignment of sequences with each other and the reference sequences. Finally, the sequences are functionally annotated allowing them to be classified based on their "function or taxonomic unit" (Thomas et al., 2012). Most reads are now 400 to 500 base pairs in length or greater. One of the challenges associated with the application of metagenomics is how to approach the analysis of the millions of sequences generated during one pyrosequencing run, of which 20–40% may need to be removed because of low quality (Lindahl et al., 2013). As a result there are several sequence quality management programs available that should be used prior to binning and alignment of sequences. Here we explore the potential of using pyrosequencing technologies in particular to evaluate the effect of biochar on rhizosphere microbial communities.

Soil samples especially those from the rhizosphere with various agricultural amendments such as biochar pose a challenge when it comes to extraction of DNA and RNA of very high quality for pyrosequencing experiments. SOM and other organic fractions interact with RNA and DNA, mineral constituents sorb these materials, and the presence of RNase and DNase contaminants lead to their degradation (Rancon-Florez et al., 2013). All extraction kits perform similarly in that cells are mechanically disrupted and solubilized to release DNA/RNA, solids are removed, and DNA/RNA is bound or precipitated then eluted and solubilized (Lindhal et al., 2013). The choice of extraction method or kit will be dependent upon the size and cell structure of a given organism, the environment from which the organism was taken (ie, soil vs water), and the types of material that could interfere with extraction chemistry. Examples include metals such as Fe, humics, and proteins, all of which can interfere with DNA and RNA extraction (Carrig et al., 2007; Leite et al., 2014; Perrine et al., 2014). The presence of biochar further complicates extraction of DNA and RNA by acting somewhat similarly to SOM (Leite et al., 2014).

Nevertheless, a number of examples exist in the literature that highlight the use of pyrosequencing technologies to evaluate the rhizosphere microbiome. Recent methods developed to isolate rhizosphere soil fractions with various affinities to the root system have enabled studying spatiotemporal distribution of microbial communities in the soil (Bulgarelli et al., 2012; Edwards et al., 2015). Taketani et al. (2013) used 16S rRNA gene pyrosequencing to verify differences in microbial communities associated with Amazonian Dark Earth soils and adjacent soils. Their work revealed distinct community shifts that have persisted, which they equated to the presence and long-term persistence of biochar. Also pyrosequencing of 16S rRNA sequences from the rhizosphere of potted pepper plants revealed that biochar addition increased and/ or maintained N-fixing rhizobium, plant growth-promoting rhizobacteria that solubilize phosphate and degrade resistant C found in biochar (Anderson et al., 2011). Peiffer et al. (2013) showed that the maize (*Zea mays*) genotype had a small but significant and heritable variation in rhizosphere bacterial diversity between genotypes. Larger variations were observed between different fields and geographical locations, but it was clear that microbiome diversity can be used as a heritable trait in plant breeding programs. The authors suggested that future studies should focus on functional groups of microbes rather than taxonomic relatedness of the microbial community.

Studies evaluating shifts in microbial communities in the rhizosphere have revealed changes in community structure because of variations in soil, climate, and time of season. For example, Verbruggen et al. (2012) used primers targeting the V7 and V6 hypervariable regions of the small-subunit SSU rRNA gene to compare the effects of growing genetically modified (GM) and non-GM corn varieties on native arbuscular mycorrhizal fungal (AMF) populations via pyrosequencing of DNA and RNA. The authors recorded few differences in AMF communities in the rhizosphere between corn varieties but substantial shifts in community structure with time of season and location. In a separate experiment, pyrosequencing verified that increases in macronutrients resulted in shifts in the AMF community structure resulting in the loss of some species (Lin et al., 2012). Indeed, biochars can impact soil systems by altering the rates and types of nutrient cycling (Warnock et al., 2007) and thus influence AMF communities.

Despite obvious benefits, DNA-based metagenomic studies are characterized by certain biases and inconsistencies. For example, some variable regions are better than others at distinguishing some species. Therefore the use of multiple variable regions of 16S rRNA in a single experiment does not always provide comparable results. Peiffer et al. (2013) used a pilot study to select the V3–V4 region as the most suitable to study the maize rhizosphere microbiome and found that the targeted variable region determined the recovered microbiome diversity. Fungal primer biases are even greater. The choice of multiple primers is dependent upon the experimental question(s) to be addressed and therefore the population(s) to be studied. The ITS primer sets have sufficient specificity in some instances to identify fungal strains in soils, the LSU region can provide similar specificity and allow for the study of phylogenetic structure, while 18S primer sets tend to be more conserved (Lindahl et al., 2013).

In addition, the presence of DNA does not verify that the organism or gene of interest within an organism is active. Microorganisms are often found in a static state and, although most DNA is degraded rapidly after an organism dies, persistent moieties could be amplified. Therefore RNA-based metagenomics can overcome such deficiencies as this nucleic acid is produced only when organisms are active. In addition, these studies provide information not only on the diversity of the microbiome, but also its activity. For example, while DNA-based metagenomics can provide species or phylogenetic information, RNA-based metagenomics provide additional gene expression information. This can help elucidate enzymatic activities and signaling mechanisms active among the rhizosphere microbial communities. In RNA-based metagenomics the use of next-generation sequencing technologies has helped overcome limitations of alternate approaches such as qPCR, for example, the need for prior knowledge of which specific gene sequences to target.

A study on the influence of three different major crop species (wheat, oat, and pea) used RNA rather than DNA to evaluate changes in the diversity, composition, and activity of the rhizosphere microbiome (Turner et al., 2013). The authors obtained environmental RNA, which was reverse transcribed into cDNA and sequenced using Illumina HiSeq. The method elucidated the microbiome structure without bias from PCR amplification. An added advantage is that RNA-based metagenomics specifically targets metabolically active microorganisms rather than dormant ones. Thus the researchers showed that pea plants influenced the rhizosphere microbiome to a significantly larger extent compared to wheat and oats. Interestingly, the pea rhizosphere was also enriched in fungi compared to oats and wheat. The study identified metabolic capabilities potentially important for rhizosphere colonization including cellulose degradation (cereals), H_2 oxidation (pea), and methylotrophy (all plants).

In addition, metaproteomics have also been utilized to understand rhizosphere microbial activities (eg, Schlaeppi et al., 2014; Yergeau et al., 2014). Unlike traditional approaches all meta-approaches evaluate the whole microbiome in an unbiased manner. Current challenges with these approaches include annotation and identification of orthologous groups of genes/proteins based on their function. Also it is crucial to correlate metagenomic information to chemical, physical, and biological properties of an environment in order to assess how microbial composition and diversity influences plant phenotype and productivity as well as provide an effective means of assessing the impacts of biochar additions on the environment. Multivariate analysis techniques are important tools that can be used to relate environmental variables to that of community structure (see also Chapter 3).

CONCLUSIONS

Accumulating evidence has indicated clearly that biochar addition can significantly alter physical and chemical characteristics of the supplemented soil and can indirectly influence microbial community composition and functional capacity. In addition, biochars can directly influence microbial communities because of their effects on microbial nutrition (eg, soil C:N ratio) as well as communication (eg, quorum sensing). However, our knowledge on the direct effects of biochar on microbial community structure and function, especially under agricultural production conditions, is fragmented. The unbiased and high-throughput nature of metagenomic technologies provides a significant opportunity to bridge this knowledge gap to enhance our understanding of the impact of biochar on rhizosphere microbial community structure and function. Such knowledge would help the development of biochar-based technologies for enhancing soil health.

References

Akiyama, K., Matsuzaki, K-I., Hayashi, H., 2005. Plant sesquiterpenes induce hyphal branching in arbuscular mycorrhizal fungi. Nature 435, 824–827.

Ameloot, N., Neve, S.D., Jegajeevagan, K., Yildiz, G., Buchan, D., Funkuin, Y.N., Prins, W., Bouckaert, L., Sleutel, S., 2013. Short-term CO_2 and N_2O emissions and microbial properties of biochar amended sandy loam soils. Soil Biol. Biochem. 57, 401–410.

Anderson, C.R., Condron, L.M., Clough, T.J., Fiers, M., Stewart, A., Hill, R.A., Sherlock, R.R., 2011. Biochar induced soil microbial community change: Implications for biogeochemical cycling of carbon, nitrogen and phosphorus. Pedobiologia 54, 309–320.

Atkinson, S., Williams, P., 2009. Quorum sensing and social networking in the microbial world. J. R. Soc. Interface 6, 959–978.

Bais, W.P., Weir, T.L., Perry, L.G., Gilroy, S., Vivanco, J.M., 2006. The role of root exudates in rhizosphere interactions with plants and other organisms. Annu. Rev. Plant Biol. 57, 233–266.

Barnes, R.T., Gallagher, M.E., Masiello, C.A., Liu, Z., Dugan, B., 2014. Biochar-induced changes in soil hydraulic conductivity and dissolved nutrient fluxes constrained by laboratory experiments. PLoS One. http://dx.doi.org/10.1371/journal.pone.0108340.

Briones, A.M., 2012. The secrets of El Dorado viewed through a microbial perspective. Front. Microbiol. 3, 239.

Brockett, B.F.T., Prescott, C.E., Grayston, S.J., 2012. Soil moisture is the major factor influencing microbial community structure and enzyme activities across seen biogeoclimatic zones in western Canada. Soil Biol. Biochem. 44, 9–20.

Bulgarelli, D., Rott, M., Schlaeppi, K., van Themaat, E.V.L., Ahmadinejad, N., Assenza, F., Rauf, P., Huettel, B., Reinhardt, R., Schmelzer, E., Peplies, J., Gloeckner, F.O., Amann, R., Eickhorst, T., SchulzeLefert, P., 2012. Revealing structure and assembly cues for Arabidopsis root-inhabiting bacterial microbiota. Nature 488, 91–95.

Carrig, C., Rice, O., Kavanagh, S., Collins, G., O'Flaherty, V., 2007. DNA extraction method affects microbial community profiles from soils and sediment. Appl. Microbiol. Biotechnol. 77, 955–964.

Cayuela, M.L., Sanchez-Monedero, M.A., Roig, A., Hanley, K., Enders, A., Lehmann, J., 2013. Biochar and denitrification in soils: when, how much and why does biochar reduce N_2O emissions? Nat. Sci. Rep. 3, 1732.

Chan, K., Van Zwieten, L., Meszaros, I., Downie, A., Joseph, S., 2007. Agronomic values of greenwaste biochar as a soil amendment. Aust. J. Soil Res. 45, 629–634.

Chaer, G.M., Fernandes, M.F., Myrold, D.D., Bottomley, P.J., 2009. Shifts in microbial community composition and physiological profiles across a gradient of induced soil degradation. Soil Sci. Soc. Am. J. 73, 1327–1334.

Chen, J.H., Liu, X.Y., Zheng, J.W., Zhang, B., Lu, H.F., Chi, Z.Z., Pan, G.X., Li, L.Q., Zheng, J.F., Zhang, X.H., Wang, J.F., Yu, X.Y., 2013. Biochar soil amendment increased bacterial but decreased fungal gene abundance with shifts in community structure in a slightly acid rice paddy from Southwest China. Appl. Soil Ecol. 71 (2013), 33–44.

Chia, C.H., Downie, A., Munroe, P., 2015. Charateristics of biochar: physical and structural properties. In: Lehmann, D.J., Joseph, S. (Eds.), Biochar for Environmental Management: Science and Technology. Earthscan Books Ltd, London & New York, pp. 89–109.

Chintala, R., Mollinedo, J., Schumacher, T.E., Malo, D.D., Papiernik, S., Clay, D.E., Kumar, S., Gulbrandson, D.W., 2013. Nitrate sorption and desorption by biochars produced from microwave pyrolysis. Microporous Mesoporous Mater. 179, 250–257.

Chintala, R., Mollinedo, J., Schumacher, T.E., Malo, D.D., Julson, J.L., 2014a. Effect of biochars on chemical properties of acidic soil. Arch. Agron. Soil Sci. 60, 393–404.

Chintala, R., Owen, R.K., Schumacher, T.E., Spokas, K.A., McDonald, L.M., Malo, D.D., Clay, D.E., Bleakley, B., 2014b. Denitrification kinetics in biomass and biochar amended soils of different landscape positions. Environ. Sci. Pollut. Res. 22, 5152–5163.

Chintala, R., Schumacher, T.E., Kumar, S., Malo, D.D., Rice, J., Bleakley, B., Chilom, G., Papiernik, S., Julson, J.L., Clay, D., Gu, Z.R., 2014c. Molecular characterization of biochar materials and their influence on microbiological properties of soil. J. Hazard. Mater. 279, 244–256.

Chintala, R., Schumacher, T.E., McDonald, L.M., Clay, D.E., Malo, D.D., Clay, S.A., Papiernik, S.K., Julson, J.L., 2014d. Phosphorus sorption and availability in biochars and soil biochar mixtures. CLEAN – Soil Air Water 42, 626–634.

Choppala, G.K., Bolan, N.S., Megharaj, M., Chen, Z., Naidu, R., 2012. The influence of biochar and black carbon on reduction and bioavailability of chromate in soils. J. Environ. Qual. 41, 1175–1184.

Clark, F.E., 1949. Soil microorganisms and plant roots. Adv. Agron 1, 241–288.

Devare, M., Londono-R, L.M., Thies, J.E., 2007. Neither transgenic Bt maize (MON863) nor tefluthrin insecticide adversely affect soil microbial activity or biomass: a 3-year field analysis. Soil Biol. Biochem. 39, 2038–2047.

Domene, X., Mattana, S., Hanley, K., Enders, A., Lehmann, J., 2014. Medium-term effects of corn biochar addition on soil biota activities and functions in a temperate soil cropped to corn. Soil Biol. Biochem. 72, 152–162.

Doran, J.W., Parkin, T.B., 1994. Defining and assessing soil quality. In: Doran, J.W., Coleman, D.C., Bezdicek, D.F., Stewart, B.A. (Eds.), Defining Soil Quality for a Sustainable Environment. SSSA Inc., Madison, Wisconsin, USA.

Dose, H.L., Fortuna, A.M., Cihacek, L.J., Norland, J., DeSutter, T.M., Clay, D.E., Bell, J., 2015. Biological indicators provide short term soil health assessment during sodic soil reclamation. Ecol. Indic. 58, 244–253.

Edwards, J., Johnson, C., Santos-Medellin, C., Lurie, E., Podishetty, N.K., Bhatnagar, S., Eisen, J.A., Sundaresan, V., 2015. Structure, variation, and assembly of the root-associated microbiomes of rice. Proc. Natl. Acad. Sci. 112, E911–E920. http://dx.doi.org/10.1073/pnas.1414592112.

Ennis, C.J., Evans, A.G., Islam, M., Ralebitso-Senior, T.K., Senior, E., 2012. Biochar: carbon sequestration, land remediation, and impacts on soil microbiology. Crit. Rev. Environ. Sci. Technol. 42, 2311–2364.

Essandoh, M., Kunwar, B., Pittman Jr., C.U., Mohan, D., Mlsna, T., 2015. Sorptive removal of slaicyclic acid and ibuprofen from aqueous solutions using pine wood fast pyrolysis biochar. Chem. Eng. J. 265, 219–227.

Fortuna, A.M., 2012. The soil biota. Nat. Educ. Knowl. 3 (10), 1. http://www.nature.com/scitable/knowledge/library/the-soil-biota-84078125.

Gomez, J.D., Denef, K., Stewart, C.E., Zheng, J., Cotrufo, M.F., 2014. Biochar addition rate influences soil microbial abundance and activity in temperate soils. Eur. J. Soil Sci. 65, 28–39.

Graber, E.R., Frenkel, O., Jaiswal, A.K., Elad, Y., 2014. How may biochar influence severityof diseases caused by soilborne pathogens? Carbon Manag. 5, 169–183.

Grayston, S.J., Campbell, C.D., Bardgett, R.D., Mawdsley, J.L., Clegg, C.D., Ritz, K., Griffths, B.S., Rodwell, J.S., Edwards, S.J., Davies, W.J., Elston, D.J., Millard, P., 2004. Assessing shifts in microbial commmunity structure across a range of grasslands of differing management intensity using CLPP, PLFA and community DNA techniques. Appl. Soil Ecol. 25, 63–84.

Gul, S., Whalen, J.K., Thomas, B.W., Sachdeva, V., Deng, H., 2015. Physico-chemical properties and microbial responses in biochar-amended soils: mechanisms and future directions. Agric. Ecosyst. Environ. 206, 46–59.

Hale, L., Luth, M., Crowley, D., 2015. Biochar characteristics relate to its utility as an alternative soil inoculum carrier to peat and vermiculite. Soil Biol. Biochem. 81, 228–235.

Hansel, C.M., Fendorf, S., Jardine, P.M., Francis, C.A., 2008. Changes in bacterial and archael community structure and functional diversity along a geochemically variable soil profile. Appl. Environ. Microbiol. 74, 1620–1633.

Hardie, M., Clothier, B., Bound, S., Oliver, G., Close, D., 2014. Does biochar influence soil physical properties and soil water availability? Plant Soil 376, 347–361.

He, X., Su, Y., Liang, Y., Chen, X., Zhu, H., Wang, K., 2012. Land reclamation and short-term cultivation change soil microbial communities and bacterial metabolic profiles. J. Sci. Food Agric. 92, 1103–1111.

Khodadad, C.L.M., Zimmerman, A.R., Green, S.J., Uthandi, S., Foster, J.S., 2011. Taxa-specific changes in soil microbial community composition induced by pyrogenic carbon amendments. Soil Biol. Biochem. 43, 385–392.

Kleber, M., Sollins, P., Sutton, R., 2007. A conceptual model of organo-mineral interactions in soils: self-assembly of organic molecular fragments into zonal structures on mineral surfaces. Biogeochemistry 85, 9–24.

Kleber, M., Hockaday, W., Nico, P.S., 2015. Charateristics of biochar: macro-molecular properties. In: Lehmann, D.J., Joseph, S. (Eds.), Biochar for Environmental Management: Science and Technology. Earthscan Books Ltd, London & New York, pp. 111–137.

Klupfel, L., Keiluweit, M., Kleber, M., Sander, M., 2014. Redox properties of plant biomass-derived black carbon (Biochar). Environ. Sci. Technol. 48, 5601–5611.

Lebeis, S.L., Paredes, S.H., Lundberg, D.S., Breakfield, N., Gehring, J., McDonald, M., Malfatti, S., del Rio, T.G., Jones, C.D., Tringe, S.G., Dangl, J.L., 2015. Salicylic acid modulates colonization of the root microbiome by specific bacterial taxa. Science 349, 860–864.

Lehmann, J., Rillig, M.C., Thies, J., Masiello, C.A., Hockaday, W.C., Crowley, D., 2011. Biochar effects on soil biota–a review. Soil Biol. Biochem. 43, 1812–1836.

Leite, D.C.A., Balieiro, F.C., Pires, C.A., Madari, B.E., Rosado, A.S., Coutinho, H.L.C., Peixoto, R.S., 2014. Comparison of DNA extraction protocols for microbial communities from soil treated with biochar. Braz. J. Microbiol. 45, 175–183.

Lim, T.J., Spokas, K.A., Feyereisen, G., Novak, J.M., 2015. Predicting the impact of biochar additions on soil hydraulic properties. Chemosphere. http://dx.doi.org/10.1016/j.chemosphere.2015.06.069.

Lim, Y.W., Kim, B.K., Kim, C., Jung, H.S., Kim, B.S., Lee, J.H., Chun, J., 2010. Assessment of soil fungal communities using pyrosequencing. J. Microbiol. 48, 284–289.

Lin, X., Feng, Y., Zhang, H., Chen, R., Wang, J., Zhang, J., Chu, H., 2012. Long-term balanced fertilization decreases arbuscular mycorrhizal fungal diversity in an arable soil in North China revealed by 454 pyrosequencing. Environ. Sci. Technol. 46, 5764–5771.

Lindahl, B.D., Nilsson, R.H., Tedersoo, L., Abarenkov, K., Carlsen, T., Kjøller, R., Kõljalg, U., et al., 2013. Fungal community analysis by high-throughput sequencing of amplified markers–a user's guide. New Phytol. 199, 288–299.

Lundberg, D.S., Lebeis, S.L., Paredes, S.H., Yourstone, S., Gehring, J., Malfatti, S., Tremblay, J., Engelbrektson, A., Kunin, V., del Rio, T.G., et al., 2012. Defining the core *Arabidopsis thaliana* root microbiome. Nature 488, 86–90.

Luo, Y., Durenkamp, M., Nobili, M.D., Lin, Q., Devonshire, B.J., Brookes, P.C., 2013. Microbial biomass growth following incorporation of biochars produced at 350 °C or 700 °C, in a silty-clay loam soil of high and low pH. Soil Biol. Biochem. 57, 513–523.

Maestrini, B., Herrmann, A.M., Nannipieri, P., Schmidt, M.W.I., Abiven, S., 2014. Ryegrass-derived pyrogenic organic matter changes organic carbon and nitrogen mineralization in a temperate forest soil. Soil Biol. Biochem. 69, 291–301.

McCormack, S.A., Ostle, N., Bardgett, R.D., Hopkins, D.W., Vanbergen, A.J., 2013. Biochar in bioenergy cropping systems: impacts on soil faunal communities and linked ecosystem processes. Glob. Change Biol. Bioenergy 5, 81–95.

Masiello, C.A., Gao, Y.C.X., Liu, S., Cheng, H-Y., Bennett, M.R., Rudgers, J.A., Wagner, D.S., Zygourakis, K., Silberg, J.J., 2013. Biochar and microbial signaling: production conditions determine effects on microbial communication. Environ. Sci. Technol. 47, 11496–11503.

Masiello, C.A., Dugan, B., Brewer, C.E., Spokas, K.A., Novak, J.M., Liu, Z., Sorrenti, G., 2015. Biochar effects on soil hydrology. In: Lehmann, D.J., Joseph, S. (Eds.), Biochar for Environmental Management: Science and Technology. Earthscan Books Ltd, London & New York (Chapter 19).

Mehari, Z.H., Meller-Harel, Y., Rav-David, D., Graber, E.R., Elad, Y., 2013. The nature of systemic resistance induced in tomato (*Solanum lycopersicum*) by biochar soil treatments. IOBC/WPRS Bull. 89, 227–230.

Meller-Harel, Y., Elad, Y., Rav-David, D., Borenstein, M., Sulchani, R., Lew, B., Graber, E.R., 2012. Biochar mediates systemic response of strawberry to foliar fungal pathogens. Plant Soil 357, 245–257.

Mello, A., Napoli, C., Murat, C., Morin, E., Marceddu, G., Bonfante, P., 2011. ITS-1 versus ITS-2 pyrosequencing: a comparison of fungal populations in truffle grounds. Mycologia 103, 1184–1193.

Miller, M.B., Bassler, B.L., 2001. Quorum sensing in bacteria. Annu. Rev. Microbiol. 55, 165–199.

Mollinedo, J., Schumacher, T.E., Chintala, R., 2015. Influence of feedstocks and pyrolysis on biochar's capacity to affect the soil water retention characteristics. J. Anal. Appl. Pyrolysis 114, 100–108.

Monnier, G., 1965. Action de matières organiques sur la stabilité structural des sols. Thèse de la faculté 140pp.

Muhammad, N., Dia, Z., Xiao, K., Meng, J., Brookes, P.C., Liu, X., Wang, H., Wu, J., Xu, J., 2014. Changes in microbial community structure due to biochars generated from different feedstocks and their relationship with soil chemical properties. Geoderma 226, 270–278.

Oh, S.-Y., Son, J.-G., Chiu, P.C., 2013. Biochar-rnediated reductive transformation of nitro herbicides and explosives. Environ. Toxicol. Chem. 32, 501–508.

Ojeda, G., Mattana, S., Avila, A., Alcaniz, J.M., Volkmann, M., Bachmann, J., 2015. Are soil-water functions affected by biochar application? Geoderma 240–250, 1–11.

Painter, T.J., 2001. Carbohydrate polymers in food preservation: an integrated view of the Maillard reaction with special reference to discoveries of preserved foods in Sphagnum-dominated peat bogs. Carbohydr. Polym. 36, 335–347.

Pankhurst, C.E., Hawke, B.G., McDonald, H.J., Kirkby, C.A., Buckerfield, J.C., Michelsen, P., O'Brien, K.A., Gupta, V.V.S.R., Doube, B.M., 1995. Evaluation of soil biological properties as potential bioindicators of soil health. Anim. Prod. Sci. 35 (7), 1015–1028.

Peiffer, J.A., Spor, A., Koren, O., Jin, Z., Tringe, S.G., Dangl, J.L., Buckler, E.S., Ley, R.E., 2013. Diversity and heritability of the maize rhizosphere microbiome under field conditions. Proc. Natl. Acad. Sci. U. S. A. 110, 6548–6553.

Perrine, C., Vigneron, A., Lucchetti-Miganeh, C., Ciron, P.E., Godfroy, A., Cambon-Bonavita, M.A., 2014. Influence of DNA extraction method, 16S rRNA targeted hypervariable regions, and sample origin on microbial diversity detected by 454 pyrosequencing in marine chemosynthetic ecosystems. Appl. Environ. Microbiol. 80, 4626–4639.

Philippot, L., Raaijmakers, J.M., Lemanceau, P., van der Putten, W.H., 2013. Going back to the roots: the microbial ecology of the rhizosphere. Nat. Rev. Microbiol. 11, 789–799.

Quilliam, R.S., Glanville, H.C., Wade, S.C., Jones, D.L., 2013. Life in the 'charosphere' – does biochar in agricultural soil provide a significant habitat for microorganisms? Soil Biol. Biochem. 65, 287–293.

Rincon-Florez, V.A., Carvalhais, L.C., Schenk, P.M., 2013. Culture-independent molecular tools for soil and rhizosphere microbiology. Diversity 5, 581–612.

Rosenzweig, R., Furman, A., Dosoretz, C., Shavit, U., 2014. Modeling biofilm dynamcs and hydraulic properties in variably saturated soils using a channel network model. Water Resour. Res. 50, 5678–5697.

Saito, M., Muramoto, T., 2002. Inoculation with arbuscular mycorrhizal fungi: the status quo in Japan and the future prospects. Plant Soil 244, 273–279.

Schlaeppi, K., Dombrowski, N., Oter, R.G., van Themaat, E.V.L., Schulze-Lefert, P., 2014. Quantitative divergence of the bacterial root microbiota in *Arabidopsis thaliana* relatives. Proc. Natl. Acad. Sci. U. S. A. 111, 585–592.

Schoch, C.L., Seifert, K.A., Huhndorf, S., Robert, V., Spouge, J.L., Levesque, C.A., Chen, W., Bolchacova, E., Voigt, K., Crous, P.W., Miller, A.N., 2012. Nuclear ribosomal internal transcribed spacer (ITS) region as a universal DNA barcode marker for Fungi. Proceedings of the National Academy of Sciences 109 (16), 6241–6246.

Seo, D.C., Delaune, R.D., 2010. Effect of redox conditions on bacterial and fungal biomass and carbon dioxide production in Louisiana coastal swamp forest sediment. Sci. Total Environ. 408, 3623–3631.

Shi, S., Richardson, A.E., O'Callaghan, M., DeAngelis, K.M., Jones, E.E., Stewart, A., Firestone, M.K., Condron, L.M., 2011. Effects of selected root exudate components on soil bacterial communities. FEMS Microbiol. Ecol. 77, 600–610.

Smith, J.L., Collins, H.P., Bailey, V.L., 2010. The effect of young biochar on soil respiration. Soil Biol. Biochem. 42, 2345–2347.

Steiner, C., Das, K.C., Garcia, M., Forster, B., Zech, W., 2008. Charcoal and smoke extract stimulate the soil microbial community in a highly weathered xanthic ferralsol. Pedobiologia 51, 359–366.

Stevenson, F.J., 1994. Humus Chemistry: Genesis, Composition, Reactions. John Wiley & Sons.

Suliman, W., Harsh, J.B., Abu-Lail, N.I., Fortuna, A.M., Dallmeyer, I., Garcia-Perez, M., 2016a. Influence of feedstock source and pyrolysis temperature on biochar bulk and surface properties. Biomass and Bioenergy 84, 37–48.

Suliman, W., Harsh, J.B., Abu-Lail, N.I., Fortuna, A.M., Dallmeyer, I., Garcia-Perez, M., 2016b. Modification of biochar surface by air oxidation: Role of pyrolysis temperature. Biomass and Bioenergy 85, 1–11.

Sun, Z., Moldrup, P., Elsgaard, L., Arthur, E., Bruun, E.W., Hauggaard-Nielsen, H., de Jonge, L.W., 2013. Direct and indirect short-term effects of biochar on physical characteristics of an arable sandy loam. Soil Sci. 178, 465–473.

Taketani, R.G., Lima, A.B., da Conceição Jesus, E., Teixeira, W.G., Tiedje, J.M., Tsai, S.M., 2013. Bacterial community composition of anthropogenic biochar and Amazonian anthrosols assessed by 16S rRNA gene 454 pyrosequencing. Antonie Leeuwenhoek 104, 233–242.

Thies, J.E., Rillig, M.C., 2009. Characteristics of biochar: biological properties. In: Lehmann, J., Joseph, S. (Eds.), Biochar for Environmental Management: Science and Technology. Earthscan Earthscan Books Ltd, London & New York, pp. 85–105.

Thies, J.E., Rillig, M.C., Graber, E.R., 2015. Biochar effects on the abundance, activity and diversity of the soil biota. In: Lehmann, D.J., Joseph, S. (Eds.), Biochar for Environmental Management: Science and Technology. Earthscan Books Ltd, London & New York, pp. 327–389.

Thomas, T., Gilbert, J., Meyer, F., 2012. Metagenomics – a guide from sampling to data analysis. Microb. Inf. Exp. 2, 1–12.

Turner, T.R., Ramakrishnan, K., Walshaw, J., Heavens, D., Alston, M., Swarbreck, D., Osbourn, A., Grant, A., Poole, P.S., 2013. Comparative metatranscriptomics reveals kingdom level changes in the rhizosphere microbiome of plants. ISME J. 7, 2248–2258.

Verbruggen, E., Kuramae, E.E., Hillekens, R., de Hollander, M., Kiers, E.T., Röling, W.F., Kowalchuk, G.A., van der Heijden, M.G., 2012. Testing potential effects of maize expressing the Bacillus thuringiensis Cry1Ab endotoxin (Bt maize) on mycorrhizal fungal communities via DNA-and RNA-based pyrosequencing and molecular fingerprinting. Appl. Environ. Microbiol. 78, 7384–7392.

Wang, Z., Li, Y., Chang, S.X., Zhang, J., Jiang, P., Zhou, G., Shen, Z., 2014. Contrasting effects of bamboo leaf and its biochar on soil CO_2 efflux and labile organic carbon in an intensively managed Chinese chestnut plantation. Biol. Fertil. Soils 50, 1109–1119.

Warnock, D.D., Lehmann, J., Kuyper, T.W., Rillig, M.C., 2007. Mycorrhizal responses to biochar in soil – concepts and mechanisms. Plant Soil 300, 9–20.

Watzinger, A., Feichtmair, S., Kitzler, B., Zehetner, F., Kloss, S., Wimmer, B., Zechmeister-Boltenstern, S., Soja, G., 2014. Soil microbial communities responded to biochar application in temperate soils and slowly metabolized ^{13}C-labelled biochar as revealed by ^{13}C PLFA analyses: results from a short-term incubation and pot experiment. Eur. J. Soil Sci. 65, 40–51.

Whalen, J.K., Sampedro, L., 2010. Soil Ecology and Management. CABI.

Woese, C.R., Fox, G.E., 1977. Phylogenetic structure of the prokaryotic domain: the primary kingdoms. Proc. Natl. Acad. Sci. U. S. A. 74, 5088–5090.

Xu, N., Zhang, B., Tan, G., Li, J., Wang, H., 2015. Influence of biochar on sorption, leaching and dissipation of bisphenol A and 17α-ethynylestradiol in soil. Environ. Sci. Process. Impacts. http://dx.doi.org/10.1039/C5EM00190K.

Yarwood, R.R., Rockhold, M.L., Niemet, M.R., Selker, J.S., Bottomley, P.J., 2006. Impact of microbial growth on water flow and solute transport in unsaturated porous media. Water Resour. Res. 42, W10405. http://dx.doi.org/10.1029/2005WR004550.

Yergeau, E., Sanschagrin, S., Maynard, C., St-Arnaud, M., Greer, C.W., 2014. Microbial expression profiles in the rhizosphere of willows depend on soil contamination. ISME J. 8, 344–358.

Zhu, D., Pignatello, J.J., 2005. Characterization of aromatic compound sorptive interactions with black carbon (charcoal) assisted by graphite as a model. Environ. Sci. Technol. 39, 2033–2041.

Zimmerman, A.R., Gao, B., Ahn, M.-Y., 2011. Positive and negative carbon mineralization priming effects among a variety of biochar-amended soils. Soil Biol. Biochem. 43, 1169–1179.

Elucidating the Impacts of Biochar Applications on Nitrogen Cycling Microbial Communities

N. Hagemann[1]*, J. Harter[1]*, S. Behrens[2]

[1]University of Tüebingen, Tüebingen, Germany; [2]University of Minnesota, Minneapolis, MN, United States and BioTechonology Institute, St. Paul, MN, United States

OUTLINE

*These authors contributed equally.

Biochar Application
http://dx.doi.org/10.1016/B978-0-12-803433-0.00007-2

163

INTRODUCTION

Nitrogen (N) is the fifth most abundant element in the solar system and the most abundant element in Earth's atmosphere. However, because it is necessary for the synthesis of major biomolecules such as proteins and nucleic acids, N availability can become a limiting growth factor in many ecosystems. Most organisms are not able to fix atmospheric nitrogen (Ehrlich, 2002). Instead, they need "fixed," organic forms of nitrogen (Canfield et al., 2010) or inorganic, "reactive" nitrogen ("N_r"; cf. Galloway et al., 2008), such as ammonium (NH_4^+) and/or nitrate (NO_3^-). Over the last 2.7 billion years a resilient nitrogen cycle evolved on Earth, which is to a large extent controlled by the activity of microorganisms (Canfield et al., 2010). During the 20th century, the so-called "Anthropocene" (as proposed by Nobel Prize laureate Paul Crutzen; cf. Zalasiewicz et al., 2011), intensification of agriculture significantly enhanced fluxes of nitrous oxide (N_2O), ammonia (NH_3), and nitrate (NO_3^-) leaching. In particular, the Haber–Bosch process and legume cultivation increased the anthropogenic input of reactive N into ecosystems. Soil biochar amendment might be a mitigation strategy to reduce undesired effects of elevated N fluxes. While many studies have quantified the effect of biochar on inorganic N leaching (eg, Laird et al., 2010; Zheng et al., 2013), ammonia volatilization (eg, Schomberg et al., 2012), and biological nitrogen fixation (eg, Güereña et al., 2015; Mia et al., 2014; Rondon et al., 2007), most research on the effect of biochar on soil microbial nitrogen cycling has focused on the reduction of N_2O emissions, first described by Yanai et al. (2007). N_2O is a precursor for ozone depleting nitric oxide (Crutzen, 1970) and an important greenhouse gas with a global warming potential 298 times that of CO_2 for a 100-year time scale (IPCC, 2007).

This chapter focuses on the soil microbial nitrogen cycle and how the physicochemical properties of biochar impact soil N cycling. The chapter summarizes the current literature on biochar research related to soil biogeochemical N cycling. The chapter concludes by presenting current knowledge gaps and proposing future research needs. The potential socioeconomic benefits and risks that biochar-based technology implementation might exert on global N cycling are also discussed.

THE MICROBIAL NITROGEN CYCLE

Many recent studies showed that biochar addition to surface mineral soils can influence soil N transformations directly. Here we review the most important nitrogen transformation processes catalyzed by microorganisms before we explore the direct and indirect effect of biochar on these transformations in the rest of the chapter.

Nitrogen Fixation

In natural systems, where no anthropogenic fertilizers are applied, nitrogen enters the microbial nitrogen cycle through the fixation of dinitrogen (N_2) from the atmosphere. Even though nitrogen is ubiquitously available as N_2 gas ($\approx 78\%$ of the atmosphere), most organisms, including plants, are not capable of N fixation and thus cannot use atmospheric N_2. In order to make nitrogen available for plants and other organisms it is of major importance that N_2-fixing microorganisms transform atmospheric N_2 into N_r (Francis et al., 2007). In soils, where the transformation of different N species is of particular interest as it regulates plant growth, N fixation is performed by free-living microorganisms or strains that form symbiotic relationships with plants (eg, legumes). Independent of their free-living or symbiotic occurrence, N_2-fixing microbes take up atmospheric N_2 and reduce it to NH_4^+ (Fig. 7.1). This process is mediated by the nitrogenase enzyme, and its catalytic subunit is encoded by the functional gene *nifH* (Reed et al., 2011). The nitrogenase enzyme is highly oxygen sensitive and therefore N fixation by free-living N_2-fixing microorganisms is usually enhanced under O_2-limited environmental conditions (Vitousek et al., 2002). Therefore, under oxic conditions, microorganisms have to protect the nitrogenase enzyme from exposure to elevated O_2 concentrations using several mechanisms that include engagement in symbiotic interactions with plants or formation of special

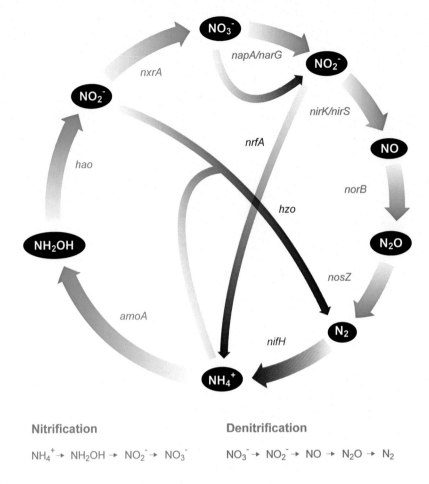

FIGURE 7.1 Schematic illustration of the microbial nitrogen cycle. *Color-coded arrows* indicate the different microbial transformation processes described in the text. Names of genes encoding for enzymes that catalyze a shown transformation process are given next to each respective *arrow*. *DNRA*, dissimilatory nitrate reduction to ammonium; *anammox*, anaerobic ammonium oxidation.

cell types (heterocysts in cyanobacteria) and subcellular compartments (Vitousek et al., 2002; Reed et al., 2011). NH_4^+ and organic nitrogen become available for other soil organisms, when free-living N_2-fixing bacteria lyse, are ingested or degraded, and when biomass from symbiotically N_2-fixing plants is mineralized (San-nai and Ming-pu, 2000; Francis et al., 2007).

Nitrification

Under oxic conditions, NH_4^+ is oxidized to NO_3^- by nitrifying bacteria and archaea (Fig. 7.1; Canfield et al., 2010). Until recently, nitrification was believed to be a two-step process carried out by two different groups of microorganisms, the ammonia oxidizers and the nitrite oxidizers (Robertson and Groffman, 2007). According to current textbook knowledge, ammonia-oxidizing bacteria (AOB) or ammonia-oxidizing archaea (AOA) oxidize NH_4^+ to hydroxylamine (NH_2OH) using the enzyme ammonia monooxygenase, which is encoded by the functional gene *amoA* (Treusch et al., 2005). Then, in another enzyme-mediated process, ammonia-oxidizing microorganisms oxidize NH_2OH to nitrite (NO_2^-) via hydroxylamine oxidoreductase encoded by the *hao* gene (Braker and Conrad, 2011). The produced NO_2^- can then be oxidized to NO_3^- by nitrite-oxidizing bacteria using the *nxrA* gene encoding a nitrite oxidoreductase (Poly et al., 2008).

Van Kessel et al. (2015) and Daims et al. (2015) independently reported the enrichment and initial characterization of three different *Nitrospira* species that encode all the enzymes necessary for NH_4^+ oxidation via NO_2^- to NO_3^- in their genomes. Both research groups used physiological experiments to show that these species are indeed capable of completely oxidizing NH_4^+ to NO_3^- to conserve energy. The ammonia monooxygenase enzymes of these newly described "comammox" (complete ammonia oxidizer) *Nitrospira* are phylogenetically distinct from currently identified bacterial AmoA and showed high similarity to a methane monooxygenase protein of *Crenothrix polyspora*, a filamentous methane oxidizing gammaproteobacterium (Stoecker et al., 2006). Similar *amoA* sequences have already been found in public sequence databases revealing that they were apparently misclassified as methane monooxygenases. Future studies on the environmental abundance, distribution, ecophysiology, and competition of complete ammonia-oxidizing microorganisms with canonical ammonia-oxidizing microorganisms will most likely change our current understanding of the microbial nitrogen cycle.

Nitrification is mainly performed by autotrophic microorganisms (Braker and Conrad, 2011). However, heterotrophic nitrification has also been documented in forest soils (Pedersen et al., 1999). Most autotrophic AOB are classified as Betaproteobacteria (eg, *Nitrosopira* spp., comprising the newly described "comammox" *Nitrospira*, and *Nitrosomonas* spp.) and only a small number of strains belong to the Gammaproteobacteria, as, for example, *Nitrosococcus* spp. (Prosser, 2007). Autotrophic nitrite-oxidizing bacteria are affiliated with Alpha-, Beta, Gamma-, and Deltaproteobacteria and the most prominent genera are *Nitrobacter* and *Nitrospira*. The relatively recently discovered AOA belong to the phylum Thaumarchaeota (Spang et al., 2010). Nitrification is also a potential source of the greenhouse gas N_2O, which can be formed by chemical decomposition of NH_2OH (Braker and Conrad, 2011). However, the amount of N_2O produced during nitrification is several orders of magnitude lower than the amounts of nitrite production from nitrification. ($\sim$$10^3$–$10^6$; Arp and Stein, 2003).

Denitrification

Under anoxic conditions, microorganisms catalyze the stepwise reduction of NO_3^- to N_2 in a process called denitrification (Fig. 7.1; Braker and Conrad, 2011). In contrast to nitrification, which is dominated by autotrophic bacteria and archaea, denitrification is mainly carried out by heterotrophic bacteria (Braker and Conrad, 2011). For the first step, the reduction of NO_3^- to NO_2^- is catalyzed by a membrane-bound nitrate reductase encoded by the functional gene *narG* or a *napA*-encoded periplasmic nitrate reductase. NO_2^- reduction to gaseous nitric oxide (NO) is performed with either a copper or a cytochrome cd1 nitrite reductase encoded by *nirK* or *nirS*, respectively (Wallenstein et al., 2006; Philippot et al., 2007). The reduction of NO to N_2O is catalyzed by a *norB*-encoded nitric oxide reductase. To date, three marginally different nitric oxide reductases are known that differ in their electron donor (Braker and Conrad, 2011). The last step of denitrification is the reduction of N_2O to N_2 mediated by the nitrous oxide reductase enzyme encoded by the functional gene *nosZ* (Fig. 7.1; Philippot et al., 2007; Braker and Conrad, 2011). Although denitrifiers have been identified in many different bacterial phyla (eg, Firmicutes, Actinobacteria, Bacteroidetes, Proteobacteria), they are also found among some fungal and archaeal taxa (Wallenstein et al., 2006). "Classical" (complete) denitrifiers are facultative aerobes and can switch from oxygen respiration to denitrification when soils become anoxic. Most of them carry a typical *nosZ* gene and are affiliated with the Proteobacteria. A number of studies provided evidence for the existence of a new clade of N_2O reducers. These particular microorganisms contain an atypical form of the *nosZ* gene. Microorganisms with an atypical N_2O reductase are physiologically more diverse, occupy a broader range of habitats, and belong to several different bacterial phyla such as Chloroflexi, Chlorobi, and Aquificae (Sanford et al., 2012; Jones et al., 2013). About half of the atypical *nosZ* gene containing N_2O reducers are not "classical" denitrifiers in the sense that they do not carry a nitrate, nitrite, and nitric oxide reductase and are only able to perform the last step of denitrification, the reduction of N_2O to N_2 (Jones et al., 2014; Orellana et al., 2014). Interestingly, atypical *nosZ* gene carrying bacteria outnumber typical *nosZ* gene containing strains in many soil environments (Jones et al., 2013). Nonetheless, their role in soil microbial nitrogen cycling, especially N_2O reduction, has not yet been studied extensively.

Depending on environmental conditions and the composition and genetic capabilities of the denitrifying microbial community, denitrification can act as both source (eg, under conditions unfavorable to N_2O reduction or in the presence of denitrifiers lacking *nosZ* genes) and sink (eg, by complete denitrification of "classical" denitrifiers and by the potential contribution of atypical N_2O reducers that only carry a *nosZ* gene) of N_2O (Braker and Conrad, 2011).

Anaerobic Ammonium Oxidation (Anammox)

For a long time, NH_4^+ was thought to be inert under anoxic conditions (Jetten, 2008). However, the discovery of anammox demonstrated the existence of an ammonium-oxidizing microbial process occurring in the absence of oxygen (Francis et al., 2007). During this process, NH_4^+ is oxidized to N_2 with NO_2^- as electron acceptor (Fig. 7.1). These findings also proved that heterotrophic denitrification is not the only N_2 forming pathway within the microbial nitrogen cycle (Zhu et al., 2011). Even though the exact pathway is not completely understood, it is thought that in the first step NO_2^- is reduced to NO by a cytochrome cd1 nitrite reductase, which then reacts with NH_4^+ to hydrazine (N_2H_2; Lam and Kuypers, 2011). The second step likely involves a hydrazine oxidoreductase, encoded by the gene *hzo*, that catalyzes the oxidation of N_2H_2 to N_2 (Schmid et al., 2008; Lam and Kuypers, 2011). The anammox process has been shown to be responsible for a substantial fraction of nitrogen losses in marine environments (Francis et al., 2007), although anammox bacteria have also been detected in freshwater environments and soils (Francis et al., 2007; Zhu et al., 2011). All anammox bacteria identified so far belong to the phylum Planctomycetes (Lam and Kuypers, 2011).

Dissimilatory Nitrate Reduction to Ammonium (DNRA)

DNRA is another microbial process that occurs under anoxic conditions in soils (Schmidt et al., 2011), but has not yet been well characterized. In this process, NO_3^- is reduced to NH_4^+ (Fig. 7.1; Braker and Conrad, 2011). It starts with the reduction of NO_3^- to NO_2^- catalyzed by a periplasmic nitrate reductase (Lam and Kuypers, 2011). In a second step, NO_2^- is reduced to NH_4^+ by a cytochrome c nitrite reductase encoded by the functional gene *nrfA* (Lam and Kuypers, 2011; Welsh et al., 2014). It has been reported that DNRA is favored over denitrification in strongly reduced and carbon-rich sediments and soils such as certain forest soils and rice paddies (Schmidt et al., 2011). Microorganisms capable of DNRA are mainly found among the Gamma-, Delta-, and Epsilonproteobacteria (Lam and Kuypers, 2011).

Mineralization and Immobilization

Nitrogen mineralization is the conversion of organic nitrogen to inorganic forms of nitrogen (N_{min}). It is part of the microbial degradation of nitrogen-rich organic matter (Robertson and Groffman, 2007). Organic nitrogen comprises about 90% of total nitrogen in most soils (Kelley and Stevenson, 1995) and includes biomass (proteins, etc.), detritus, and humics. The most important enzyme classes involved in nitrogen mineralization are deaminases, O-glycosidases, and acetyl hydrolases (Mengel, 1996). The reverse process of mineralization is immobilization. During nitrogen immobilization, microorganisms assimilate inorganic nitrogen for the synthesis of proteins and other nitrogen-containing organic compounds (Janssen, 1996; Mengel, 1996). This process occurs frequently when nitrogen-poor organic matter is decomposed (Janssen, 1996; Robertson and Groffman, 2007). Both processes, nitrogen mineralization and immobilization, occur simultaneously in soil. Therefore it is of great importance to distinguish between gross and net mineralization and immobilization. When gross mineralization exceeds immobilization the N_{min} content of the soil increases. Net immobilization occurs when gross immobilization rates are higher than mineralization rates, resulting in a decreased N_{min} content (Robertson and Groffman, 2007). The balance between nitrogen mineralization and immobilization is mainly controlled by the C:N ratio of the organic material being decomposed (Stevenson and Cole, 1999; Robertson and Groffman, 2007). As long as C and N are available from compounds with similar degradation rates, a C:N ratio of below 20 usually leads to net mineralization. Net immobilization occurs mainly when the C:N ratio of the decomposed organic matter exceeds 30 (Stevenson and Cole, 1999).

Factors Controlling Nitrogen Cycling in Soil

Microbial nitrogen cycling in soils is strongly controlled by environmental factors. The most important parameter is oxygen availability, because it determines which processes dominate. Soils with high oxygen concentrations promote nitrification, while oxygen-limited and anoxic conditions stimulate microbial denitrification, DNRA, anammox, and nitrogen fixation by free-living nitrogen-fixing bacteria (Braker and Conrad, 2011; Lam and Kuypers, 2011; Reed et al., 2011). Typically, heterotrophic organic matter decomposition, nitrogen mineralization, and immobilization rates are higher under oxic conditions (redox potentials $E_h > 300\,mV$; Mengel, 1996; Robertson and Groffman, 2007).

In soils, oxygen supply is often linked directly to soil moisture content because this determines the number of air-filled and water-filled soil pores. A high number of water-filled pores results quickly in the depletion of dissolved oxygen concentrations by heterotrophic microbial respiration, which consequently promotes anaerobic processes (Drenovsky et al., 2004). Apart from oxygen partial pressure and water content, many other environmental factors have been suggested to control the occurrence and activity of the various microbial nitrogen-transforming processes. Among them, the most important parameters are pH, soil organic carbon content, and N availability (Davidsson and Stahl, 2000; Hsu and Buckley, 2009; Nicol et al., 2008; Robertson and Groffman, 2007; Stevens et al., 1998; Strous et al., 1999; Verhagen and Laanbroek, 1991).

BIOCHAR'S PHYSICOCHEMICAL PROPERTIES AFFECTING SOIL MICROBIAL NITROGEN TRANSFORMATIONS

The physical and chemical properties of a soil are strong determinants of the nature and extent of its microbial nitrogen cycling. Since biochar addition changes the physicochemical properties of soils, in many ways the consequences for the composition and functioning of the inhabiting nitrogen-transforming microbial community can be multifarious. For example, through sorption or by changing the soil water regime, soil biochar amendment will affect the availability of many nutrients and other compounds such as carbon substrates, electron acceptors, trace elements, and toxins that determine microbial growth and activity. Thus Gul et al. (2015) made a distinction between "direct and microscale" and "indirect and larger-scale" effects of biochar amendment. The latter are mediated by improved plant growth in biochar-amended soil that results in increased input of carbon (litter, root exudates) and stimulates microbial activity in the rhizosphere. On the other hand, improved plant growth also causes higher plant nutrient uptake and thus potentially lower availability of N in the soil (Saarnio et al., 2013). However, these indirect effects of biochar on soil microbial community structure and function are difficult to tease apart and are out of the scope of this chapter. Therefore we will focus on the direct effects of biochar on the composition and activity of nitrogen-cycling soil microbial communities and attempt to distinguish between bulk and microsite (μm- to nm-scale) processes in what Quilliam et al. (2013b) defined as the "charosphere." So far, most studies have focused on bulk effects, while microsite impacts seem to have gained more attention only recently and will be addressed later.

Soil Water Content and Soil Structure

With respect to soil water conditions, two aspects seem to be most relevant for soil microbial nitrogen cycling. First, changes in the soil water retention by biochar will alter the transport (ie, leaching) of nitrogen compounds and other nutrients (Clough et al., 2013). Second, the water-filled pore space (WFPS) largely determines the extent and relative proportions of the different nitrogen transformation processes in a soil because the amount of water-filled pores affects oxygen concentrations in the soil (Braker and Conrad, 2011).

That biochar increases soil water retention has been shown for silty and loamy soils (Laird et al., 2010; Karhu et al., 2011) and sandy soils (Kammann et al., 2011; Uzoma et al., 2011). However, clay addition can be as effective as biochar in increasing soil water retention in sandy soils

(Dempster et al., 2012). Furthermore, it has been reported that biochar can increase soil hydraulic conductivity, for example, in highly weathered clay soils, because the relatively large particle size of the biochar compared to the soil can increase the flow rate of water (Kameyama et al., 2012). However, this effect depends on the particle size distribution of the respective biochar.

The soil WFPS is a key variable that determines the availability of nutrients in soil. The amount of WFPS affects diffusion as well as the availability of oxygen in soils. Denitrification is the dominating metabolism at a WFPS of 70% or higher because water mobilizes soil organic carbon, which stimulates heterotrophic microbial growth and reduces the availability of oxygen (Bateman and Baggs, 2005). A WFPS of 60% seems to be optimal for autotrophic nitrification, because the transport and diffusion of substrate becomes a limiting factor for microbial growth and activity at dryer water regimes. The relative contribution of denitrification increases again for a WPFS < 20% as a consequence of the nonhomogeneous distribution of water in the soil (Smith, 1980). Small water-filled pores that are still large enough to host microbial cells can provide anoxic microenvironments and become local "hotspots" for microbial denitrification even in relatively dry soils (eg, WPFS < 20%).

The water-filled pore space can be calculated as follows (Yanai et al., 2007):

$$\text{WFPS} = \frac{w_g \times \rho_{bulk}}{\rho_{H_2O} \times \left(1 - \frac{\rho_{bulk}}{\rho_{particle}}\right)}$$

Gravimetric water content (w_g) is determined by drying the soil at 105°C until it reaches a constant weight. Particle density ($\rho_{particle}$) is assumed to be 2.65 g cm^{-3} for mineral soil and 1.5–2.0 g cm^{-3} for most biochars (Chia et al., 2015). Based on these two values the $\rho_{particle}$ of any soil biochar mixture can be approximated accordingly. For laboratory-scale experiments, soil is compacted to a desired bulk density whereas in field experiments, 100 cm^3 of soil are sampled with a metal cylinder and dried at 105°C to calculate water content and bulk density. Biochars tend to reduce WFPS as most of them have a higher porosity $\left(1 - \frac{\rho_{bulk}}{\rho_{particle}}\right)$ than soil. However, this effect can be negligible because of the increased water-holding capacity of biochar-amended soils as exemplified by Troy et al. (2013) who reported a slightly increased WFPS after soil biochar addition in a column experiment. Nevertheless, field data on the effects of biochar on WFPS is scarce and inconsistent. Thus Quin et al. (2014) reported that biochar addition to a clay soil in the field increased porosity and pore connectivity indicating a lower WFPS and improved soil aeration. Also Saarnio et al. (2013) found that biochar increased N_2O emissions from soil mesocosms when both the control and biochar treatments were irrigated equally aiming at a water content of 20–30 vol%. Under these dry conditions, biochar might locally increase water content and thus promote denitrification at conditions otherwise favorable for nitrification. That biochar soil mixing decreases the WFPS, causing better soil aeration, has been suggested previously as a potential mechanism to explain the reduction of soil N_2O emissions upon biochar addition (van Zwieten et al., 2010c). However, Case et al. (2012) demonstrated that increased soil aeration by biochar amendment might only partially explain the repeatedly observed reductions in N_2O emission.

Biochar and Soil pH

Most biochars are alkaline and increase soil pH (eg, van Zwieten et al., 2010a,b); however, their effects on this important parameter are complex and have not been systematically investigated to date, although some studies have highlighted the influence of soil pH on microbial nitrogen transformations. Čuhel et al. (2010) reported that lowering soil pH increased the molar $\left(\dfrac{N_2O}{N_2O + N_2} \right)$ ratio with constant N_2O production and might decrease nitrifying activity. Also Baggs et al. (2010) reported an increase in N_2O emission at circumneutral pH and 50% WFPS after soil liming, suggesting a higher contribution of microbial ammonia oxidation to overall N_2O production. For the biochar context, van Zwieten et al. (2014) attributed the observed reduction in N_2O emissions from soil biochar mixtures at 85% WFPS to an increased soil pH and an elevated abundance of nitrous oxide reductase (*nosZ*) gene copy numbers. Also Obia et al. (2015) showed that the addition of NaOH to acidic soil slurries reduced the net N_2O emission in the same way as unweathered biochar.

Unlike N_2O, NH_3 emissions are affected directly by soil pH because of the NH_4^+–NH_3 chemical equilibrium. Thus increase in soil pH caused by biochar addition has been shown to increase N losses via ammonia volatilization (Schomberg et al., 2012).

In order to characterize the alkalinity of biochars, most studies measure biochar pH in water or a diluted $CaCl_2$ solution. As an additional parameter, van Zwieten et al. (2013) suggested to determine the acid-neutralizing capacity of biochars as %$CaCO_3$ equivalent. The acid-neutralizing capacity of biochars is an important parameter because biochars do not only have a high pH by themselves, but also tend to remove protons from solution, which increases pH in solution. This effect is caused by negatively charged surface functional groups (carboxylic, phenolic, etc.) as well as silicates and carbonates bound to the biochar surface (Gul et al., 2015). The contact of biochar with soil solution should result in slow saturation of biochar's surface functional groups over weeks and months. In agreement with these assumptions, Spokas (2013) as well as Heitkötter and Marschner (2015) found a reduction of biochar pH with aging in soil.

Biochar Redox Properties

Biochars are redox active and can reversibly accept and donate up to 2 mmol electrons per gram of char (Kluepfel et al., 2014a). These authors showed that biochars produced at 400–700°C had the highest capacities to accept and donate electrons. They attributed the electron-donating and electron-accepting capacities of different biochars to their relative pool size of phenolic moieties, quinone moieties, and polycondensed aromatic sheets. Consequently, Chen et al. (2014) demonstrated that biochar can promote direct interspecies electron transfer in a manner similar to that previously reported for granular-activated carbon. Similar to dissolved and solid-phase humic substances (Kappler et al., 2004; Roden et al., 2010), biochar can also stimulate anaerobic microbial respiration of insoluble electron acceptors such as iron(oxy)hydroxide minerals, by functioning as an "electron shuttle" between the microbial cell and the extracellular electron acceptor (Kappler et al., 2014). Furthermore, Joseph et al. (2015a) described biochar as a reductant and suggested that iron oxides on its surface might play an important role in abiotic nitrogen transformations. Notwithstanding these, published research on direct or indirect microbial catalyzed redox reactions with biochar and their effect on microbial (respiratory) activities is still very limited. In their critical review article, Cayuela et al. (2013) speculated that biochar might reduce denitrification rates because microorganisms use biochar as an alternate terminal electron acceptor instead of nitrate. They suggested further that the combination of biochar "electron shuttling" properties and effect on soil pH (liming) could promote nitrous oxide reductase (NosZ) activity explaining why many, but not all, biochars seem to be capable of reducing soil N_2O emissions.

Nitrogen Leaching

Laird and Rogovska (2015) suggested eight mechanisms describing the interaction of biochar and nutrients as: (1) physical retention of water (and thus nutrients as discussed earlier); (2) cation exchange capacity; (3) anion exchange capacity; (4) nutrient release by biochar; (5) liming effect of biochar (pH-dependent nutrient solubility); (6) adsorption of organic nutrients (eg, organic forms of nitrogen); (7) microbial activity affected by biochar; and (8) coprecipitation of nutrients with or onto biochar. Also Conte et al. (2013) suggested that the formation of nonconventional water bonds on the surface of biochar could lock hydrated nitrate molecules into its nanopores, thus affecting nitrate availability and desorption. This might also explain the slow desorption of nitrate from co-composted biochar as reported by Kammann et al. (2015).

Overall, biochar seems to reduce both NH_4^+ and NO_3^- leaching (eg, Zheng et al., 2013) by an interplay of the mechanisms discussed previously. However, implications for N availability to microbes need further research. NO_3^- sorption by biochar has been described with Freundlich isotherms by Chintala et al. (2013), who found that 20–75% of the initial N could be desorbed in an aqueous extraction. Also increased leaching rates have been reported for fresh soil biochar mixtures (Singh et al., 2010). Dempster et al. (2012) showed that leaching of dissolved organic nitrogen was not affected by biochar, while the same biochar reduced mineral nitrogen leaching. Taghizadeh-Toosi et al. (2012) reported that ammonia sorbed onto biochar must be bioavailable because they found ^{15}N in plant biomass after growth on biochar that was exposed to $^{15}N-NH_3$ for 1 week.

Biochar Carbon and Soil Carbon Interactions

Biochars can both supply and remove labile organic carbon and natural organic matter from soil solution. Bioavailable organic carbon can be a substrate for heterotrophic microbial growth while nonbiodegradable organic matter can affect microbial activities in soil, for example, as signaling molecules or electron shuttles. Most biochars contain volatiles that accumulate on the biochar surfaces during the cooling or quenching of biochar after pyrolysis. These organic compounds can usually be removed from biochar surfaces by rinsing and are therefore available to microbes following soil biochar mixing (Smith et al., 2010; Spokas et al., 2014). Organic condensates have been suggested to influence soil microbial nutrient cycling because they might contain molecules that have a toxic effect on microbial activity. However, elevated polycyclic aromatic hydrocarbon (PAH, here: naphthalene, phenanthrene, and pyrene) concentrations could not explain the reduction of N_2O emission observed after the addition of a pine wood biochar produced at 350 and 550°C to haplic Calcisol soil (Alburquerque et al., 2015).

Spokas et al. (2010) described the release of ethylene by biochar with the subsequent increase in soil ethylene production following biochar addition. While the mechanisms of ethylene production in biochar-amended soils are still unclear, it is, for example, known that ethylene originating from calcium carbide can reduce the formation of N_2O and NO_3^- in fertilized soils. Thus Spokas et al. (2010) suggested biochar as a potential nitrification inhibitor. However, Fulton et al. (2013) showed that biochar does not emit measurable amounts of ethylene 7–90 days after production when stored under ambient conditions.

Although only limited information exists about how biochar might interfere or possibly enhance the activity of microbial signaling compounds, Masiello et al. (2013) showed that their sorption onto biochar does affect microbial community composition. Given the strong sorption affinity of organic matter to biochar surfaces, it is anticipated that biochar will interfere with signaling between microorganisms, generally, and then between plant roots and rhizosphere microbial communities (Lehmann et al., 2011), as a specific unique ecosystem. DeLuca et al. (2006) suggested that biochar (wildfire charcoal) might sorb compounds like monoterpenes, which have been shown to inhibit nitrification (White, 1994). Hale et al. (2015) confirmed this by demonstrating that the monoterpenes α-pinene and limonene sorbed onto biochar at a similar rate as onto peat. Nevertheless, the relevance of the effects of biochar on cell-to-cell communication and sorption of signaling compounds has not been empirically investigated yet.

Biochar Nitrogen Content

It has previously been thought that the N contained in plant-derived biochar is not bioavailable because it is part of heterocyclic carbon structures (Knicker and Skjemstad, 2000). However, de la Rosa and Knicker (2011) showed in a ^{15}N study that nitrogen in plant-derived biochar is bioavailable and can be found in new synthesized biomass after incubation in soil for 72 days. The manure-derived biochars studied by Revell et al. (2012) and van Zwieten et al. (2013) had N contents of >3% and thus were a relevant source of N in their investigated ecosystems. Also biochar produced from coffee grounds or defatted coffee grounds after biodiesel production showed an elevated N content of up to 4.3% (Vardon et al., 2013). Generally, low temperature biochars made from manures and other biosolids contain more easily hydrolyzable organic N forms such as amino acids compared to high-temperature plant biomass-derived biochars. The mineralization and release of bioavailable forms of N from biochar will generally depend on how recalcitrant the biochar is to microbial degradation, its C:N ratio, and the soil N pool. In order to further understand the effect of biochar on immobilization and mineralization of N in soils, more studies using ^{15}N-labeled biochar should be performed.

Biochar Porosity and Microbial Microsite Formation

As also described by Ennis et al. (2012), biochar not only affects the bulk soil, but also provides novel microsites in the soil environment. As a result the authors suggested the need for a closer analysis of these unique microgeographical ecosystems, for which Quilliam et al. (2013b) introduced the term "charosphere." Thus, using electron microscopy to image the surface of soil-aged biochar, Quilliam et al. (2013b) explored these biochar microenvironments and showed that microorganisms attach to the surface of biochar and colonize accessible pores. So biochar does serve as a habitat for soil microorganisms. Compared to surrounding soil the charosphere as a microbial habitat is characterized by a higher pH, increased availability of labile organic matter (including potential toxins), elevated concentrations of reactive nitrogen species (including N_2O; Cornelissen et al., 2013) and other nutrients (such as trace elements), and potentially a less fluctuating water regime.

van Zwieten et al. (2009) suggested that these "charosphere" microenvironments can lead to the formation of "anoxic microsites" when the available oxygen is consumed rapidly through the activity of heterotrophic microorganisms. This might create local "hotspots" for anaerobic microbial processes with consequences for soil microbial nitrogen transformation, for example, through stimulation of microbial denitrification and nitrogen fixation. Also soil biochar amendment will enhance soil aggregation with consequences for microbial biochar colonization as well as microbial community structure and function (Gul et al., 2015; see also Chapter 11). The formation of soil organomineral complexes depends on the biochar's physicochemical characteristics, soil organic matter availability, and soil texture. It is therefore important to consider biochar particle size, porosity, and surface area because they will largely determine the bioavailability of nutrients and pore accessibility for microbial colonization.

BIOCHAR EFFECTS ON SOIL MICROBIAL NITROGEN CYCLING

Nitrogen Fixation

Several laboratory and field studies have revealed positive effects of biochar amendment on microbial nitrogen fixation in soil. Thus stable isotope and acetylene approaches, as well as molecular biological techniques such as quantitative polymerase chain reaction (qPCR) and DNA sequencing, suggested biochar-associated increases in nitrogenase activities, root nodule numbers, proportion of soil nitrogen input derived from microbial N fixation, and abundance of nitrogen-fixing microorganisms (Anderson et al., 2011; Quilliam et al., 2013a; Harter et al., 2014; Mia et al., 2014; Rondon et al., 2007). Specifically, Ducey et al. (2013) and Harter et al. (2014) demonstrated that biochar can increase *nifH* gene copy numbers in controlled soil microcosm experiments. Anderson et al. (2011) performed pot experiments with different biochar application rates and used terminal-restriction fragment length polymorphism (T-RFLP) in combination with 16S rRNA gene amplicon sequencing to detect shifts in the general microbial community. They showed that soil biochar amendment increased the relative abundance of many known nitrogen-fixing bacterial families such as Bradyrhizobiaceae, Frankiaceae, and Rhizobiaceae. A biochar-induced increase in relative sequence abundance of the Bradyrhizobiaceae-affiliated genus *Bradyrhizobium* was also reported in a rice paddy study (Chen et al., 2015). Taken together, these studies demonstrated that the addition of biochar to soil can have stimulating effects on the relative abundance of microorganisms capable of nitrogen fixation and, with that, the functional genetic potential for nitrogen fixation in soil microbial communities (see also Chapter 7).

Also studies focusing on symbiotic nitrogen fixation in legume plants using stable isotope techniques revealed similar trends. In a pot experiment, Rondon et al. (2007) showed that the addition of wood-derived biochar to an oxisol with a very low fertility could improve symbiotic nitrogen fixation by common beans (*Phaseolus vulgaris* L.). Biochar application considerably increased both plant biomass production and the proportion of nitrogen derived from nitrogen fixation. Güereña et al. (2015) confirmed these results in a greenhouse experiment with a broad range of biochars produced from different feedstocks under various charring conditions. In this study, biochar not only increased the proportion of nitrogen derived from nitrogen fixation but also increased nodule biomass and nodule numbers on common legume plants. Similarly, Mia et al. (2014) and Oram et al. (2014) used microcosm experiments and showed that biochar considerably increased biological nitrogen fixation and the number of plant root nodules of red clover (*Trifolium pratense* L.). In contrast, Quilliam et al. (2013a) reported that aged biochar did not increase the amount of root nodules in clover (*Trifolium repens* L.) in a pot experiment but they showed that individual nodules exhibited a significantly increased nitrogenase activity. Despite the observation of higher nitrogenase activity in individual root nodules, biochar had no impact on the nitrogen fixation rate in the whole root system nor did it affect above-ground green biomass N contents. Also a field experiment by Van de Voorde et al. (2014), showed that biochar had no effect on symbiotic nitrogen fixation rates of red clover plants.

In general, the potential explanations for the underlying mechanisms by which biochar affects symbiotic nitrogen fixation are as diverse as the results of these studies. According to Rondon et al. (2007) the reason for the higher symbiotic nitrogen fixation in legumes in the presence of biochar is an increased availability of the trace nutrients boron and molybdenum. However, biochar-induced changes in symbiotic nitrogen fixation might also be affected, to some extent, by decreased N availability, increased availability of K, Ca, and P, as well as higher pH and lower aluminum content in the investigated tropical soils (Rondon et al., 2007). Güereña et al. (2015) described that an increased P uptake by plants might be responsible for the observed stimulation of symbiotic nitrogen fixation. They also showed that changes in soil pH and the mineral nutrient content of biochar were not the dominating driving forces for the increased fraction of plant N derived from symbiotic N fixation. In experiments with red clover the addition of phosphorous or micronutrients (such as B and Mo) together with biochar did not further increase biological nitrogen fixation (Mia et al., 2014; Oram et al., 2014). Both studies explained the increased nitrogen fixation and root nodule numbers with an elevated availability of potassium. Until now the mechanisms responsible for higher abundances of nitrogen-fixing microorganisms and increased microbial nitrogen fixation by leguminous plants in the presence of biochar have not been systematically investigated. However, it is known from biochar-free studies that among other geochemical parameters, pH, oxygen availability, nitrogen availability, and the C:N ratio of soil organic matter can have strong effects on microbial nitrogen fixation (Cusack et al., 2009; Pereira e Silva et al., 2011; Roper, 1985; Roper and Smith, 1991). As suggested in Harter et al. (2014), the high C:N ratio of biochar and the formation of anoxic microsites in biochar-amended soils might favor the growth of free-living as well as plant-associated nitrogen-fixing microorganisms. Furthermore, limited availability of mineral nitrogen caused by adsorption onto biochar particles, and an increased pH in the charosphere, might provide a competitive advantage to microorganisms capable of nitrogen fixation.

Mineralization (Ammonification) and Immobilization

For N mineralization and ammonification in soil, contrasting reports on the effect of biochar exist. For example, Nelissen et al. (2012) reported a twofold increase of gross N mineralization immediately after biochar amendment in ^{15}N tracer incubation experiments. Using similar isotope tracer experiments, Prommer et al. (2014) found that soil organic N transformations decreased in a field trial 4–18 months after biochar amendment. Since they also found an increase in gross nitrification rates, the researchers suggested a "de-coupling of the intrinsic organic and inorganic soil N cycles" during the first 18 months after biochar amendment. Castaldi et al. (2011) reported that the initial increase of net N mineralization in biochar field experiments declined over 1 year. A similar effect was described by Nelissen et al. (2015).

Decreased rates of ammonification have been reported also from short-term incubation studies (Sarkhot et al., 2012; Tammeorg et al., 2012). However, only bulk N_{min} measurements were performed in both these experiments and no additional microbiological or stable isotope data have been collected to verify the observed decrease in ammonification rates. Thus it is unclear if the mineralized N pool was low because of a decline in nitrogen mineralization or because N_{min} is immediately immobilized again by microbes as soon as it became available (Tammeorg et al., 2012). This interpretation is also supported by the data of Ippolito et al. (2012), who observed an increase in microbial nitrogen immobilization upon soil biochar amendment.

Nitrification (Ammonia Oxidation, Nitrite Oxidation)

Several studies reported that soil biochar amendment increased nitrification in arable soils from different climate zones. This has been shown using: conventional N_{min} measurements in a subtropical acidic soil (Zhao et al., 2014); ^{15}N stable isotope approaches in a temperate loamy sand (eg, Nelissen et al., 2012); a combination of ^{15}N stable isotope probing and PCR-based quantification of the ammonia monooxygenase gene (*amoA*) in a temperate Austrian soil (Prommer et al., 2014); and using *amoA*-targeted qPCR and clone libraries in a temperate Chinese soil (Song et al., 2014). The latter two studies reported increases of both archaeal and bacterial *amoA* gene copy numbers per gram soil and correlated archaeal *amoA* gene abundance to ammonia oxidation and gross nitrification rates. Bacterial *amoA* gene abundance correlated with gross nitrification rates in the study by Prommer et al. (2014). The *amoA* gene clone libraries in the study by Song et al. (2014) revealed that biochar amendment seemed to decrease the diversity of AOA while the AOB diversity increased. Ducey et al. (2013) did not find a statistically significant biochar effect on the abundance of bacterial *amoA* gene copy numbers over a period of 6 months in their soil incubation experiments. However, they also did not find any correlation of *amoA* gene abundance and nitrogen speciation. Also Xu et al. (2014) did not observe any biochar-induced differences in bacterial *amoA* gene or transcript copy numbers in a pot experiment with rape (*Brassica napus*). Nonetheless, they reported that the observed increase in nitrification rates correlated with an increase in archaeal *amoA* transcript copy numbers in their biochar-amended setup. Interestingly, the effect was more pronounced in the pot experiments without plants.

Castaldi et al. (2011) reported no significant changes in nitrification in a field experiment, although they applied $60 t ha^{-1}$ biochar. Also Nelissen et al. (2015) questioned the time-dependent effect of biochar on aerobic nitrogen transformations in the field. They found a 13% increase in gross nitrification along with a 34% increase in gross N mineralization right after biochar amendment, but did not observe any difference in nitrogen transformations between biochar and control field plots 1 year after they started the experiment. They attributed the initial changes in gross nitrification and N mineralization to stimulation of heterotrophic nitrification and ammonification by labile, biodegradable carbon (volatiles, condensates) that became available to the nitrifying microbial community directly after biochar amendment.

Microbial nitrification can also be an important source of soil N_2O. However, Xu et al. (2014) recorded reduced N_2O emissions in the presence of biochar under oxic conditions (50% water-holding capacity) in a subtropical acidic soil, although biochar increased overall nitrification rates. Sánchez-García et al. (2014) hypothesized that for soils in which microbial N_2O production is dominated by nitrification, biochar addition might further increase N_2O emissions, while biochar addition to soils dominated by microbial denitrification could lead to decreased N_2O emissions. According to Sánchez-García et al. (2014) this might explain why some soils show increased N_2O emissions after biochar addition, although the same type of biochar effectively mitigates emissions in other soils. The lack of experimental data precludes further speculation, hence additional systematic experiments are necessary to elucidate the effect of biochar on soil nitrification and nitrous oxide emission in more detail.

Denitrification

In order to study the effect of biochar on microbial denitrification, a wide range of different molecular and analytical techniques comprising acetylene inhibition and stable isotope tracer tests have been used. Anderson et al. (2011) investigated the impact of a pine wood biochar on microbial community composition in a pot experiment with soil from a pasture field. Using T-RFLP and 16S rRNA gene sequencing, the authors identified an increased relative abundance of microbial families known to be capable of complete denitrification (eg, Bradyrhizobiaceae) in the presence of biochar. These findings were also confirmed by Chen et al. (2015), who recorded an increased relative sequence abundance of the genus *Bradyrhizobium* in a biochar-amended rice paddy. A follow-up study by Anderson and colleagues with samples taken directly from the pasture field verified the biochar-induced increase in Bradyrhizobiaceae abundance (Anderson et al., 2014). However, functional marker genes for microbial denitrification (*nirK*, *nirS*, and *nosZ*) quantified using qPCR were not altered by soil biochar amendment. Dicke et al. (2015) quantified *nosZ* gene copy numbers of typical denitrifiers and atypical N_2O reducers in a field trial amended with different biochars and biochar–digestate mixtures. They also did not observe any significant changes in *nosZ* gene copy numbers in the presence of biochar or biochar–digestate.

On the other hand, soil microcosm experiments with soil water regimes favoring microbial denitrification revealed strong impacts of biochar on the abundance of functional marker genes for this pathway (Ducey et al., 2013; Harter et al., 2014; van Zwieten et al., 2014; Xu et al., 2014). Thus Ducey et al. (2013) showed that switchgrass-derived biochar increased *nirS*, *nirK*, and *nosZ* gene copy numbers, indicating higher abundances of NO_2^- and N_2O reducers. In a biochar co-composting experiment with pig manure Wang et al. (2013a) reported a biochar-induced decrease in *nirK* gene copies and increases in *nirS* and *nosZ* gene copy numbers. Higher *nosZ* gene copy numbers in the presence of biochars have also been found in a sandy soil (Tenosol; van Zwieten et al., 2014) and a loamy sand (calcaric leptosol; Harter et al., 2014). In addition, Harter et al. (2014) and Xu et al. (2014) also reported a significant increase in *nosZ* transcript copies, suggesting a higher activity of N_2O-reducing microorganisms. Besides transcript copy quantification, Xu et al. (2014) quantified the activity of denitrifying enzymes with the denitrification enzyme activity (DEA) assay and showed that biochar considerably increased DEA rates. These results were in agreement with the findings of Castaldi et al. (2011), who reported significantly higher DEA rates in biochar-amended field plots 3 months after biochar application.

Based on the literature findings summarized previously, biochar seems to affect microbial denitrification. It remains unclear, however, whether the observed effects are a consequence of direct microbe–biochar interactions or if the sum of multifactorial alterations of the soil physicochemical characteristics upon biochar addition results in conditions that affect microbial denitrification indirectly. Although current evidence from the summarized DEA and qPCR studies cannot answer this question, several studies have suggested that biochar stimulates "complete" denitrification. Because denitrification is a major source of N_2O in many soil environments (Braker and Conrad, 2011; Philippot et al., 2009), most publications related decreased N_2O emissions to biochar-induced changes in microbial denitrification. Interestingly, increased gene and transcript abundance of the nitrous oxide reductase (encoded by the gene *nosZ*) that catalyzes the reduction of N_2O to N_2 might provide a potential explanation for the observed decrease in soil N_2O emissions in the presence of biochar. These findings also seem to be supported by the results of Cayuela et al. (2013), who performed short-term soil microcosm experiments under soil water conditions promoting microbial denitrification. The soil microcosms were amended with ^{15}N-enriched NO_3^- and the authors quantified $^{15}N-N_2O$ and $^{15}N-N_2$ in the headspace of their soil microcosms. Biochar addition decreased the $N_2O:(N_2O + N_2)$ ratio for all 15 agricultural soils, which suggested that the supplementation could have stimulated complete denitrification and promoted the biotic reduction of N_2O to N_2. This has also been suggested by several other studies (Case et al., 2015; Taghizadeh-Toosi et al., 2011; van Zwieten et al., 2010c; Yanai et al., 2007; Zhang et al., 2012a). Cayuela et al. (2014) summarized the findings in a meta-analysis and demonstrated that the majority of field- and laboratory-based studies performed until 2014 confirmed the capability of biochar to mitigate N_2O emissions from soils. Notwithstanding, it should not be neglected that, in contrast to this general trend, some field studies that have been performed under conditions favoring nitrification or in the presence of manure-derived biochars, reported no effects of biochar addition on N_2O emissions (Scheer et al., 2011; Suddick and Six, 2013; Verhoeven and Six, 2014).

Several mechanisms that might explain the decreased N_2O emissions from biochar-amended soils have been proposed. Most frequently, decreased N_2O emissions were thought to be linked to decreased denitrification rates (Kammann et al., 2011; Yanai et al., 2007); enhanced N_2O reduction activities resulting in complete denitrification (Cayuela et al., 2013; Harter et al., 2014; Xu et al., 2014); or a combination of both (Case et al., 2015; Kammann et al., 2012; Khan et al., 2013; van Zwieten et al., 2010c; Zhang et al., 2012b). Lower denitrification rates in biochar-amended soils might occur because of increased soil aeration or decreased substrate availability (eg, by NO_3^- immobilization; van Zwieten et al., 2009). Also an enhanced N_2O reduction activity might result from the effect of biochar addition on soil pH. It is known that a low pH can negatively affect microbial N_2O reduction (Bakken et al., 2012). N_2O reduction activity might also be elevated when the number of anoxic microsites in soil increases because N_2O reductases are highly oxygen sensitive (McKenney et al., 1994; van Zwieten et al., 2009). Cornelissen et al. (2013) showed that biochar can sorb significantly more N_2O than any other soil constituent such as iron oxides. However, these experiments were conducted in an artificial, water-free environment and need to be repeated under conditions more relevant for biochars in soil. It has also been speculated that biochars can serve as electron shuttles for microorganisms (Cayuela et al., 2013, 2015). Microorganisms are able to donate or receive electrons from biochar depending on the redox potential of the biochar and the overall soil redox state, similar to microbial redox interactions with dissolved and solid state humic substances as also discussed previously (Kappler et al., 2004, 2014; Roden et al., 2010; Kluepfel et al., 2014a,b). However, how this affects microbial denitrification and N_2O reduction has not been investigated in detail so far. Biochar has also been shown to release toxic compounds like PAHs, which inhibit nitrification or denitrification (Spokas et al., 2010; Wang et al., 2013b). On the contrary, Alburquerque et al. (2015) showed that high concentrations of PAHs (naphthalene, phenanthrene, and pyrene) did not inhibit biochar's capacity to reduce soil N_2O emissions.

Based on the current state of knowledge of the effects of biochar on microbial denitrification, a potential working hypothesis that explains the frequently observed decrease in N_2O emissions from biochar-amended soil might be described as follows. After heavy rainfall, water saturates the soil and transports labile organic carbon and NO_3^- (eg, from fertilization) through the soil matrix. Biochar has been shown to absorb and retain water (Atkinson et al., 2010; Yu et al., 2013), NO_3^- (Clough et al., 2013), and organic carbon (Pignatello et al., 2006). Consequently, biochar

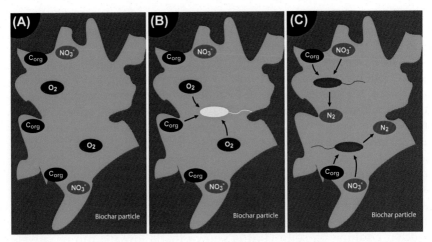

FIGURE 7.2   The proposed mechanism responsible for decreased N_2O emissions via enhanced N_2O reduction in biochar pores. (A) A biochar pore filled with water, organic carbon compounds, and NO_3^- after a rain event. (B) The transition from oxic to anoxic conditions where heterotrophic microorganisms (yellow) consume oxygen by aerobically respiring organic carbon. (C) The situation when oxygen is depleted and denitrifiers (purple) perform complete denitrification by reducing NO_3^- to N_2 using organic carbon as electron donor.

surfaces will accumulate NO_3^- and labile organic carbon as biochar pores soak up rainwater (Fig. 7.2A). Because of aerobic microbial respiration, oxygen concentrations inside the water-saturated biochar pores will decline quickly and anoxic microsites will form (Fig. 7.2B). Anoxia and the high pH microenvironment of biochar pores provide excellent conditions for the activity of N_2O-reducing microorganisms shifting the denitrification end product from N_2O to N_2 (Fig. 7.2C). Furthermore, it is conceivable that reduced biochar surface functional groups act as electron donors (Kappler et al., 2014; Kluepfel et al., 2014a) for microbial N_2O reduction when organic carbon becomes limited. N_2O released by denitrifiers not capable of N_2O reduction (Philippot et al., 2011) will remain in soils entrapped in water-saturated biochar pores or sorbed onto biochar surfaces as suggested by Cornelissen et al. (2013). Atypical N_2O reducers that only carry a *nosZ* gene but no other denitrification genes rely on the supply of N_2O from other microorganisms (Jones et al., 2014). They might benefit from the dissolved N_2O entrapped in biochar pores or sorbed onto biochar surfaces and further reduce it to N_2. Consequently, less N_2O will partition into the gas phase and ultimately be emitted from the soil. In order to verify or falsify this hypothesis, N_2 emissions from soil should be quantified following soil microcosm incubations with $^{15}N-NO_3^-$ stable isotope tracers.

KNOWLEDGE GAPS AND FUTURE RESEARCH

This chapter showed that there is evidence for multiple, mostly desirable effects of biochar on microbial nitrogen cycling in agroecosystems, including increased soil nitrogen retention and reduction of N_2O emissions. Despite the rising number of published studies on this topic, the underlying mechanisms of the interactions of biochar and the nitrogen-cycling microbial communities remain largely unknown or speculative. It is, however, necessary and critical to understand these mechanisms to be able to design biochars and biochar postproduction treatments for specific purposes and ultimately predict and prevent potentially undesired outcomes of biochar soil amendment. Therefore in our discussion that follows we suggest general quality standards for biochar research and identify the aspects of biochar's interactions with the microbial nitrogen cycle that have not yet received adequate attention:

1. Most studies focusing on, for example, nitrogen mineralization and nitrification in biochar-amended soils do not quantify N_2O and, vice versa, many studies on N_2O emission under conditions favorable for denitrification do not characterize fully the N_{min} content and associated microbial nitrogen transformation reactions. These data are essential to understand the underlying mechanisms.
2. Kappler et al. (2014) demonstrated that experiments with pure cultures in defined culture media can provide further insights into the electron-shuttling mechanisms of biochar to insoluble iron(oxy) hydroxides. To our knowledge, there is no similar published data with respect to other anaerobic microbial respirations such as denitrification.
3. Some effects of biochar seem to decline over time. However, the effect of soil aging on the physicochemical characteristics of biochar have not been systematically investigated and linked to soil microbial nitrogen cycling. Therefore both soil incubation studies and pure culture experiments should be performed with aged biochar recovered from field experiments, as exemplified by Spokas (2013). Spectroscopy methods including scanning (SEM) or transmission electron microscopy (TEM) that can be combined with electron energy loss spectroscopy (EELS) and energy dispersive X-ray spectroscopy (EDS), Fourier-transformed infrared spectroscopy (FTIR), X-ray photoelectron spectroscopy (XPS), nuclear magnetic resonance spectroscopy (NMR), or near edge X-ray absorption fine structures (NEXAFS) can provide further insights into biochar's physical and chemical surface properties. So far there are studies that characterized pristine (fresh) biochar spectroscopically, for example: (1) NEXAFS, FTIR, and X-ray diffraction of two thermosequences of

pristine biochars from herbaceous and woody feedstock (Keiluweit et al., 2010); and (2) a correlation of germination rates and plant biomass production after supplementation to NMR and SEM-EDS spectroscopy on pristine biochars from different feedstock (de la Rosa et al., 2014). Few studies applied these methods on aged biochar, including (1) a combination of SEM, TEM-EELS, and EDS as well as Raman spectroscopy on carbonaceous *Terra Preta* particles (Jorio et al., 2012); (2) the characterization of carbon functional groups on pristine and laboratory soil-aged biochars using XPS and NEXAFS (Singh et al., 2014); and (3) the characterization of organomineral complexes on biochar after soil aging in the field with TEM and other techniques (Joseph et al., 2010; Lin et al., 2012).

4. To date, N_2O emissions have been quantified by various techniques in the gas phase above soils. These measurements should be extended and/or complemented with N_2O quantification in the solution (soil pore water), using (micro)electrodes, for example.

Biochar is a complex material that requires some specific considerations when used in applied research, hence the following should be considered:

1. Biochar consists of solid pyrogenic carbon (also fixed carbon), ash, and volatile organic carbon (VOC; Keiluweit et al., 2010). To preserve the VOC, biochar should not be dried at temperatures higher than 40°C if it is subsequently used for analytical purposes. Contact with water alters the structure of the ash via dissolution and reprecipitation, and can cause physical disintegration of biochar (Spokas et al., 2014). Thus all preparatory steps need to be reported carefully, especially, but not only, for microscopic and spectroscopic analyses.

2. Biochars should be characterized according to the standards identified by the European Biochar Certificate (EBC, 2012) or International Biochar Initiative (IBI, 2014). This practice will ensure that the results of any biochar study can be interpreted within the context of global biochar research and facilitate the compilation of meaningful reviews and meta-analysis.

3. Soil water conditions should be reported in % WFPS as discussed previously.

4. Adequate and appropriate control experiments need to be designed and conducted for each biochar investigation. Most studies typically use an unfertilized and/or nonamended soil as a control that receives the same management as the biochar treatment. For studies in soil, amendments with the original feedstock that was used to produce the biochar should be considered. This should be considered especially when working with manure or sewage sludge chars where, for example, a $1\,t\,ha^{-1}$ sewage sludge biochar can be compared to $2\,t\,ha^{-1}$ sewage sludge amendment, if the mass loss during pyrolysis is

50%. Alternately, biochar and respective feedstock can be applied normalized to total carbon (van Zwieten et al., 2013), nitrogen, or phosphorous input (Zhu et al., 2014) or at the same application rate (Paz-Ferreiro et al., 2012). If biochar received any postproduction treatment or was aged in soil or compost, then the control treatment should receive the equivalent amount of nutrients. For example, bovine urine activated biochar should be compared to a bovine urine only control. Composted or soil-aged biochars typically contain water or 2 M KCl extractable NH_4^- and NO_3^-, hence equivalent amounts of nutrients should be added to the control treatments. When biochar effects are to be studied in defined cultivation media, then the control experiments should be amended with an inert or nonreactive solid, for example, graphite or microporous aluminum silicates such as zeolites (with comparable pore volumes).

5. A similar approach should be applied when different biochars are compared within one experiment. Different normalization strategies to better compare the different treatments should be considered carefully. The most common procedure is to normalize on biochar dry matter content. Although this is suitable for many experiments, other options including organic carbon added with the biochar (or nitrogen, phosphorus), the dry matter weight of the feedstock of the respective biochars, and the cation exchange capacity or the electron accepting and donating capacity of the biochars might be alternatives to consider. Depending on the research question asked, one or more of these alternative normalization options might provide the most meaningful set of controls.

GLOBAL IMPACT OF BIOCHAR ON SOIL NITROGEN CYCLING

The concept of biochar originates from research on the *Terra Preta do Indio*, also called Amazonian Dark Earth (ADE). First biogeochemical studies (Glaser et al., 2000; Lehmann et al., 2003; Glaser, 2007) highlighted ADE as a model for present-day agriculture with respect to climate change mitigation and sustainable intensification of agriculture. Understanding the effects of biochar on the microbial nitrogen cycle is key for our endeavor toward a sustainable, carbon-negative agriculture using biochar. The fate of nitrogen, and thus the rate of soil-borne N_2O emissions and plant availability of the most important macronutrient, is largely governed by microbial activity. Thus biochar application as a means to counteract climate change is not only about carbon sequestration, its total mitigation potential also originates from its effects on soil nitrogen transformation reactions.

In addition, the anthropogenic influence on the nitrogen cycle by the Haber–Bosch process, the burning of both biomass and fossil fuels, as well as the cultivation of leguminous plant exceeds the suggested planetary boundary and shows a high risk to endanger the stability of the Earth system (Rockstrom et al., 2009; Steffen et al., 2015). From this perspective, biochar could also benefit nonagricultural ecosystems by reducing the demand for reactive nitrogen as fertilizers and thus reducing global eutrophication.

Until soil biochar amendment can be considered as a risk-free application to mitigate climate change and improve soil quality, we should follow a key precautionary principle, ie, we need a detailed understanding of all mechanisms of biochar's effects in different soils. However, this needs to be communicated carefully following experiments with appropriate controls and baselines. When we as scientists report that biochar alters the microbial community structure, we must mention that plowing, fertilization, and other traditional agricultural practices also have these effects. Additionally, we should be aware that biochar is actually not a novel technology, but a global traditional phenomenon. It is comparable to the production of alcoholic beverages from fruits and cereals or the preservation of food such as kimchi (Korea), dongbei suancai (China), or sauerkraut (Germany) by lactic acid fermentation that have been developed independently by various cultures. So far there is evidence for the use of charcoal in agriculture for North America (Allen, 1847), Europe (Wiedner et al., 2015), Africa (Frausin et al., 2014), and Australia (Downie et al., 2011). This can be looked at as a general proof of concept for biochar.

Furthermore, to mainstream biochar we not only need to understand the underlying mechanisms that determine its reactivity in soils, we must also learn how to build integrated systems using the material. Laird (2008) shared his "charcoal vision" and described a "win–win–win scenario" for biochar that comprises energy production by pyrolysis, carbon sequestration by soil amendment, and increased environmental quality of both soil and ground water. Today, further benefits and applications of biochar are suggested, for example, in "55 uses of biochar" compiled by Schmidt and Wilson (2014). According to these authors and in line with Laird's vision, suitable applications of biochar should be arranged as cascades with the potential to create both agroeconomic benefits and relevant research opportunities. In this context, Joseph et al. (2015b) set up a cascade using biochar as cow feed additive with subsequent soil application of the resulting biochar-enriched manure as a high N fertilizer. Although economic benefits to the farmer, improved soil fertility, and increased pasture growth were proven, the authors highlight the need for further research on cow gut microbiology, soil macrobiota, and the long-term effect of gut-aged biochar following soil application.

These practices are relevant as the costs of biochar, for example, US\$222 to 584 t^{-1} in the United Kingdom (Shackley et al., 2011), will probably frustrate the application of high levels of biochar in arable farming. Instead, cascade use and concentrated root zone application of biochar-based substrates in horticulture resulting in application rates of less than 1 t ha^{-1} (Schmidt et al., 2015) are already feasible today and more likely to become the dominating biochar application practices. As such practices are especially relevant for organic farming, and most studies on biochar focus on conventional agricultural practices such as the application of mineral N fertilizer (Hagemann and Potthast, 2015), there is a particular need for more specific biochar research in this context. To address these challenges and to design appropriate research, we need a "charcoal vision—revisited" that is generated in a participatory manner (Gebhard et al., 2015) and shared by researchers from various disciplines and practitioners to develop, informatively, the future *Terra Preta*.

Acknowledgments

NH is funded by a BMBF PhD scholarship provided by Rosa-Luxemburg Foundation, Germany. We thank Lukas van Zwieten (New South Wales Department for Primary Industries, Australia) for valuable discussions and his input on control experiments during the 3rd International Biochar Training Course (Nanjing, China). We also would like to thank Andreas Kappler (Geomicrobiology, University of Tuebingen, Germany) for generous support and inspiring discussions during the writing of this chapter.

References

Alburquerque, J.A., Sánchez-Monedero, M.A., Roig, A., Cayuela, M.L., 2015. High concentrations of polycyclic aromatic hydrocarbons (naphthalene, phenanthrene and pyrene) failed to explain biochar's capacity to reduce soil nitrous oxide emissions. Environ. Pollut. 196, 72–77.

Allen, R.L., 1847. Brief Compend of American Agriculture, third ed. C.M. Saxton, New York.

Anderson, C.R., Condron, L.M., Clough, T.J., Fiers, M., Stewart, A., Hill, R.A., Sherlock, R.R., 2011. Biochar induced soil microbial community change: implications for biogeochemical cycling of carbon, nitrogen and phosphorus. Pedobiologia 54, 309–320.

Anderson, C.R., Hamonts, K., Clough, T.J., Condron, L.M., 2014. Biochar does not affect soil n-transformations or microbial community structure under ruminant urine patches but does alter relative proportions of nitrogen cycling bacteria. Agric. Ecosyst. Environ. 191, 63–72.

Arp, D.J., Stein, L.Y., 2003. Metabolism of inorganic n compounds by ammonia-oxidizing bacteria. Crit. Rev. Biochem. Mol. Biol. 38 (6), 471–495.

Atkinson, C., Fitzgerald, J., Hipps, N., 2010. Potential mechanisms for achieving agricultural benefits from biochar application to temperate soils: a review. Plant Soil 337 (1), 1–18.

Baggs, E., Smales, C., Bateman, E., 2010. Changing pH shifts the microbial source as well as the magnitude of N_2O emission from soil. Biol. Fertil. Soils 46 (8), 793–805.

Bakken, L.R., Bergaust, L., Liu, B., Frostegård, Å., 2012. Regulation of denitrification at the cellular level: a clue to the understanding of N_2O emissions from soils. Phil. Trans. R. Soc. B 367 (1593), 1226–1234.

Bateman, E.J., Baggs, E.M., 2005. Contributions of nitrification and denitrification to N_2O emissions from soils at different water-filled pore space. Biol. Fertil. Soils 41 (6), 379–388.

Braker, G., Conrad, R., 2011. Diversity, structure, and size of N_2O-producing microbial communities in soils-what matters for their functioning? Adv. Appl. Microbiol. 75, 33–70.

Canfield, D.E., Glazer, A.N., Falkowski, P.G., 2010. The evolution and future of earth's nitrogen cycle. Science 330 (6001), 192–196.

Case, S.D.C., Mcnamara, N.P., Reay, D.S., Whitaker, J., 2012. The effect of biochar addition on N_2O and CO_2 emissions from a sandy loam soil – the role of soil aeration. Soil Biol. Biochem. 51 (0), 125–134.

Case, S.D.C., Mcnamara, N.P., Reay, D.S., Stott, A.W., Grant, H.K., Whitaker, J., 2015. Biochar suppresses N_2O emissions while maintaining n availability in a sandy loam soil. Soil Biol. Biochem. 81, 178–185.

Castaldi, S., Riondino, M., Baronti, S., Esposito, F.R., Marzaioli, R., Rutigliano, F.A., Vaccari, F.P., Miglietta, F., 2011. Impact of biochar application to a mediterranean wheat crop on soil microbial activity and greenhouse gas fluxes. Chemosphere 85 (9), 1464–1471.

Cayuela, M.L., Sánchez-Monedero, M.A., Roig, A., Hanley, K., Enders, A., Lehmann, J., 2013. Biochar and denitrification in soils: when, how much and why does biochar reduce N_2O emissions? Sci. Rep. 3.

Cayuela, M.L., van Zwieten, L., Singh, B.P., Jeffery, S., Roig, A., Sánchez-Monedero, M.A., 2014. Biochar's role in mitigating soil nitrous oxide emissions: a review and meta-analysis. Agric. Ecosyst. Environ. 191, 5–16.

Cayuela, M.L., Jeffery, S., van Zwieten, L., 2015. The molar h:Corg ratio of biochar is a key factor in mitigating N_2O emissions from soil. Agric. Ecosyst. Environ. 202, 135–138.

Chen, S., Rotaru, A.-E., Shrestha, P.M., Malvankar, N.S., Liu, F., Fan, W., Nevin, K.P., Lovley, D.R., 2014. Promoting interspecies electron transfer with biochar. Sci. Rep. 4, 5019.

Chen, J.H., Liu, X.Y., Li, L.Q., Zheng, J.W., Qu, J.J., Zheng, J.F., Zhang, X.H., Pan, G.X., 2015. Consistent increase in abundance and diversity but variable change in community composition of bacteria in topsoil of rice paddy under short term biochar treatment across three sites from south China. Appl. Soil Ecol. 91, 68–79.

Chia, C.H., Downie, A., Munroe, P., 2015. Characteristics of biochar: physical and structural properties. In: Lehmann, J., Joseph, S.D. (Eds.), Biochar for Environmental Management: Science, Technology and Implementation, second ed. Taylor and Francis, Florence, KY, USA, pp. 89–109.

Chintala, R., Mollinedo, J., Schumacher, T.E., Papiernik, S.K., Malo, D.D., Clay, D.E., Kumar, S., Gulbrandson, D.W., 2013. Nitrate sorption and desorption in biochars from fast pyrolysis. Micropor. Mesopor. Mater. 179, 250–257.

Clough, T., Condron, L., Kammann, C., Müller, C., 2013. A review of biochar and soil nitrogen dynamics. Agronomy 3 (2), 275–293.

Conte, P., Marsala, V., De Pasquale, C., Bubici, S., Valagussa, M., Pozzi, A., Alonzo, G., 2013. Nature of water-biochar interface interactions. GCB Bioenergy 5 (2), 116–121.

Cornelissen, G., Rutherford, D.W., Arp, H.P.H., Dörsch, P., Kelly, C.N., Rostad, C.E., 2013. Sorption of pure N_2O to biochars and other organic and inorganic materials under anhydrous conditions. Environ. Sci. Technol. 47 (14), 7704–7712.

Crutzen, P.J., 1970. The influence of nitrogen oxides on the atmospheric ozone content. Q. J. R. Meteorol. Soc. 96 (408), 320–325.

Čuhel, J., Šimek, M., Laughlin, R.J., Bru, D., Chèneby, D., Watson, C.J., Philippot, L., 2010. Insights into the effect of soil pH on N_2O and N_2 emissions and denitrifier community size and activity. Appl. Environ. Microbiol. 76 (6), 1870–1878.

Cusack, D.F., Silver, W., Mcdowell, W.H., 2009. Biological nitrogen fixation in two tropical forests: ecosystem-level patterns and effects of nitrogen fertilization. Ecosystems 12, 1299–1315.

Daims, H., Lebedeva, E.V., Pjevac, P., Han, P., Herbold, C., Albertsen, M., Jehmlich, N., Pala-tinszky, M., Vierheilig, J., Bulaev, A., Kirkegaard, R.H., Bergen, M.V., Rattei, T., Bendinger, B., Nielsen, P.H., Wagner, M., 2015. Complete nitrification by Nitrospira bacteria. Nature 528 (7583), 504–509 Advance online publication.

Davidsson, T.E., Stahl, M., 2000. The influence of organic carbon on nitrogen transformations in five wetland soils. Soil Sci. Soc. Am. J. 64, 1129–1136.

Deluca, T.H., Mackenzie, M.D., Gundale, M.J., Holben, W.E., 2006. Wildfire-produced char-coal directly influences nitrogen cycling in ponderosa pine forests. Soil Sci. Soc. Am. J. 70 (2), 448–453.

Dempster, D.N., Jones, D.L., Murphy, D.V., 2012. Clay and biochar amendments decreased inorganic but not dissolved organic nitrogen leaching in soil. Soil Res. 50 (3), 216–221.

Dicke, C., Andert, J., Ammon, C., Kern, J., Meyer-Aurich, A., Kaupenjohann, M., 2015. Effects of different biochars and digestate on N_2O fluxes under field conditions. Sci. Total Envi-ron. 524–525, 310–318.

Downie, A.E., van Zwieten, L., Smernik, R.J., Morris, S., Munroe, P.R., 2011. Terra preta australis: reassessing the carbon storage capacity of temperate soils. Agric. Ecosyst. Environ. 140 (1–2), 137–147.

Drenovsky, R.E., Vo, D., Graham, K.J., Scow, K.M., 2004. Soil water content and organic carbon availability are major determinants of soil microbial community composition. Microb. Ecol. 48, 424–430.

Ducey, T.F., Ippolito, J.A., Cantrell, K.B., Novak, J.M., Lentz, R.D., 2013. Addition of activated switchgrass biochar to an aridic subsoil increases microbial nitrogen cycling gene abun-dances. Appl. Soil Ecol. 65 (0), 65–72.

Ebc, 2012. European Biochar Foundation - European Biochar Certificate - Guidelines for a Sustainable Production of Biochar. Version 6.1 of June 19, 2015. European Biochar Foun-dation (EBC), Arbaz, Switzerland.

Ehrlich, H.L., 2002. Geomicrobiology. Marcel Dekker Inc., New York.

Ennis, C.J., Evans, A.G., Islam, M., Ralebitso-Senior, T.K., Senior, E., 2012. Biochar: carbon sequestration, land remediation, and impacts on soil microbiology. Crit. Rev. Environ. Sci. Technol. 42 (22), 2311–2364.

Francis, C.A., Beman, J.M., Kuypers, M.M.M., 2007. New processes and players in the nitro-gen cycle: the microbial ecology of anaerobic and archaeal ammonia oxidation. ISME J. 1, 19–27.

Frausin, V., Fraser, J., Narmah, W., Lahai, M., Winnebah, T.A., Fairhead, J., Leach, M., 2014. "God made the soil, but we made it fertile": gender, knowledge, and practice in the forma-tion and use of african dark earths in liberia and sierra leone. Hum. Ecol. 42 (5), 695–710.

Fulton, W., Gray, M., Prahl, F., Kleber, M., 2013. A simple technique to eliminate ethylene emissions from biochar amendment in agriculture. Agron. Sustain. Dev. 33 (3), 469–474.

Galloway, J.N., Townsend, A.R., Erisman, J.W., Bekunda, M., Cai, Z., Freney, J.R., Martinelli, L.A., Seitzinger, S.P., Sutton, M.A., 2008. Transformation of the nitrogen cycle: recent trends, questions, and potential solutions. Science 320 (5878), 889–892.

Gebhard, E., Hagemann, N., Hensler, L., Schweizer, S., Wember, C., 2015. Agriculture and food 2050: visions to promote transformation driven by science and society. J. Agric. Environ. Ethics 28 (3), 497–516.

Glaser, B., Balashov, E., Haumaier, L., Guggenberger, G., Zech, W., 2000. Black carbon in density fractions of anthropogenic soils of the Brazilian Amazon region. Org. Geochem. 31 (7–8), 669–678.

Glaser, B., 2007. Prehistorically modified soils of central Amazonia: a model for sustainable agriculture in the twenty-first century. Phil. Trans. R. Soc. B 362 (1478), 187–196.

Güereña, D.T., Lehmann, J., Thies, J.E., Enders, A., Karanja, N., Neufeldt, H., 2015. Partition-ing the contributions of biochar properties to enhanced biological nitrogen fixation in common bean (Phaseolus vulgaris). Biol. Fertil. Soils 51, 479–491.

Gul, S., Whalen, J.K., Thomas, B.W., Sachdeva, V., Deng, H.Y., 2015. Physico-chemical properties and microbial responses in biochar-amended soils: mechanisms and future directions. Agric. Ecosyst. Environ. 206, 46–59.

Hagemann, N., Potthast, T., 2015. Necessary new approaches towards sustainable agriculture – innovations for organic agriculture. In: Dumitras, D.E., Jitea, I.M., Aerts, S. (Eds.), Know Your Food-Food Ethics and Innvoation. Wageningen Academic Publishers, Wageningen, pp. 107–113.

Hale, S.E., Endo, S., Arp, H.P.H., Zimmerman, A.R., Cornelissen, G., 2015. Sorption of the monoterpenes α-pinene and limonene to carbonaceous geosorbents including biochar. Chemosphere 119, 881–888.

Harter, J., Krause, H.M., Schuettler, S., Ruser, R., Fromme, M., Scholten, T., Kappler, A., Behrens, S., 2014. Linking N$_2$O emissions from biochar-amended soil to the structure and function of the n-cycling microbial community. ISME J. 8 (3), 660–674.

Heitkötter, J., Marschner, B., 2015. Interactive effects of biochar ageing in soils related to feedstock, pyrolysis temperature, and historic charcoal production. Geoderma 245, 56–64.

Hsu, S.F., Buckley, D.H., 2009. Evidence for the functional significance of diazotroph community structure in soil. ISME J. 3, 124–136.

Ibi, 2014. International Biochar Initiative - Standardized Product Definition and Product Testing Guidelines for Biochar that Is Used in Soil (Aka Ibi Biochar Standards) Version 2.0.

Ipcc, 2007. Climate Change 2007: The Physical Science Basis. Contribution of Working Group I to the Fourth Assessment Report of the Intergovernmental Panel on Climate Change. Cambridge University Press, Cambridge, United Kingdom and New York, NY, USA.

Ippolito, J.A., Novak, J.M., Busscher, W.J., Ahmedna, M., Rehrah, D., Watts, D.W., 2012. Switchgrass biochar affects two aridisols. J. Environ. Qual. 41 (4), 1123–1130.

Janssen, B.H., 1996. Nitrogen mineralization in relation to C:N ratio and decomposability of organic materials. Plant Soil 181, 39–45.

Jetten, M.S.M., 2008. The microbial nitrogen cycle. Environ. Microbiol. 10, 2903–2909.

Jones, C.M., Graf, D.R., Bru, D., Philippot, L., Hallin, S., 2013. The unaccounted yet abundant nitrous oxide-reducing microbial community: a potential nitrous oxide sink. ISME J. 7, 417–426.

Jones, C.M., Spor, A., Brennan, F.P., Breuil, M.C., Bru, D., Lemanceau, P., Griffiths, B., Hallin, S., Philippot, L., 2014. Recently identified microbial guild mediates soil N$_2$O sink capacity. Nat. Clim. Chang. 4, 801–805.

Jorio, A., Ribeiro-Soares, J., Cançado, L.G., Falcão, N.P.S., Dos Santos, H.F., Baptista, D.L., Martins Ferreira, E.H., Archanjo, B.S., Achete, C.A., 2012. Microscopy and spectroscopy analysis of carbon nanostructures in highly fertile Amazonian anthrosoils. Soil Till. Res. 122, 61–66.

Joseph, S.D., Camps-Arbestain, M., Lin, Y., Munroe, P., Chia, C.H., Hook, J., van Zwieten, L., Kimber, S., Cowie, A., Singh, B.P., Lehmann, J., Foidl, N., Smernik, R.J., Amonette, J.E., 2010. An investigation into the reactions of biochar in soil. Soil Res. 48 (7), 501–515.

Joseph, S., Husson, O., Graber, E.R., van Zwieten, L., Taherymoosavi, S., Thomas, T., Nielsen, S., Ye, J., Pan, G., Chia, C., 2015a. The electrochemical properties of biochars and how they affect soil redox properties and processes. Agronomy 5 (3), 322–340.

Joseph, S., Pow, D., Dawson, K., Mitchell, D.R.G., Rawal, A., Hook, J., Taherymoosavi, S., van Zwieten, L., Rust, J., Donne, S., Munroe, P., Pace, B., Graber, E., Thomas, T., Nielsen, S., Ye, J., Lin, Y., Pan, G., Li, L., Solaiman, Z.M., 2015b. Feeding biochar to cows: an innovative solution for improving soil fertility and farm productivity. Pedosphere 25 (5), 666–679.

Kameyama, K., Miyamoto, T., Shiono, T., Shinogi, Y., 2012. Influence of sugarcane bagasse-derived biochar application on nitrate leaching in calcaric dark red soil. J. Environ. Qual. 41 (4), 1131–1137.

Kammann, C.I., Linsel, S., Gößling, J., Koyro, H.-W., 2011. Influence of biochar on drought tolerance of *Chenopodium quinoa* Willd and on soil–plant relations. Plant Soil 345 (1–2), 195–210.

Kammann, C., Ratering, S., Eckhard, C., Müller, C., 2012. Biochar and hydrochar effects on greenhouse gas (carbon dioxide, nitrous oxide, and methane) fluxes from soils. J. Environ. Qual. 41 (4), 1052–1066.

Kammann, C.I., Schmidt, H.-P., Messerschmidt, N., Linsel, S., Steffens, D., Müller, C., Koyro, H.-W., Conte, P., Stephen, J., 2015. Plant growth improvement mediated by nitrate capture in co-composted biochar. Sci. Rep. 5.

Kappler, A., Benz, M., Schink, B., Brune, A., 2004. Electron shuttling via humic acids in microbial iron(iii) reduction in a freshwater sediment. FEMS Microbiol. Ecol. 47 (1), 85–92.

Kappler, A., Wuestner, M.L., Ruecker, A., Harter, J., Halama, M., Behrens, S., 2014. Biochar as an electron shuttle between bacteria and Fe(iii) minerals. Environ. Sci. Technol. Let. 1 (8), 339–344.

Karhu, K., Mattila, T., Bergström, I., Regina, K., 2011. Biochar addition to agricultural soil increased ch4 uptake and water holding capacity – results from a short-term pilot field study. Agric. Ecosyst. Environ. 140 (1–2), 309–313.

Keiluweit, M., Nico, P.S., Johnson, M.G., Kleber, M., 2010. Dynamic molecular structure of plant biomass-derived black carbon (biochar). Environ. Sci. Technol. 44 (4), 1247–1253.

Kelley, K.R., Stevenson, F.J., 1995. Forms and nature of organic n in soil. Fertil. Res. 42, 1–11.

Khan, S., Chao, C., Waqas, M., Arp, H.P., Zhu, Y.G., 2013. Sewage sludge biochar influence upon rice (Oryza sativa l) yield, metal bioaccumulation and greenhouse gas emissions from acidic paddy soil. Environ. Sci. Technol. 47, 8624–8632.

Kluepfel, L., Keiluweit, M., Kleber, M., Sander, M., 2014a. Redox properties of plant biomass-derived black carbon (biochar). Environ. Sci. Technol. 48 (10), 5601–5611.

Kluepfel, L., Piepenbrock, A., Kappler, A., Sander, M., 2014b. Humic substances as fully regenerable electron acceptors in recurrently anoxic environments. Nat. Geosci. 7 (3), 195–200.

Knicker, H., Skjemstad, J.O., 2000. Nature of organic carbon and nitrogen in physically protected organic matter of some Australian soils as revealed by solid-state 13C and 15N nmr spectroscopy. Soil Res. 38 (1), 113–128.

Laird, D.A., Rogovska, N., 2015. Biochar effects on nutrient leaching. In: Lehmann, J., Joseph, S.D. (Eds.), Biochar for Environmental Management: Science, Technology and Implementation, second ed. Taylor and Francis, Florence, KY, USA, pp. 512–542.

Laird, D., Fleming, P., Wang, B., Horton, R., Karlen, D., 2010. Biochar impact on nutrient leaching from a midwestern agricultural soil. Geoderma 158 (3–4), 436–442.

Laird, D.A., 2008. The charcoal vision: a win–win–win scenario for simultaneously producing bioenergy, permanently sequestering carbon, while improving soil and water quality. Agron. J. 100 (1), 178–181.

Lam, P., Kuypers, M.M.M., 2011. Microbial nitrogen cycling processes in oxygen minimum zones. Ann. Rev. Mar. Sci. 3, 317–345.

Lehmann, J., Pereira Da Silva, J., Steiner, C., Nehls, T., Zech, W., Glaser, B., 2003. Nutrient availability and leaching in an archaeological anthrosol and a ferralsol of the central Amazon basin: fertilizer, manure and charcoal amendments. Plant Soil 249 (2), 343–357.

Lehmann, J., Rillig, M.C., Thies, J., Masiello, C.A., Hockaday, W.C., Crowley, D., 2011. Biochar effects on soil biota – a review. Soil Biol. Biochem. 43 (9), 1812–1836.

Lin, Y., Munroe, P., Joseph, S., Kimber, S., van Zwieten, L., 2012. Nanoscale organo-mineral reactions of biochars in ferrosol: an investigation using microscopy. Plant Soil 357 (1–2), 369–380.

Masiello, C.A., Chen, Y., Gao, X., Liu, S., Cheng, H.-Y., Bennett, M.R., Rudgers, J.A., Wagner, D.S., Zygourakis, K., Silberg, J.J., 2013. Biochar and microbial signaling: production conditions determine effects on microbial communication. Environ. Sci. Technol. 47 (20), 11496–11503.

Mckenney, D.J., Drury, C.F., Findlay, W.I., Mutus, B., Mcdonnell, T., Gajda, C., 1994. Kinetics of denitrification by pseudomonas fluorescens: oxygen effects. Soil Biol. Biochem. 26 (7), 901–908.

Mengel, K., 1996. Turnover of organic nitrogen in soils and its availability to crops. Plant Soil 181, 83–93.

Mia, S., Van Groenigen, J.W., Van De Voorde, T.F.J., Oram, N.J., Bezemer, T.M., Mommer, L., Jeffery, S., 2014. Biochar application rate affects biological nitrogen fixation in red clover conditional on potassium availability. Agric. Ecosyst. Environ. 191, 83–91.

Nelissen, V., Rütting, T., Huygens, D., Staelens, J., Ruysschaert, G., Boeckx, P., 2012. Maize biochars accelerate short-term soil nitrogen dynamics in a loamy sand soil. Soil Biol. Biochem. 55, 20–27.

Nelissen, V., Rütting, T., Huygens, D., Ruysschaert, G., Boeckx, P., 2015. Temporal evolution of biochar's impact on soil nitrogen processes – a 15N tracing study. GCB Bioenergy 7 (4), 635–645.

Nicol, G.W., Leininger, S., Schleper, C., Prosser, J.I., 2008. The influence of soil pH on the diversity, abundance and transcriptional activity of ammonia oxidizing archaea and bacteria. Environ. Microbiol. 10, 2966–2978.

Obia, A., Cornelissen, G., Mulder, J., Dörsch, P., 2015. Effect of soil pH increase by biochar on NO, N_2O and N_2 production during denitrification in acid soils. PLoS ONE 10 (9), e0138781.

Oram, N.J., Van De Voorde, T.F.J., Ouwehand, G.-J., Bezemer, T.M., Mommer, L., Jeffery, S., Groenigen, J.W.V., 2014. Soil amendment with biochar increases the competitive ability of legumes via increased potassium availability. Agric. Ecosyst. Environ. 191, 92–98.

Orellana, L.H., Rodriguez-R, L.M., Higgins, S., Chee-Sanford, J.C., Sanford, R.A., Ritalahti, K.M., Loffler, F.E., Konstantinidis, K.T., 2014. Detecting nitrous oxide reductase (nosz) genes in soil metagenomes: method development and implications for the nitrogen cycle. mBio 5 e01193-01114.

Paz-Ferreiro, J., Gascó, G., Gutiérrez, B., Méndez, A., 2012. Soil biochemical activities and the geometric mean of enzyme activities after application of sewage sludge and sewage sludge biochar to soil. Biol. Fertil. Soils 48 (5), 511–517.

Pedersen, H., Dunkin, K.A., Firestone, M.K., 1999. The relative importance of autotrophic and heterotrophic nitrification in a conifer forest soil as measured by 15N tracer and pool dilution techniques. Biogeochemistry 44, 135–150.

Pereira E Silva, M.C., Semenov, A.V., Van Elsas, J.D., Salles, J.F., 2011. Seasonal variations in the diversity and abundance of diazotrophic communities across soils. FEMS Microbiol. Ecol. 77, 57–68.

Philippot, L., Hallin, S., Schloter, M., 2007. Ecology of denitrifying prokaryotes in agricultural soil. Adv. Agron. 96, 249–305.

Philippot, L., Čuhel, J., Saby, N.P.A., Chèneby, D., Chroňáková, A., Bru, D., Arrouays, D., Martin-Laurent, F., Šimek, M., 2009. Mapping field-scale spatial patterns of size and activity of the denitrifier community. Environ. Microbiol. 11 (6), 1518–1526.

Philippot, L., Andert, J., Jones, C.M., Bru, D., Hallin, S., 2011. Importance of denitrifiers lacking the genes encoding the nitrous oxide reductase for N_2O emissions from soil. Glob. Change Biol. 17, 1497–1504.

Pignatello, J.J., Kwon, S., Lu, Y., 2006. Effect of natural organic substances on the surface and adsorptive properties of environmental black carbon (char): attenuation of surface activity by humic and fulvic acids. Environ. Sci. Technol. 40, 7757–7763.

Poly, F., Wertz, S., Brothier, E., Degrange, V., 2008. First exploration of nitrobacter diversity in soils by a pcr cloning-sequencing approach targeting functional gene nxra. FEMS Microbiol. Ecol. 63, 132–140.

Prommer, J., Wanek, W., Hofhansl, F., Trojan, D., Offre, P., Urich, T., Schleper, C., Sassmann, S., Kitzler, B., Soja, G., Hood-Nowotny, R.C., 2014. Biochar decelerates soil organic nitrogen cycling but stimulates soil nitrification in a temperate arable field trial. PLoS ONE 9 (1), e86388.

Prosser, J.I., 2007. The ecology of nitrifying bacteria. In: Newton, W.E., Bothe, H., Ferguson, S.J. (Eds.), Biology of the Nitrogen Cycle. Elsevier, Amsterdam, pp. 223–243.

Quilliam, R.S., Deluca, T.H., Jones, D.L., 2013a. Biochar application reduces nodulation but increases nitrogenase activity in clover. Plant Soil 366 (1–2), 83–92.

Quilliam, R.S., Glanville, H.C., Wade, S.C., Jones, D.L., 2013b. Life in the 'charosphere' – does biochar in agricultural soil provide a significant habitat for microorganisms? Soil Biol. Biochem. 65 (0), 287–293.

Quin, P.R., Cowie, A.L., Flavel, R.J., Keen, B.P., Macdonald, L.M., Morris, S.G., Singh, B.P., Young, I.M., van Zwieten, L., 2014. Oil mallee biochar improves soil structural properties—a study with X-ray micro-ct. Agric. Ecosyst. Environ. 191, 142–149.

Reed, S.C., Cleveland, C.C., Townsend, A.R., 2011. Functional ecology of free-living nitrogen fixation: a contemporary perspective. In: Futuyma, D.J., Shaffer, H.B., Simberloff, D. (Eds.), Annual Review of Ecology, Evolution and SystematicsPalo Alto, Annual Reviews, 42, pp. 489–512.

Revell, K.T., Maguire, R.O., Agblevor, F.A., 2012. Influence of poultry litter biochar on soil properties and plant growth. Soil Sci. 177 (6), 402–408.

de la Rosa, J.M., Knicker, H., 2011. Bioavailability of n released from n-rich pyrogenic organic matter: an incubation study. Soil Biol. Biochem. 43 (12), 2368–2373.

de La Rosa, J.M., Paneque, M., Miller, A.Z., and Knicker, H., 2014. Relating physical and chemical properties of four different biochars and their application rate to biomass production of lolium perenne on a calcic cambisol during a pot experiment of 79 days. Sci. Total Environ. 499, 175–184.

Robertson, G.P., Groffman, P.M., 2007. Nitrogen transformation. In: Paul, E.A. (Ed.), Soil Microbiology, Biochemistry and Ecology. Springer, New York, New York, pp. 341–364.

Rockstrom, J., Steffen, W., Noone, K., Persson, A., Chapin, F.S., Lambin, E., Lenton, T.M., Scheffer, M., Folke, C., Schellnhuber, H.J., Nykvist, B., De Wit, C.A., Hughes, T., Van Der Leeuw, S., Rodhe, H., Sorlin, S., Snyder, P.K., Costanza, R., Svedin, U., Falkenmark, M., Karlberg, L., Corell, R.W., Fabry, V.J., Hansen, J., Walker, B., Liverman, D., Richardson, K., Crutzen, P., Foley, J., 2009. Planetary boundaries: exploring the safe operating space for humanity. Ecol. Soc. 14 (2), 33.

Roden, E.E., Kappler, A., Bauer, I., Jiang, J., Paul, A., Stoesser, R., Konishi, H., Xu, H., 2010. Extracellular electron transfer through microbial reduction of solid-phase humic substances. Nat. Geosci. 3 (6), 417–421.

Rondon, M.A., Lehmann, J., Ramírez, J., Hurtado, M., 2007. Biological nitrogen fixation by common beans (Phaseolus vulgaris l.) increases with bio-char additions. Biol. Fertil. Soils 43 (6), 699–708.

Roper, M.M., Smith, N.A., 1991. Straw decomposition and nitrogenase activity (C_2H_2 reduction) by free-living microorganisms from soil: effects of pH and clay content. Soil Biol. Biochem. 23, 275–283.

Roper, M.M., 1985. Straw decomposition and nitrogenase activity (C_2H_2 reduction) - effects of soil-moisture and temperature. Soil Biol. Biochem. 17, 65–71.

Saarnio, S., Heimonen, K., Kettunen, R., 2013. Biochar addition indirectly affects N_2O emissions via soil moisture and plant n uptake. Soil Biol. Biochem. 58, 99–106.

Sánchez-García, M., Roig, A., Sanchez-Monedero, M.A., Cayuela, M.L., 2014. Biochar increases soil N_2O emissions produced by nitrification-mediated pathways. Front. Environ. Sci. 2.

Sanford, R.A., Wagner, D.D., Wu, Q., Chee-Sanford, J.C., Thomas, S.H., Cruz-García, C., Rodríguez, G., Massol-Deyá, A., Krishnani, K.K., Ritalahti, K.M., Nissen, S., Konstantinidis, K.T., Löffler, F.E., 2012. Unexpected nondenitrifier nitrous oxide reductase gene diversity and abundance in soils. Proc. Natl. Acad. Sci. U. S. A. 109 (48), 19709–19714.

San-Nai, J., Ming-Pu, Z., 2000. Nitrogen transfer between N_2-fixing plant and non-N_2-fixing plant. J. For. Res. 11, 75–80.

Sarkhot, D.V., Berhe, A.A., Ghezzehei, T.A., 2012. Impact of biochar enriched with dairy manure effluent on carbon and nitrogen dynamics. J. Environ. Qual. 41 (4), 1107–1114.

Scheer, C., Grace, P.R., Rowlings, D.W., Kimber, S., van Zwieten, L., 2011. Effect of biochar amendment on the soil-atmosphere exchange of greenhouse gases from an intensive sub-tropical pasture in northern new South Wales, Australia. Plant Soil 345 (1–2), 47–58.

Schmid, M.C., Hooper, A.B., Klotz, M.G., Woebken, D., Lam, P., Kuypers, M.M.M., Pommerening-Roeser, A., Op Den Camp, H.J.M., Jetten, M.S.M., 2008. Environmental detection of octahaem cytochrome c hydroxylamine/hydrazine oxidoreductase genes of aerobic and anaerobic ammonium-oxidizing bacteria. Environ. Microbiol. 10, 3140–3149.

Schmidt, H.-P., Wilson, K., 2014. The 55 uses of biochar. Biochar J. www.biochar-journal.org/en/ct/2.

Schmidt, C.S., Richardson, D.J., Baggs, E.M., 2011. Constraining the conditions conducive to dissimilatory nitrate reduction to ammonium in temperate arable soils. Soil Biol. Biochem. 43, 1607–1611.

Schmidt, H.-P., Pandit, B., Martinsen, V., Cornelissen, G., Conte, P., Kammann, C., 2015. Four-fold increase in pumpkin yield in response to low-dosage root zone application of urine-enhanced biochar to a fertile tropical soil. Agriculture 5 (3), 723.

Schomberg, H.H., Gaskin, J.W., Harris, K., Das, K.C., Novak, J.M., Busscher, W.J., Watts, D.W., Woodroof, R.H., Lima, I.M., Ahmedna, M., Rehrah, D., Xing, B., 2012. Influence of biochar on nitrogen fractions in a coastal plain soil. J. Environ. Qual. 41 (4), 1087–1095.

Shackley, S., Hammond, J., Gaunt, J., Ibarrola, R., 2011. The feasibility and costs of biochar deployment in the UK. Carbon Manag. 2 (3), 335–356.

Singh, B.P., Hatton, B.J., Balwant, S., Cowie, A.L., Kathuria, A., 2010. Influence of biochars on nitrous oxide emission and nitrogen leaching from two contrasting soils. J. Environ. Qual. 39 (4), 1224–1235.

Singh, B., Fang, Y., Cowie, B.C.C., Thomsen, L., 2014. Nexafs and xps characterisation of carbon functional groups of fresh and aged biochars. Org. Geochem. 77, 1–10.

Smith, J.L., Collins, H.P., Bailey, V.L., 2010. The effect of young biochar on soil respiration. Soil Biol. Biochem. 42 (12), 2345–2347.

Smith, K.A., 1980. A model of the extent of anaerobic zones in aggregated soils and its potential application to estimates of denitrification. J. Soil Sci. 31 (2), 263–277.

Song, Y., Zhang, X., Ma, B., Chang, S., Gong, J., 2014. Biochar addition affected the dynamics of ammonia oxidizers and nitrification in microcosms of a coastal alkaline soil. Biol. Fertil. Soils 50 (2), 321–332.

Spang, A., Hatzenpichler, R., Brochier-Armanet, C., Rattei, T., Tischler, P., Spieck, E., Streit, W., Stahl, D.A., Wagner, M., Schleper, C., 2010. Distinct gene set in two different lineages of ammonia-oxidizing archaea supports the phylum Thaumarchaeota. Trends Microbiol. 18, 331–340.

Spokas, K., Baker, J., Reicosky, D., 2010. Ethylene: potential key for biochar amendment impacts. Plant Soil 333 (1–2), 443–452.

Spokas, K.A., Novak, J.M., Masiello, C.A., Johnson, M.G., Colosky, E.C., Ippolito, J.A., Trigo, C., 2014. Physical disintegration of biochar: an overlooked process. Environ. Sci. Tec. Let. 1 (8), 326–332.

Spokas, K.A., 2013. Impact of biochar field aging on laboratory greenhouse gas production potentials. GCB Bioenergy 5 (2), 165–176.

Steffen, W., Richardson, K., Rockström, J., Cornell, S.E., Fetzer, I., Bennett, E.M., Biggs, R., Carpenter, S.R., De Vries, W., De Wit, C.A., Folke, C., Gerten, D., Heinke, J., Mace, G.M., Persson, L.M., Ramanathan, V., Reyers, B., Sörlin, S., 2015. Planetary boundaries: guiding human development on a changing planet. Science 347 (6223).

Stevens, R.J., Laughlin, R.J., Malone, J.P., 1998. Soil pH affects the processes reducing nitrate to nitrous oxide and di-nitrogen. Soil Biol. Biochem. 30 (8–9), 1119–1126.

Stevenson, F.J., Cole, M.A., 1999. The internal cycle of nitrogen in soil. In: Cole, M.A., Stevenson, F.J. (Eds.), Cycles of Soils: Carbon, Nitrogen, Phosphorus, Sulfur, Micronutrients. John Wiley & Sons, USA, pp. 191–229.

Stoecker, K., Bendinger, B., Schöning, B., Nielsen, P.H., Nielsen, J.L., Baranyi, C., Toenshoff, E.R., Daims, H., Wagner, M., 2006. Cohn's crenothrix is a filamentous methane oxidizer with an unusual methane monooxygenase. Proc. Natl. Acad. Sci. U. S. A. 103 (7), 2363–2367.

Strous, M., Kuenen, J.G., Jetten, M.S.M., 1999. Key physiology of anaerobic ammonium oxidation. Appl. Environ. Microbiol. 65, 3248–3250.

Suddick, E.C., Six, J., 2013. An estimation of annual nitrous oxide emissions and soil quality following the amendment of high temperature walnut shell biochar and compost to a small scale vegetable crop rotation. Sci. Total Environ. 465, 298–307.

Taghizadeh-Toosi, A., Clough, T.J., Condron, L.M., Sherlock, R.R., Anderson, C.R., Craigie, R.A., 2011. Biochar incorporation into pasture soil suppresses in situ nitrous oxide emissions from ruminant urine patches. J. Environ. Qual. 40 (2), 468–476.

Taghizadeh-Toosi, A., Clough, T., Sherlock, R., Condron, L., 2012. Biochar adsorbed ammonia is bioavailable. Plant Soil 350 (1–2), 57–69.

Tammeorg, P., Brandstaka, T., Simojoki, A., Helenius, J., 2012. Nitrogen mineralisation dynamics of meat bone meal and cattle manure as affected by the application of softwood chip biochar in soil. Trans. Earth. Sci. 103 (01), 19–30.

Treusch, A.H., Leininger, S., Kletzin, A., Schuster, S.C., Klenk, H.P., Schleper, C., 2005. Novel genes for nitrite reductase and amo-related proteins indicate a role of uncultivated mesophilic crenarchaeota in nitrogen cycling. Environ. Microbiol. 7, 1985–1995.

Troy, S.M., Lawlor, P.G., O'Flynn, C.J., Healy, M.G., 2013. Impact of biochar addition to soil on greenhouse gas emissions following pig manure application. Soil Biol. Biochem. 60, 173–181.

Uzoma, K.C., Inoue, M., Andry, H., Zahoor, A., Nishihara, E., 2011. Influence of biochar application on sandy soil hydraulic properties and nutrient retention. J. Food Agr. Environ. 9 (3–4), 1137–1143.

Van De Voorde, T.F.J., Bezemer, T.M., Van Groenigen, J.W., Jeffery, S., Mommer, L., 2014. Soil biochar amendment in a nature restoration area: effects on plant productivity and community composition. Ecol. Appl. 24, 1167–1177.

Van Kessel, M.a.H.J., Speth, D.R., Albertsen, M., Nielsen, P.H., Op Den Camp, H.J.M., Kartal, B., Jetten, M.S.M., Lücker, S., 2015. Complete nitrification by a single microorganism. Nature 528 (7583), 555–559 Advance online publication.

Vardon, D.R., Moser, B.R., Zheng, W., Witkin, K., Evangelista, R.L., Strathmann, T.J., Rajagopalan, K., Sharma, B.K., 2013. Complete utilization of spent coffee grounds to produce biodiesel, bio-oil, and biochar. ACS Sustain. Chem. Eng. 1 (10), 1286–1294.

Verhagen, F.J.M., Laanbroek, H.J., 1991. Competition for ammonium between nitrifying and heterotrophic bacteria in dual energy-limited chemostats. Appl. Environ. Microbiol. 57, 3255–3263.

Verhoeven, E., Six, J., 2014. Biochar does not mitigate field-scale N_2O emissions in a northern California vineyard: an assessment across two years. Agric. Ecosyst. Environ. 191, 27–38.

Vitousek, P.M., Cassman, K., Cleveland, C., Crews, T., Field, C.B., Grimm, N.B., Howarth, R.W., Marino, R., Martinelli, L., Rastetter, E.B., Sprent, J.I., 2002. Towards an ecological understanding of biological nitrogen fixation. Biogeochemistry 57, 1–45.

Wallenstein, M.D., Myrold, D.D., Firestone, M., Voytek, M., 2006. Environmental controls on nitrifying communities and denitrification rates: insights from molecular methods. Ecol. Appl. 16 (6), 2143–2152.

Wang, C., Lu, H., Dong, D., Deng, H., Strong, P.J., Wang, H., Wu, W., 2013a. Insight into the effects of biochar on manure composting: evidence supporting the relationship between N_2O emission and denitrifying community. Environ. Sci. Technol. 47 (13), 7341–7349.

Wang, Z., Zheng, H., Luo, Y., Deng, X., Herbert, S., Xing, B., 2013b. Characterization and influence of biochars on nitrous oxide emission from agricultural soil. Environ. Pollut. 174, 289–296.

Welsh, A., Chee-Sanford, J.C., Connor, L.M., Loffler, F.E., Sanford, R.A., 2014. Refined nrfa phylogeny improves pcr-based nrfa gene detection. Appl. Environ. Microbiol. 80, 2110–2119.

White, C.S., 1994. Monoterpenes: their effects on ecosystem nutrient cycling. J. Chem. Ecol. 20 (6), 1381–1406.

Wiedner, K., Schneeweiß, J., Dippold, M.A., Glaser, B., 2015. Anthropogenic dark earth in northern Germany – the nordic analogue to *Terra preta* de índio in Amazonia? Catena 132, 114–125.

Xu, H.J., Wang, X.H., Li, H., Yao, H.Y., Su, J.Q., Zhu, Y.G., 2014. Biochar impacts soil microbial community composition and nitrogen cycling in an acidic soil planted with rape. Environ. Sci. Technol. 48, 9391–9399.

Yanai, Y., Toyota, K., Okazaki, M., 2007. Effects of charcoal addition on N_2O emissions from soil resulting from rewetting air-dried soil in short-term laboratory experiments. Soil Sci. Plant Nutr. 53 (2), 181–188.

Yu, O.-Y., Raichle, B., Sink, S., 2013. Impact of biochar on the water holding capacity of loamy sand soil. Intern. J. Energy Environ. Eng. 4, 44.

Zalasiewicz, J., Williams, M., Haywood, A., Ellis, M., 2011. The anthropocene: a new epoch of geological time? Phil Trans. Math. Phys. Eng. Sci. 369 (1938), 835–841.

Zhang, A., Bian, R., Pan, G., Cui, L., Hussain, Q., Li, L., Zheng, J., Zheng, J., Zhang, X., Han, X., Yu, X., 2012a. Effects of biochar amendment on soil quality, crop yield and greenhouse gas emission in a chinese rice paddy: a field study of 2 consecutive rice growing cycles. Field Crop Res. 127, 153–160.

Zhang, A.F., Liu, Y.M., Pan, G.X., Hussain, Q., Li, L.Q., Zheng, J.W., Zhang, X.H., 2012b. Effect of biochar amendment on maize yield and greenhouse gas emissions from a soil organic carbon poor calcareous loamy soil from central China plain. Plant Soil 351, 263–275.

Zhao, X., Wang, S., Xing, G., 2014. Nitrification, acidification, and nitrogen leaching from subtropical cropland soils as affected by rice straw-based biochar: laboratory incubation and column leaching studies. J. Soils Sediments 14 (3), 471–482.

Zheng, H., Wang, Z., Deng, X., Herbert, S., Xing, B., 2013. Impacts of adding biochar on nitrogen retention and bioavailability in agricultural soil. Geoderma 206 (0), 32–39.

Zhu, G., Wang, S., Wang, Y., Wang, C., Risgaard-Petersen, N., Jetten, M.S., Yin, C., 2011. Anaerobic ammonia oxidation in a fertilized paddy soil. ISME J. 5, 1905–1912.

Zhu, K., Christel, W., Bruun, S., Jensen, L.S., 2014. The different effects of applying fresh, composted or charred manure on soil N_2O emissions. Soil Biol. Biochem. 74, 61–69.

van Zwieten, L., Singh, B., Joseph, S.D., Kimber, S., Cowie, A., Chan, K.Y., 2009. Biochar and the emissions of non-CO_2 greenhouse gases from soil. In: Lehmann, J., Joseph, S. (Eds.), Biochar for Environmental Management, Science and Technology. Earthscan, London.

van Zwieten, L., Kimber, S., Downie, A., Morris, S., Petty, S., Rust, J., Chan, K.Y., 2010a. A glasshouse study on the interaction of low mineral ash biochar with nitrogen in a sandy soil. Soil Res. 48 (7), 569–576.

van Zwieten, L., Kimber, S., Morris, S., Chan, K.Y., Downie, A., Rust, J., Joseph, S., Cowie, A., 2010b. Effects of biochar from slow pyrolysis of papermill waste on agronomic performance and soil fertility. Plant Soil 327 (1–2), 235–246.

van Zwieten, L., Kimber, S., Morris, S., Downie, A., Berger, E., Rust, J., Scheer, C., 2010c. Influence of biochars on flux of N_2O and CO_2 from ferrosol. Soil Res. 48 (7), 555–568.

van Zwieten, L., Kimber, S.W.L., Morris, S.G., Singh, B.P., Grace, P.R., Scheer, C., Rust, J., Downie, A.E., Cowie, A.L., 2013. Pyrolysing poultry litter reduces N_2O and CO_2 fluxes. Sci. Total Environ. 465, 279–287.

van Zwieten, L., Singh, B.P., Kimber, S.W.L., Murphy, D.V., Macdonald, L.M., Rust, J., Morris, S., 2014. An incubation study investigating the mechanisms that impact N_2O flux from soil following biochar application. Agric. Ecosyst. Environ. 191, 53–62.

Microbial Ecology of the Rhizosphere and Its Response to Biochar Augmentation

C.H. Orr, T. Komang Ralebitso-Senior, S. Prior

Teesside University, Middlesbrough, United Kingdom

Biochar Application
http://dx.doi.org/10.1016/B978-0-12-803433-0.00008-4

199

INTRODUCTION

Currently the need for sustainable food provision grows for a global population exceeding 7 billion, which is projected to increase to 9.5 billion by 2050. The unabated demand for increased agricultural output, which often results in degraded soil quality, concomitant with increasing awareness for sustainable environmental management and reduced carbon footprints, mandate innovative and diverse agricultural practices. Biochar, char (charcoal) applied to land, has the potential to close the circle of enhanced yield (rhizosphere microbial stimulation) with reduced energy output (lower fertilizer application), minimized carbon emissions (carbon sequestration), and decreased demand for water (improved soil moisture-holding capacity). The realization of these elements must be underpinned by understanding the fundamentals of soil microbial community response to biochar supplementation as well as the relationships between biogeochemical cycles, such as the nitrogen cycle (see also Chapter 7), soil amendments, and functional bacterial groups.

Biochar

Intense interest in biochar is stimulated by two potential benefits: stabilized storage of photosynthetic carbon, which, in turn, addresses carbon pool loss; and soil improvement (Lal, 2009; Molina et al., 2009; Woolf et al., 2010; Kolton et al., 2011). Biochar has unique physical, chemical, and biological characteristics, including high water-holding capacity, large internal surface area, and a wide range of pore sizes (see critical reviews, eg, Atkinson et al., 2010; Ennis et al., 2012). These can be tailored through choice of source material and production conditions to produce bespoke biochar, which is ideal for holding soluble organic matter and nutrients and therefore creating an excellent habitat/shelter for many different soil microorganisms (Pietikainen et al., 2000). Historically, biochar was used by old civilizations, especially in Brazil's Amazon region, to improve crop yield. Centuries later, these soils are still darker and more fertile than similar ones without biochar. Specifically, *Terra Preta* (modified Brazilian oxisols) improved fertility and this was recognized in the 16th century. Today, biochar could: produce comparable soils; sequester carbon ($\geq 12\% v/v$ of anthropogenic greenhouse gases) (Woolf et al., 2010) to mitigate effectively climate change (Molina et al., 2009); and redress organic pool carbon loss (78 ± 12 Gt since 1850) (Lal, 2009). Although the efficacy of porous media such as biochar to promote plant growth and remediate contaminated soil is recognized (du Plessis et al., 1998; Kolton et al., 2011), few definitive studies have been reported.

Soil bacteria play an essential role in the cycling of important nutrients such as organic carbon and nitrogen that contribute to plant growth and a range of associated ecosystem services and functions (Zhao et al., 2014; Fitter et al., 2005). Perturbations to soil can have significant effects on the abundance and distribution of many groups of soil bacteria. Biochar, specifically, provides a medium for biofilm formation and hence supports the growth of robust, dynamic, and diverse soil microbial communities. Also it has the potential to improve crop yield through in situ nitrogen fixation. As nitrogen is often the key limiting growth factor, understanding the effects of biochar addition on nitrogen fixation dynamics in (agronomic) soils is critical. Although the nitrogen cycle is being investigated increasingly in relation to greenhouse gas emissions from char-augmented soils, very few studies have explored the microbial ecology of nitrogen fixation in this context (eg, Clough et al., 2010; Lin et al., 2015; Orr and Ralebitso-Senior, 2014). Since less than 5% of all microorganisms have been cultured in the laboratory, advances in molecular microbial ecology have provided the opportunity to study key functional groups in soil at a meaningful resolution and so offer an ecological shorthand to elucidate the functions and dynamics of soil systems.

The Importance of Studying Biochar Supplementation in Agriculture

Sales of biochar have tripled annually since 2008 as the potential benefits, particularly to small farm holders in developing countries, have been publicized (Cernansky, 2015). The benefits can be environmental, in terms of carbon sequestration, reduced greenhouse gas emissions, and reduced energy emissions, and economical, in terms of increased crop yield. Because of its large surface area and porosity, biochar has the potential to improve soil moisture-holding capacity, decrease irrigation requirements, and hence reduce agronomic carbon footprints (eg, as reviewed by Atkinson et al., 2010; Anderson et al., 2011; Ennis et al., 2012; Lehmann et al., 2011). Also the addition of biochar to agricultural soil has been recommended to decrease possibly the amounts and application frequencies of chemical (nitrogen-based) fertilizers with the potential to then reduce nitrogen leaching to groundwater, emissions of greenhouse gases, and energy input (eg, Orr and Ralebitso-Senior, 2014). Also biochar's ability to remediate pollution by immobilizing metals and binding toxic chemicals (Cernansky, 2015) has the potential to sorb excess chemical pesticides, which may cause environmental concerns (Chapters 9 and 10). Therefore seminal studies are being implemented to elucidate and/or demonstrate the advantages and disadvantages of the addition of biochar.

Biochar has previously been shown to enhance crop output in terms of yield (Alburqueque et al., 2013), biomass (Soja, 2013), and overall soil productivity (Liu et al., 2013). Specifically, increases in soil productivity have been recorded when biochar was applied to agronomic settings because of reduced soil acidity, improved cation exchange capacity, increased water-holding capacity, and increased microbial community diversity (Shackley et al., 2009). Thus meta-analysis of the literature by Jeffery et al. (2011) showed that, although the effects vary depending on biochar feedstock and pyrolysis conditions, and soil type and cropping regime, there is an average 10% increase in crop productivity caused by biochar addition.

Often the highest increases in crop yield result when biochar is applied in combination with a nitrogen source. For example, Alberqueque et al. (2013) supplemented durum wheat (*Triticum durum*) with two different agricultural waste-derived biochars, wheat straw and olive tree pruning, and different mineral fertilizer concentrations. Biochar had no affect when applied on its own but led to an increase of 20–30% grain yield when applied with mineral fertilizer, potentially because of increased available P but decreased N and Mn. Similar trends were recorded in other cropping systems in the Traismauer Austrian field experiments (Soja, 2013). In this study the biochar and N fertilizer combination increased the growth of sunflower (*Helianthus* sp.; 0.4%), winter wheat (*Triticum aestivum*; 0.2%), corn (4.1%), and maize (*Zea mays*; 6.6%) and significantly increased the growth of spring barley (10.4%) compared to N fertilizer application alone.

As reflected in the literature, increases in crop yield are often correlated with increases in P and N availability (Filiberto and Gaunt, 2013). Although no statistically significant changes in N immobilization have been recorded because of biochar application, there is evidence that soil pH, organic C, and total N can increase, while bulk density decreases (Laufer and Tomlinson, 2013; Zhang et al., 2010). Also changes in pH have reportedly increased soil K, P, Ca, and Mg availability, which have consequently been found in higher concentrations in plants grown in the presence of biochar compared to nonbiochar controls (Gaskin et al., 2010). In these studies the biochars were not nutrient rich but released base cations into the surrounding soil instead, unlocking nutrients for plant uptake (Gaskin et al., 2010). Biochar is frequently associated with soil "liming" effect caused by the presence of alkaline earth metals. Therefore, while soil pH is usually fairly stable, the addition of biochars has been observed to increase pH by 0.5 pH units (Hodgson et al., 2012). As pH is the key driver of the soil microbial community (Fierer and Jackson, 2006), changes in this parameter can also result in changed plant growth.

By comparing findings from 16 studies, Jeffery et al. (2011) suggested conditions that can potentially result in positive responses to biochar application. Thus the highest improvements in crop productivity resulted for acidic to neutral soils with a coarse to medium texture. Also they surmised that possibilities for positive effects were increased for soils that are characterized by low pH, low organic carbon, balanced N supply, periodic drought, compacted structure, or (historical) contamination. Furthermore, the use of poultry litter as a feedstock had the strongest positive effect

(Jeffery et al., 2011) potentially because of increased concentrations of N, P, and K (Filiberto and Gaunt, 2013). An Australian study reported that poultry litter biochar was the most economical when the increase in crop yield and reduced dependence on nitrogen fertilizers was considered (Laufer and Tomlinson, 2013). Therefore this would be a viable/promising feedstock given the waste management issues associated with poultry litter from farms, particularly in the United States.

A critical and comprehensive literature interrogation by Liu et al. (2013) analyzed 119 laboratory- and field-scale experiments and summarized that increased crop productivity differed with crop type where legumes, vegetables, and grasses recorded larger increases compared to cereals, wheat, and rice. Notwithstanding this, maize appeared to be a crop that can be affected significantly by biochar application. Thus a summary by the International Biochar Institute revealed that increases in maize yield ranged from 20% to 140% compared with control plots (Laufer and Tomlinson, 2013). In general, all the meta-analyses carried out between 2011 and 2015 (Jeffery et al., 2011; Liu et al., 2013) emphasized the short-term nature of the majority of studies and consequently highlighted the need for longer-term investigations.

Furthermore, while there is ample evidence that the presence of biochar can lead to increased crop yield, the underlying mechanisms remain largely unknown. Presumably the changes in nutrient, carbon, nitrogen, and water availability to the plant facilitate growth. Typically the interactions between the plant root and rhizosphere soil are investigated and manipulated toward increase crop yield. It is in this zone that vital nutrient transformations and uptake occur, which ultimately controls crop productivity (Shen et al., 2012).

Despite the emerging knowledge and understanding, comprehensive investigations on changes in nitrogen cycling in particular are required to underpin subsequent critical reassessments of biochar as a soil improvement medium. For example, many studies have considered how biochar addition may affect bulk soil and plant growth, but few have studied the impact biochar may have within the rhizosphere. Therefore the remainder of this chapter will explore this often overlooked area to further explain the change biochar addition can have on the soil environment.

What Is the Rhizosphere?

The rhizosphere is the narrow zone of soil that contacts plant roots and provides a unique ecological niche in which a microbial community responds to specific plant interactions and root exudates (Ahemad and Kibret, 2014). Several organic substances released from plant roots promote a more active and diverse microbial community than the surrounding bulk soil. Also rhizospheric microbial genes far outnumber plant genes while bacterial counts exceed those of bulk soil by factors between 10 and 1000 (Lugtenberg and Kamilova, 2009; Mendes et al., 2013; Li et al., 2014). In turn, as reviewed by Mendes et al. (2013), the rhizosphere microbiome can promote: plant growth and development; nutrient acquisition; protection against pathogens; immune response; and tolerance to abiotic stress. Generally these benefits are categorized as: promotion of plant growth via resource acquisition; modulation of plant hormones; and reduction of inhibitory pathogenic effect (Ahemad and Kibret, 2014). Growth promotion was exemplified by the work of Li et al. (2014) who used a GeoChip gene array to detect functional gene abundance within the rhizosphere of maize and reported a predominance of Proteobacteria, Firmicutes, Bacteroidetes, and Cyanobacteria with increased functional capacity for carbon, nitrogen, phosphorus, and sulfur compound transformations compared with bulk soil.

Plant root release of rhizodeposits is the key process by which C is transferred to the soil and can account for up to 40% (w/w) of the net carbon (Singh et al., 2004). These nutrient-rich exudates can then stimulate the indigenous microbial communities (Cook et al., 1995; Mendes et al., 2013). Many specific groups of rhizosphere microorganisms, such as nitrogen-fixing bacteria, plant growth-promoting rhizobacteria, and mycorrhizal fungi, have been studied for their beneficial effects on plant growth (Mendes et al., 2013). According to Singh et al. (2004), the rhizosphere nitrogen-fixing rhizobia and mycorrhizal fungi in particular receive the root exudates-based carbon and, in turn, enhance plant nutrient availability by providing nitrogen, potassium, and phosphorus via phosphate solubilization, so promoting plant growth. As a result the species have been described as biofertilizers (Lugtenberg and Kamilova, 2009). Nitrogen-fixing bacteria have also been described as phytostimulators where soils have been inoculated with species such as *Azotobacter paspali* to increase plant growth specifically through the release of auxins such as indole-3-acetic acid, gibberellins, and cytokinins (Lugtenberg and Kamilova, 2009). Ultimately this has led to symbiotic (eg, *Rhizobium*, *Bradyrhizobium*, and *Mesorhizobium*) and nonsymbiotic (eg, *Pseudomomanas*, *Azotobacter*, and *Azospirillum*) microorganisms being used as plant growth promoting bioinoculants (Ahemad and Kibret, 2014).

Plants have also been shown to consolidate symbiotic bacteria within their rhizosphere for pathogenic resistance in a process similar to human microbiome protection (Mendes et al., 2011, 2013). Mendes et al. (2011) used, for example, metagenomic analysis and showed that, when attacked by fungal pathogens, plants such as sugar beet can co-opt soil microbial communities, particularly members of the Gammaproteobacteria, that release fungicides. Similarly, Lugtenberg and Kamilova (2009) suggested that rhizobacteria such as the naphthalene degrading *Pseudomonas putida* PCL1444 can protect plants by catabolizing potentially phytotoxic pollutants. Also many bacteria can ameliorate plant stress through siderophore production. For example, rhizobia release siderophores, which chelate environmental iron making it bioavailable to the microorganisms and plants. These siderophores can also form stable complexes with potentially toxic heavy metals such as Cu, Pb, and Zn (Ahemad and Kibret, 2014). Since plant disease accounts annually for >€200 billion of crop loss, this biological control potential could be economically beneficial (Lugtenberg and Kamilova, 2009).

Although the key determinant of rhizosphere microbial community structure appears to be the amount and composition of carbon-based plant root exudates, other plant and soil augmentations, made often during standard farm management practices, can also affect structure and activity (Marscher et al., 2004). The researchers used 16S rDNA-based polymerase chain reaction-denaturing gradient gel electrophoresis to investigate the importance of soil augmentation and plant species on the rhizosphere bacterial community structure. They found that soil pH and the type of P fertilization and N addition were the major determinants. Also the rhizosphere community varied for different plants and for the same plant in different soils. Furthermore, Nallanchakravarthula et al. (2014) used pyrosequencing to show that strawberry plant rhizosphere fungal community structure changed considerably between cultivars and between soils with different C, N, and P concentrations.

Fertilizer is added to agricultural soil to increase nutrient concentrations and therefore enhance plant growth. Such additions have been shown to change inadvertently soil pH and nutrient status, and, in turn, soil microbial communities (Lauber et al., 2008; Orr et al., 2012). Jack et al. (2011) applied terminal restriction fragment length polymorphism (T-RFLP) analysis on different composts used to propagate tomatoes and found that compost type not only influenced strongly plant quality but also led to changes in rhizosphere bacterial communities long after transplantation to field. They attributed this to changes in pH, total N, NO_3–N, NH_4–N, P, K, and organic C.

Parallel with fertilizer addition, the efficacy of biochar supplementation of agricultural soil is gaining more interest to farmers and researchers through its potential to reduce both irrigation and the carbon footprint, while increasing crop yield (Arbestain et al., 2014; Atkinson et al., 2010; Ippolitto et al., 2012). Depending on the feedstock and production conditions, biochar addition has, however, been shown to alter several soil bio-/physicochemical parameters including N and C dynamics, pH, P availability, and heavy metal availability (Ippolitto et al., 2012; Spokas et al., 2012). For example, a study by Prendergast-Miller et al. (2014) reported that biochar augmentation increased phosphorus and mineral N, by a factor of two for the latter, compared to control soils. Although biochar C is largely considered recalcitrant and thus unavailable for microbial catabolism, it can introduce metabolically labile C compounds (Anderson et al., 2011). Overall, since these changes influence rhizosphere communities, it is possible that biochar addition would have further effects and, in turn, probably affect plant growth.

Previous Studies of the Rhizosphere and the Influence of Biochar

Although many studies have been made of the effects of biochar on soil microbiota, few have focused specifically on changes within the rhizosphere (Lehmann et al., 2008). For bulk soil, biochar supplementation can lead to changes in microbial biomass as revealed by biochemical (Hockaday et al., 2007; O'Neill et al., 2009; Steinbeiss et al., 2009) and molecular-based (Khodadad et al., 2011) analyses. Kolton et al. (2011) used T-RFLP, denaturing gradient gel electrophoresis (DGGE) and pyrosequencing to investigate rhizospheric microbial community dynamics of greenhouse sweet pepper plants (*Capsicum annuum* L.) 3 months after 3% (w/w) citrus wood-derived biochar addition. The workers recorded distinct community fingerprint clustering that separated the rhizosphere and bulk soil of the biochar and nonbiochar control. Specifically the pyrosequencing data revealed a considerable community structure shift where the relative abundance of rhizosphere Bacteroidetes increased (12–30%) while that of Proteobacteria decreased (72–48%) in response to biochar. Also the operational taxonomic units (OTU) that were affected significantly were *Flavobacterium*, *Chitinopaga*, and *Cellvibrio* genera. Therefore the researchers suggested that the recognized abilities of the component species to inhibit bacterial and fungal growth, produce secondary metabolites including antibiotics, and act as biocontrol agents could potentially explain why biochar can mitigate plant disease. These results contrasted those of Khodadad et al. (2011) who reported increases in Actinobacteria and Gemmatimonadetes in bulk soil supplemented with laurel oak heartwood and eastern gamagrass biochars, thus highlighting the effects of feedstock and pyrolysis conditions.

Together with biochar-promoted pepper plant growth, Graber et al. (2010) recorded more abundant rhizosphere culturable bacteria. Since the plant nutrient contents were comparable with the controls, they concluded that the biochar increased the plant growth-promoting bacteria or fungi. Studies that investigated rhizosphere biochemical changes following biochar addition have shown that nitrate-N concentration increased coincident with root length increase in wheat plants but not plant biomass or N content (Prendergast-Miller et al., 2011). This possibly explains why some studies have shown a reduction in nitrogen-fixing *Bradyrhizobium* within the rhizosphere (Anderson et al., 2011).

Anderson et al. (2011) used T-RFLP and 454-pyrosequencing to investigate the effects of a *Pinus radiata* biochar on the rhizosphere and bulk soil communities of rye grass. They found that the pH within the rhizosphere was raised while the soil moisture content was reduced. Also the rhizosphere community structure was more diverse. For both the bulk soil and rhizosphere, the Streptosporangineae and Thermomonosporaceae numbers increased following biochar addition. These promotions could prove positive since these families contain species that degrade cellulose and lignocellulose residues, and reduce nitrate and so facilitate increased turnover of soil organic matter (Anderson et al., 2011).

To test the effects of biochar on plant root growth Prendergast-Miller et al. (2014) cultivated spring barley (*Hordeum vulgare* L.) seedlings in rhizobox mesocosms with and without biochar. They found that plants with biochar had larger rhizosphere zones than the controls and the roots gravitated toward the biochar particles. In the study the biochar retained N and supplied the growing plant with P.

If biochar particles contact root systems, root hairs can penetrate pores or bond to the surfaces allowing root organic compounds to be sorbed (Joseph et al., 2010). As the rhizosphere forms a redox interface, which allows reduction and oxidation of various compounds and results in the transformations of C, N, and S, the sorptive potential of biochar may interfere (Joseph et al., 2010). Additionally the volatile compounds of some biochars may effect plant root growth and promote mycorrhizal fungi (Amonette and Joseph, 2009b). Following biochar addition, Jin (2010) reported increases in rhizosphere microbial abundance compared to bulk soil. Specifically, significant increases in cellulose-degrading Zygomycota and mycorrhizal-forming Glomeromycota were found possibly as a result of sorbed organic C (Lehmann et al., 2011). Warnock et al. (2007) proposed several mechanisms by which biochar could affect mycorrhizal fungi including: changes in nutrient availability; changes in abundance of phosphate-solubilizing bacteria; interference of plant–fungal signaling; and refuge from predators (Anderson et al., 2011).

Potential Mechanisms of Biochar's Effects on Rhizosphere Communities

As discussed previously and widely in the literature, rhizosphere and bulk soil conditions may differ markedly. For example, Daly et al. (2015) reported that rhizosphere soil is potentially more acidic while the cyclic action of water uptake and release by roots can lead to increased and decreased soil moisture content during the day and night, respectively. It is imperative therefore to research in particular whether such physicochemical differences dictate microbial community changes. First, biochar may contain compounds that are readily soluble in moist soil and so potentially alter the local pH. This may be pronounced in high ash biochars, in particular because of surface oxidation (Nguyen and Lehmann, 2009; Joseph et al., 2010). Therefore any cyclic changes in rhizosphere soil moisture content could potentially lead to pH shifts with subsequent changes in microbial community composition, diversity, and activity as proposed by Fierer and Jackson (2006) and Fernández-Calvino and Bååth (2010). This suggests that rhizosphere pH changes, in regions immediately surrounding biochar particles, could lead to microbiogeographies or "charospheres" (Quilliam et al., 2013) with distinct microbial communities. Furthermore, the rhizosphere night moisture content could possibly also result in the release of soluble organics from the biochar, potentially stimulating seed germination and fungal growth. These complex reactions, which may occur on the biochar surface, are probably intensified in interactions with roots/root hairs and so affect rhizosphere microbial activity (Joseph et al., 2010). Also the combined effects of increased plant root-derived organic substrates and the added biochar could stimulate the indigenous microbial communities.

Although largely recalcitrant, biochar is a potential source of labile carbon. In particular, high-temperature biochars contain elevated concentrations of aromatic, bioavailable C (Joseph et al., 2010). Differences in carbon availability affect soil quality, structure, and microbial biomass with elevated soil concentrations, generally facilitating proliferation of fast-growing, opportunistic bacteria (Azeez, 2009; Sarathchandra et al., 2001). A study assessing the effects of different C:N ratios on rhizosphere microbial community dynamics revealed that increased labile carbon enhanced soil respiration and microbial DNA content with rhizosphere microorganisms recording higher rates than the bulk soil species (Blagodatskaya et al., 2014).

Although dependent on feedstock and pyrolysis temperature, electrical conductivity and redox potential have been shown to be altered in soil immediately surrounding biochar particles (Joseph et al., 2010) (see also Chapter 2). Furthermore, changes in electrical conductivity and redox potential have been reported to also alter microbial biomass carbon (Mavi and Marschner, 2013), potentially confounding the effects discussed previously.

AN ILLUSTRATIVE STUDY

Rationale

Surpassing energy production and water provision, carbon sequestration is now the single greatest challenge facing the scientific community. Within the United Kingdom, farming techniques such as minimum tillage, increased tree coverage, and improved grassland management have been encouraged to address this. An alternative emerging option is to stabilize photosynthetically fixed carbon by the production of charcoal for subsequent application to soil as biochar. Further and major economic benefits can, however, be gained through agricultural application to improve plant productivity.

Biochar is a charcoal soil supplement that has the potential to address several important environmental and agricultural issues. Current biochar applications include: (1) carbon sequestration; (2) organic and inorganic compound adsorbent; (3) ecosystem restoration/reclamation; (4) structure modifier (bulking agent; moisture, nutrient, and microbiota retention); and (5) fertilizer. When applied to agricultural soils, it has influenced demonstrably many key factors including: long-term carbon storage; pesticide adsorption; and reduced nutrient leaching. Previous research has also shown that it can affect soil indigenous microbial populations probably through physicochemical parameter changes (Mia et al., 2014). Nevertheless, considerable paucity of work on improved crop yield necessitates essential study of microbial diversity, structure, and functional responses to char type and application strategy. This key knowledge gap must be addressed to enable success of all bespoke biochar applications in sustainable agricultural and environmental management.

As bioavailable nitrogen is a key limiting factor of crop yield, industrially produced fertilizers are often applied to increase soil concentrations. There are, however, large energy costs and potentially negative environmental impacts from fertilizer production processes. As a result the impacts of biochar on N-cycling microbial communities, whose functional activities can increase soil nitrogen capital, have been studied with different results (Orr and Ralebitso-Senior, 2014). For example, increased copies of nitrogen-fixing (*nifH*) and denitrifying (*nirK*, *nirS*, and *nosZ*) genes have been reported after biochar augmentation. Also decreases in nitrous oxide emissions were recorded in some studies and highlighted the potential of biochar to mitigate inimical agricultural environmental impacts (Ducey et al., 2013). In contrast some studies have reported no net effects while others (Troy et al., 2013; Wang et al., 2013; Nelissen et al., 2014) have recorded increased nitrous oxide emissions. Furthermore, agricultural soil is a major source of nitrogen-based greenhouse gas emissions, in particular nitrous oxide, with a contribution of approximately 60% to the total anthropogenic output. This results from several pathways, including the incomplete reduction of nitrate to dinitrogen gas by denitrifying soil bacterial communities, and is exacerbated when carbon is readily available as an energy source. Understanding the effects of soil perturbations, particularly those that can alter C dynamics, may be crucial to elucidate nitrogen cycling.

Although the combined and enhanced role of biochar and soil microbial populations in the rhizosphere are recognized, limited research exists of microbial diversity and functional response to the approach. The distinctive physico- and biochemical properties of biochar, including high water-holding capacity, large internal surface area, cation exchange capacity, elemental composition, and wide ranges of pore size/volume/distribution, effect its recognized benefits. Current information suggests that biochar can act as N additive and reduce harmful N-based emissions. This research, however, often does not examine specifically its effects on the N-cycling bacterial communities.

Generally these different findings highlight the existing considerable knowledge gap of the effects of biochar on organic or unfertilized soil.

Therefore this study was designed to determine the impacts of biochar on the diversity and taxa richness of nitrogen-fixing communities of unfertilized soil.

Materials and Methods

Soil (20 kg) was dug from a well-secured site at Framwellgate Moor, County Durham, UK (lat. 53.15°N, long. 1.59°W) and stored in a sterilized 25-L airtight bucket prior to sieving (ASTM-standard soil sieve No. 10; 2 mm mesh) to ensure homogeneity. The soil and locally produced biochar were analyzed as described in Chapter 4 (Table 4.1; Table 8.1).

The experimental protocol consisted of seedling trays with randomized triplicate treatments (Fig. 8.1) of: soil only (110 g fresh weight, S-); soil + biochar (5% w/w, S+); soil + clover (0.1 g *Trifolium pratense* seeds, P-); and soil + biochar + clover (P+) with moisture maintained via capillary action with deionized water. Thus the clover seeds were planted in study week 2 and allowed to germinate for 2 weeks with germination recorded as study week 0. Triplicate cells were then sampled destructively to collect bulk (controls) and rhizosphere soil every 2 weeks up to week 6. The plants were

TABLE 8.1   Physicochemical characteristics of the soil and mixed Broadleaf forestry-derived biochar

Parameter	Soil
Clay (%)	26
Silt (%)	21
Sand (%)	53
Al (g kg^{-1})	13
Ca (g kg^{-1})	2.2
Mg (g kg^{-1})	1.1
K (g kg^{-1})	1.8
Na (g kg^{-1})	0.25
Electrical conductivity (μS cm^{-1})	140
Calorific value (MJ kg^{-1})	2.5
Total organic carbon (%)	4.1
Total S (%)	0.03
P (mg kg^{-1})	1.2
Nitrate aqueous extract as NO_3 (mg L^{-1})	4.6
pH	6.3

FIGURE 8.1   Week 1 randomized planting for ambient temperature incubation.

harvested and shaken to dislodge the bulk soil while the roots were washed subsequently in sterile saline to collect the rhizosphere soil. The rinsates were centrifuged (10,000×g for 10 min), the supernatants discarded, and 0.5 g of the pelleted soils used for DNA extraction (FastDNA SPIN Kit for Soil, MP Biomedicals) with the extracts stored at −80°C.

The *nifH* gene was amplified with a PolF/PolR primer set (Ducey et al., 2013) where the reaction mixtures (25 μL) consisted of 12.5 μL PCR master mix (Promega), 0.5 μL PolF and PolR (10 μM), 1.25 μL BSA (10 mg mL⁻¹), 8.25 μL RNase/DNase-free water, and 2 μL DNA templates. The thermal cycling conditions (Primus 96 Plus, MWG-Biotech, Ebersberg, Germany) were: 94°C for 5 min; 30 cycles of 94°C for 1 min, 55°C for 1 min, 72°C for 2 min; and 72°C for 5 min. The DNA and amplification products (5 μL + 1 μL 6X loading buffer) were visualized by electrophoresis in 1.5% (w/v) agarose gels stained with SYBR Safe DNA Gel Stain (Molecular Probes, Eugene, USA). Amplicons (20 μL) were then separated by DGGE on 10% (w/v) acrylamide gel (acrylamide/bisacrylamide 37.5:1) with a 40–70% denaturant gel, with parameters described by Dias et al. (2012). The gel was stained with SYBR Gold (Invitrogen, 2 μL of 10,000X in 20 mL TAE) and visualized (AlphaImager HP, Alpha Innotech, Braintree, UK) under UV light.

The DGGE images were quantified by Phoretix 1D software (TotalLab, Newcastle, UK) and cluster analysis was made with unweighted pair-group using arithmetic average (UPGMA). The number, presence and absence, and relative abundance (intensity) of operational taxonomic units (OTUs) were used to determine the ecological indices for Shannon–Wiener diversity (H′), richness (S), and evenness (E) as described by Olakanye et al. (2014). All data were evaluated statistically using general linear model ANOVA with Tukey's comparison testing (Minitab 16.2.4, Minitab Inc.).

Results and Discussion

Biochar properties depend critically on feedstock and pyrolysis conditions. Characteristically the feedstock determines its chemical composition while process temperature determines the surface area, pore size, volume, distribution, sorption, and partitioning of the biochar (eg, Atkinson et al., 2010; Amonette and Joseph, 2009a). Therefore a cost-effective biochar, produced from sustainable and locally relevant agricultural waste, was used.

The presence of plants in a soil environment can alter significantly the microbial community. Specifically, clover is planted often as a ley crop to return nitrogen to the soil. Nitrogen-fixing bacteria then form a symbiotic relationship with the clover. Laboratory-based microcosms were therefore established each with a single biochar addition (1:5 w/w) to compare its effects on bulk and rhizosphere soil planted with clover (*T. pratense*). Generally the addition of biochar statistically significantly increased the total plant biomass as determined by total average height ($p < 0.001$) and weight ($p < 0.019$) of the red clover plants per triplicate treatment.

The results of the ecological indices and UPGMA clustering are shown in Fig. 8.2 and Fig. 8.3, respectively. The *T. pratense* rhizosphere diversity

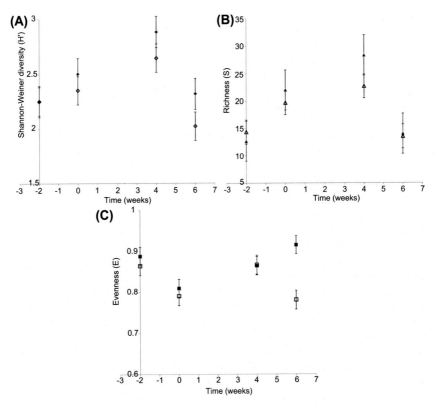

FIGURE 8.2   Changes in: (A) Shannon–Wiener diversity (H′), (B) taxa richness (S), and (C) evenness (E) of *Trifolium pratense* rhizosphere *nifH* communities in the presence (*solid symbols*) and absence (*open symbols*) of biochar.

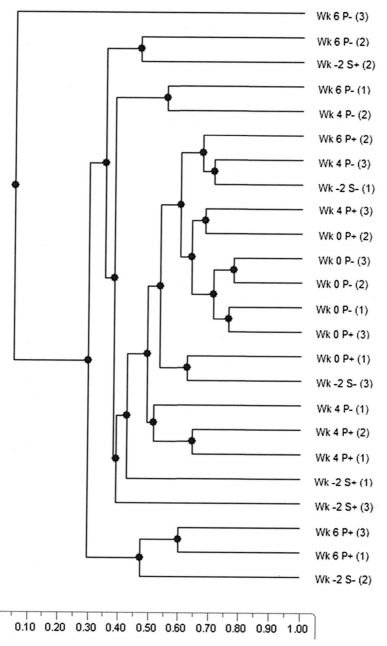

FIGURE 8.3 Unweighted pair-group using arithmetic average (UPGMA) clustering of *nifH* DGGE profiles with (P) and without (S) clover and in the presence (+) and absence (−) of biochar during weeks (Wk) −2, 0, 4, and 6. Replicates are shown in *brackets*.

(H′) of the nitrogen-fixing community and the abundance of *nifH* OTU showed an upward progression up to week 4 with different increases in the presence of biochar (H′ = 2.486 ± 0.347; richness = 19.25 ± 7.899; evenness = 0.869 ± 0.058) in contrast to the control (H′ = 2.342 ± 0.301; richness = 16.5 ± 7.681; evenness = 0.869 ± 0.052). Also evenness was the only index that changed statistically significantly between the treatments. In week 6, the H′ and S decreased in the experimental and control microcosms, but the diversity (H′) remained higher (Fig. 8.2) in the biochar-supplemented soil than the control. Ducey et al. (2013) attributed rises in diversity and richness to increased microbial activity and growth, possibly because of enhanced biochar-based water retention. The decrease could be explained by this factor, with increased activity dependent on nutrients that were not replaced by the deionized water, or lost by leaching because of the mobility of some elements in soil (Mia et al., 2014). Also some clover dieback occurred in week 6 in one of triplicate treatments, which could have also impacted the average microbial community dynamics indices.

The UPGMA clustering (Fig. 8.3) showed that mostly two of three replicates clustered together at 76% (week 0P−), 64% (week 4P+), and 60% (week 6P+). Also, with the exception of P+(1), all the week 0 triplicate soils with 2-week seedlings recorded a 63% similarity independent of the biochar augmentation. Finally, the week 6 profiles then diverged and showed only a 30% similarity between the soils. Nonetheless, no distinct relationship was recorded between rhizosphere OTU occurrence and the presence or absence of biochar, or the physical location of the different treatments in the seed trays. Similar observations were recorded by Ducey et al. (2013) where the additions (1%, 2%, 10% w/w) of a 350°C *Panicum virgatum* (switchgrass)-derived biochar to coarse-silty subsoil, with subsequent quantitative polymerase chain reaction (qPCR) analysis of nitrogen-cycling gene abundances, showed increases in the nitrogen-fixing community abundance from the first of the 6-month study but with statistically significant divergences recorded only from month 2.

In summary, biochar did not affect community structure but increased the functional diversity and abundance of nitrogen-fixing bacteria in rhizosphere soil. This upward trend was not confirmed by universal statistically significant differences for the 6-week study, therefore extended experimentation and more frequent sampling are required for confirmation. Also definitive pilot, long-term, and in situ research are required to explore the impacts of biochar on these communities before recommendations for agricultural soil applications can be made. For example, Rutigliano et al. (2014) supplemented (30 and 60 t ha⁻¹) durum wheat (*T. durum* L., cultivar Neolatino) crop, in situ, twice with wood biochar and determined substrate-induced respiration as a microbial community activity indicator, with functional diversity evaluated by catabolic fingerprinting and 16S rRNA gene DGGE profiling. The authors reported significant increases in microbial functional diversity 3 months but not 14 months after the augmentation. Additionally, on the basis of focused bacterial community 16S rDNA analysis for the control and biochar-treated soils, the researchers attributed the recorded differences for month 14 to nutrient availability, agricultural treatments, and the wheat growth cycle rather than biochar supplementation per se.

Study Conclusions and Knowledge Gaps

The need for high nitrogen concentrations in arable soil and farmland can result in some farmers applying industrially produced fertilizers with, at times, inimical environmental impacts and/or increased costs of their produce. Biochar has shown potential to mitigate many contemporary agricultural and environmental concerns because of its ability to alter the physicochemical and biological properties of soil. Currently there is, however, only limited knowledge of the effects of biochar particularly on nitrogen-cycling microorganisms, and whether augmentation could be a viable alternative to ammonia-based fertilizers. Therefore the current study focused on assessing the impacts of biochar on nitrogen-fixing communities in unfertilized soil and provided some preliminary community dynamics data that showed increased *nifH* gene diversity and richness particularly in biochar-augmented rhizosphere soil. Therefore the positive trends recorded in this laboratory-based study justify further and protracted in situ investigations with other functional N-cycle genes and communities targeted.

The current case study, and that discussed in Chapter 7, highlight some of the emergent and developed knowledge, and key gaps. Therefore they justify further investigation with larger soil microcosms and in situ pilot studies prior to large-scale tests. Specifically, microcosms with and without biochar, with some planted with clover and/or other plant models, can be used to determine bacterial interactions in the rhizosphere compared with the bulk soil. Different biomolecules, such as DNA, RNA, and fatty acids/lipids, can be extracted from bulk and rhizosphere soil with the enriched bacterial communities compared first with 16S rRNA gene primers to gain an overall understanding of the total bacterial community structure and composition. Primers specific for genes that encode enzymes involved in nitrogen cycling (*nifH* for nitrogen fixation, *nirK*/*nirS* for nitrate reduction, and *nosZ* for nitrous oxide reduction to N_2) can be used in parallel to gain an understanding of how biochar affects N cycling in particular, and measure how specific elements in the cycle respond. Typically these phylogenetic and functional analyses could include the application of semiquantitative (DNA-based DGGE; T-RFLP; phospholipid-derived fatty acids) and quantitative (qPCR; metagenomic sequencing) techniques.

Microcosm experiments enable preliminary investigations to be made on the effects of biochar on the indigenous soil microbial communities. They do, however, provide a largely controlled scenario that is unrealistic.

As a result, in situ studies should be made to determine the effects of biochar on bulk and rhizosphere soil in actual agronomic contexts to characterize the total and nitrogen-fixing bacterial and archaeal communities.

CONCLUSION

It is well accepted that plant productivity and health are affected by interactions, processes, and mechanisms of root-associated microbiomes. For example, a seminal study by Ofek-Lalzar et al. (2014) measured genetic potential and functional expression using metagenomic and metatranscriptomic analysis, respectively, and recorded differences in root-associated microbial communities of different agricultural crops, *Triticum turgidum* (wheat) and *Cucumis sativus* (cucumber), grown in the same soil. Thus host- and niche-based functional differentiation resulted in the presence of wheat versus cucumber and between their bulk and rhizoplane soils, respectively. As also discussed in Chapters 7, 9 and 10, in particular, biochar has a recognized potential for exploitation in agricultural and environmental contexts because of its ability to influence the physicochemical and biological properties of the augmented soil. Considerable knowledge gaps do, however, exist regarding its effects on nitrogen-cycling microbial communities, particularly in agronomic contexts. Therefore the need to understand the impacts of biochar augmentation on plant-specific rhizosphere N-cycling populations in similar and different soil types is critical. In particular, the effects of biochar addition on the structure and composition of total and nitrogen-fixing bacterial communities in bulk soils and the rhizosphere are essential if the potential to reduce fertilizer inputs and agronomic carbon footprints is to be realized fully.

Previous discourse (eg, Ennis et al., 2012; Orr and Ralebitso-Senior, 2014; Chapters 3, 4, 5, and 6) have highlighted the strengths and limitations of different microbial ecology techniques in biochar-related contexts. Therefore the methods that have been used so far to explore the current knowledge gaps relative to the N-cycle dynamics have been summarized and included in Chapter 1, Table 1.1. These have then been expanded to deliberate on potential alternative/additional techniques and highlight plausible new key drivers for knowledge development including, but not restricted to, impacts of biochar specifically on the nitrogen-cycling genes in the rhizosphere. For example, high-throughput platforms such as the PhyloChip and GeoChip have been proposed for microbial phylogenetic and functional profiling following biochar augmentation (Ennis et al., 2012). Although not in a biochar context, the potential for knowledge transfer has been illustrated by Li et al. (2014) who applied GeoChip 3.0 to characterize the functional bulk soil and rhizosphere microbiome of maize (*Z. mays*). The researchers recorded statistically significant differences in gene structure between the two microbiomes. In particular, ~50%

of the detected 5777 rhizosphere genes, which were largely from bacteria, included those responsible for carbon fixation and degradation, nitrogen fixation, denitrification, and ammonification, and recorded significant increases ($p < 0.05$). Also these differences resulted despite similar total N concentrations in the bulk and maize rhizosphere soils. In summary, this study included a comprehensive data figure showing the percent expression of key N-cycling genes as detected by a cutting-edge high-throughput platform. This could therefore be applied directly to biochar-impacted ecosystems, and the rhizosphere in particular.

Acknowledgments

We would like to thank: Jordan Stansfield for gifting and preparing the study soil; Christopher Schroeter for helping set up the experiment; and the Society for Applied Microbiology Students into Work Scheme and Teesside University Research Fund, for supporting this work.

References

Ahemad, M., Kibret, M., 2014. Mechanisms and applications of plant growth promoting rhizobacteria: current perspective. J. King Saud Univ. Sci. 26 (1), 1–20.

Alburquerque, J.A., Salazar, P., Barron, V., Torrent, J., Carmen del Campillo, M., Gallardo, A., Villar, R., 2013. Enhanced wheat yield by biochar addition under different mineral fertilization levels. Agron. Sustain. Dev. 33 (3), 475–484.

Amonette, J.E., Joseph, S., 2009a. Characteristics of biochar e micro-chemical properties. In: Lehmann, J., Joseph, S. (Eds.), Biochar for Environmental Management: Science and Technology. Earthscan, London, pp. 13–32.

Amonette, J.E., Joseph, S.D., 2009b. Characteristics of biochar: microchemical properties. In: Lehmann, J., Joseph, S.D. (Eds.), Biochar for Environmental Management. Earth Scan, pp. 33–53.

Anderson, C.R., Condron, L.M., Clough, T.J., Fiers, M., Stewart, A., Hill, R.A., Sherlock, R.R., 2011. Biochar induced soil microbial community change: implications for biogeochemical cycling of carbon, nitrogen and phosphorus. Pedobiologia 54 (5–6), 309–320.

Arbestain, M.C., Saggar, S., Leifeld, J., 2014. Environmental benefits and risks of biochar application to soil. Agric. Ecosyst. Environ. 191, 1–4.

Atkinson, C.J., Fitzgerald, J.D., Hipps, N.A., 2010. Potential mechanisms for achieving agricultural benefits from biochar application to temperate soils: a review. Plant Soil 337 (1–2), 1–18.

Azeez, G., 2009. Soil Carbon and Organic Farming. Available at: http://www.soilassociation.org (accessed 12.01.16.).

Blagodatskaya, E., Blagodatsky, S., Anderson, T.H., Kuzyakov, Y., 2014. Microbial growth and carbon use efficiency in the rhizosphere and root-free soil. PLos One 9, 4

Cernansky, R., 2015. Agriculture: state-of-the-art soil. Nature 517, 258–260.

Clough, T.J., Bertram, J.E., Ray, J.L., Condron, L.M., O'Callaghan, M., Sherlock, R.R., Wells, N.S., 2010. Unweathered wood biochar impact on nitrous oxide emissions from a bovine-urine-amended pasture soil. Soil Sci. Soc. Am. J. 74, 852–860.

Cook, R.J., Thomashow, L.S., Weller, D.M., Fujimoto, D., Mazzola, M., Bangera, G., Kim, D.S., 1995. Molecular mechanisms of defense by rhizobacteria against root disease. Proc. Natl. Acad. Sci. U.S.A. 92 (10), 4197–4201.

Daly, K.R., Mooney, S.J., Bennett, M.J., Crout, N.M.J., Roose, T., Tracy, S.R., 2015. Assessing the influence of the rhizosphere on soil hydraulic properties using X-ray computed tomography and numerical modelling. J. Exp. Bot. 66, 2305.

Dias, A.C.F., Silva, M.C.P., Cotta, S.R., Dini-Andreote, F., Soares Jr., F.L., Salles, J.F., Azevedo, J.L., van Elsas, J.D., Andreote, F.D., 2012. Abundance and genetic diversity of nifH gene sequences in anthopogenically affected Brazilian mangrove sediments. App. Environ. Microbiol. 78, 7960–7967.

du Plessis, C.A., Senior, E., Hughes, J.C., 1998. Growth kinetics of microbial colonisation of porous media. S. Afr. J. Sci. 94, 33–38.

Ducey, T.F., Ippolito, J.A., Cantrell, K.B., Novak, J.M., Lentz, R.D., 2013. Addition of activated switchgrass biochar to an acidic subsoil increases microbial nitrogen cycling gene abundances. Appl. Soil Ecol. 65, 65–72.

Ennis, C.J., Evans, G.A., Islam, M., Ralebitso-Senior, T.K., Senior, E., 2012. Biochar: carbon sequestration, land remediation and impacts on soil microbiology. Crit. Rev. Environ. Sci. Technol. 42 (22), 2311–2364.

Fernández-Calvino, D., Bååth, E., 2010. Growth Repsonse of the bacterial community to pH in soils differing in pH. FEMS Microbiol. Ecol. 73, 149–156.

Fierer, N., Jackson, R.B., 2006. The diversity and biogeography of soil bacterial communities. PNAS 105 (3), 628–631.

Filiberto, D.M., Gaunt, J.L., 2013. Practicality of Biochar Additions to Enhance Soil and Crop Productivity. Agriculture 3, 715–725.

Fitter, A.H., Gilligan, C.A., Hollingworth, K., Kleczkowski, A., Twyman, R.M., Pitchford, J.W., 2005. Biodiversity and ecosystem function in soil. Funct. Ecol. 19, 369–377.

Gaskin, J.W., Speir, R.A., Harris, K., Das, K.C., Lee, R.D., Morris, L.A., Fisher, D.W., 2010. Effect of peanut hull and pine chip biochar on soil nutrients, corn nutrient status, and yield. Agron. J. 102, 623–633.

Graber, E.R., Harel, Y.M., Kolton, M., Cytryri, E., Silber, A., David, D.R., Tsechansky, L., Borenshtein, M., Elad, Y., 2010. Biochar impact on development and productivity of pepper and tomato grown in fertigated soilless media. Plant Soil 337, 481–496.

Hockaday, W.C., Grannas, A.M., Kim, S., Hatcher, P.G., 2007. The transformation and mobility of charcoal in a fire-impacted watershed. Geochim. Cosmochim. Acta 71 (14), 3432–3445.

Hodgson, E., Bevan, A., Farrar, K., 2012. Bichar for Sustainable Soil, Agriculture and Energy Systems. Available at: https://www.aber.ac.uk/en/media/departmental/ibers/innovations/innovations4/04.pdf (accessed 17.12.15.).

Ippolito, J.A., Laird, D.A., Busschner, W.J., 2012. Environmental benefits of biochar. J. Environ. Qual. 41, 967–972.

Jeffery, S., Verheijen, F.G.A., van der Velde, M., Bastos, A.C., 2011. A quantitative review of the effects of biochar application to soils on crop productivity using meta-analysis. Agric. Ecosyst. Environ. 144, 175–187.

Jack, A.L.H., Rangarajan, A., Culman, S.W., Sooksa-Nguan, R., Thies, J.E., 2011. Choice of organic amendments in tomato transplants has lasting effects on bacterial rhizosphere communities and crop performance in the field. Appl. Soil Ecol. 48, 94–101.

Jin, H., 2010. Characterization of microbial life colonizing biochar and biochar amended soils. PhD Dissertation. Cornell University, Ithaca, NY.

Joseph, S.D., Camps-Arbestain, M., Lin, Y., Munroe, P., Chia, C.H., Hook, J., van Zwieten, L., Kimber, S., Cowie, A., Singh, B.P., Lehmann, J., Foidl, N., Smernik, R.J., Amonette, J.E., 2010. An investigation into the reactions of biochar in soil. Aust. J. Soil Res. 48, 501–515.

Khodadad, C.L.M., Zimmerman, A.R., Green, S.J., Uthandi, S., Foster, J.S., 2011. Taxa-specific changes in soil microbial community composition induced by pyrogenic carbon amendments. Soil Biol. Biochem. 43 (2), 385–392.

Kolton, M., Harel, Y.M., Pasternak, Z., Graber, E.R., Elad, Y., Cytryri, E., 2011. Impact of biochar application to soil on the root-associated bacterial community structure of fully developed greenhouse pepper plants. Appl. Environ. Microbiol. 77 (14), 4924–4930.

Lal, R., 2009. Challenges and opportunities in soil organic matter research. Eur. J. Soil Sci. 60, 158–169.

Lauber, C.L., Strickland, M.S., Bradford, M.A., Fierer, N., 2008. The influence of soil properties on the structure of bacterial and fungal communities across land-use types. Soil. Biol. Biochem. 40, 2407–2415.

Laufer, J., Tomlinson, T., 2013. Biochar Field Studies: An IBI Research Summary. International Biochar Initiative. Available at: http://www.biochar-international.org/publications/IBI.

Lehmann, J., Solomon, D., Kinyangi, J., Dathe, L., Wirick, S., Jacobsen, C., 2008. Spatial complexity of soil organic matter forms at nanometre scales. Nature Geosci. 1, 238–242.

Lehmann, J., Rillig, M.C., Thies, J., Masiello, C.A., Hockaday, W.C., Crowley, D., 2011. Biochar effects on soil biota: a review. Soil Biol. Biochem. 43, 1812–1836.

Li, X., Rui, J., Xiong, J., Li, J., He, Z., Zhou, J., Yannarell, A.C., Mackie, R.I., 2014. Functional potential of soil microbial communities in the maize rhizosphere. PLoS One 9 (11), e112609.

Lin, X.W., Xie, Z.B., Zheng, J.Y., Liu, Q., Bei, Q.C., Zhu, J.G., 2015. Effects of biochar application on greenhouse gas emissions, carbon sequestration and crop growth in coastal saline soil. Eur. J. Soil Sci. 66 (2), 329–338.

Liu, X., Zhang, A., Ji, C., Joseph, S., Bian, R., Li, L., Pan, G., Paz-Ferreiro, J., 2013. Biochar's effect on crop productivity and the dependence on experimental conditions – a meta-analysis of literature data. Plant Soil 373, 583–594.

Lugtenberg, B., Kamilova, F., 2009. Plant-growth-promoting rhizobacteria. Annu. Rev. Microbiol. 63 (1), 541–556.

Marschner, P., Crowley, D., Yang, C.H., 2004. Development of specific rhizosphere bacterial communities in relation to plant species, nutrition and soil type. Plant Soil 261, 199–208.

Mavi, M., Marschner, P., 2013. Salinity affects the response of soil microbial activity and biomass to addition of carbon and nitrogen. Soil Res. 51, 68–75.

Mendes, R., Kruijt, M., de Bruijn, I., Dekkers, E., van der Voort, M., Schneider, J.H.M., Piceno, Y.M., DeSantis, T.Z., Andersen, G., Bakker, P.A.H.M., Raaijmakers, J.M., 2011. Deciphering the rhizosphere microbiome for disease-suppressive bacteria. Science 332, 1097–1100.

Mendes, R., Garbeva, P., Raaijmakers, J.M., 2013. The rhizosphere microbiome: significance of plant beneficial, plant pathogenic, and human pathogenic microorganisms. FEMS Microbiol. Rev. 37 (5), 634–663.

Mia, S., van Groenigen, J.W., van de Voorde, T.F.J., Oram, N.J., Bezemer, T.M., Mommer, L., Jeffery, S., 2014. Biochar application rate affects biological nitrogen fixation in red clover conditional on potassium availability. Agric. Ecosyst. Environ. 191, 83–91.

Molina, M., Zaelke, D., Sarma, K.M., Andersen, S.O., Ramanathan, V., Kaniaru, D., 2009. Reducing abrupt climate change risk using the Montreal protocol and other regulatory actions to complement cuts in CO_2 emissions. Proc. Natl. Acad. Sci. U. S. A. 106, 20616–20621.

Nallanchakravarthula, S., Mahmood, S., Alström, S., Finlay, R.D., 2014. Influence of soil type, cultivar and *Verticillium dahliae* on the structure of the root and rhizosphere soil fungal microbiome of strawberry. PLoS One 9 (10), e111455.

Nelissen, V., Saha, B.P., Ruysschaert, G., Boeckx, P., 2014. Effect of different biochar and fertilizer types on N_2O and NO emissions. Soil Biol. Biochem. 70, 244–255.

Nguyen, B., Lehmann, J., 2009. Black carbon decomposition under varying water regimes. Org. Geochem. 40, 846–853.

Ofek-Lazar, M., Sela, N., Goldman-Voronov, M., Green, S.J., Hadar, Y., Minz, D., 2014. Niche and host-associated functional signatures of the root surface microbiome. Nat. Comm. 5, 4950. http://dx.doi.org/10.1038/ncomms5950.

Olakanye, A.O., Thompson, T., Ralebitso-Senior, T.K., 2014. Changes to soil bacterial profiles as a result of *Sus scrofa domesticus* decomposition. Forensic. Sci. Int. 245, 101–106.

O'Neil, B., Grossman, J., Tsai, M.T., Gomes, J.E., Lehmann, J., Peterson, J., Neves, E., Thies, J.E., 2009. Bacterial community composition in Brazilian Anthrosols and adjacent soils characterized using culturing and molecular identification. Microb. Ecol. 58 (1), 23–35.

Orr, C.H., Leifert, C., Cummings, S.P., Cooper, J.M., 2012. Impacts of Organic and Conventional Crop Management on Diversity and Activity of Free-Living Nitrogen Fixing Bacteria are Subsidiary to Temporal Effects. PLos One 7, 12.

Orr, C.H., Ralebitso-Senior, T.K., January 18, 2014. Tracking N-cycling genes in biochar-supplemented ecosystems: a perspective. OA Microbiol. 2 (1), 1.

Pietikainen, J., Kiikkila, O., Fritze, H., 2000. Charcoal as a habitat for microbes and its effects on the microbial community of the underlying humus. Oikos 89, 231–242.

Prendergast-Miller, M.T., Duvall, M., Sohi, S.P., 2011. Localisation of nitrate in the rhizosphere of biochar-amended soils. Soil Biol. Biochem. 43 (11), 2243–2246.

Prendergast-Miller, M.T., Duvall, M., Sohi, S.P., 2014. Biochar–root interactions are mediated by biochar nutrient content and impacts on soil nutrient availability. Eur. J. Soil Sci. 65 (1), 173–185.

Quilliam, R.S., Glanville, H.C., Wade, S.C., Jones, D.L., 2013. Life in the 'charosphere' – does biochar in agricultural soil provide a significant habitat for microorganisms? Soil Biol. Biochem. 65, 287–293.

Rutigliano, F.A., Romano, M., Marzaioli, R., Baglivo, I., Baronti, S., Miglietta, F., Castaldi, S., 2014. Effect of biochar addition on soil microbial community in a wheat crop. Eur. J. Soil Biol. 60, 476–481.

Sarathchandra, S.U., Ghani, A., Yeates, G.W., Burch, G., Cox, N.R., 2001. Effect of nitrogen and phosphate fertilisers on microbial and nematode diversity in pasture soils. Soil Biol. Biochem. 33, 953–964.

Shackley, S., Sohi, S., Haszeldine, S., Manning, D., Mašek, O., 2009. Biochar, Reducing and Removing CO$_2$ while Improving Soils: A Significant and Sustainable Response to Climate Change. Evidence Submitted to the Royal Society Geo-Engineering Climate Enquiry. Available at: http://www.geos.ed.ac.uk/home/homes/sshackle/WP2.pdf (accessed 17.12.15.).

Shen, J., Li, C., Mi, G., Li, L., Yuan, L., Jiang, R., Zhang, F., 2012. Maximizing root/rhizosphere efficiency to improve crop productivity and nutrient use efficiency in intensive agriculture of China. J. Exp. Bot. 64 (5), 1181–1192.

Singh, B.K., Millard, P., Whiteley, A.S., Murrell, J.C., 2004. Unravelling rhizosphere–microbial interactions: opportunities and limitations. Trends Microbiol. 12 (8), 386–393.

Soja, G., Zehetner, F., Kitzler, B., 2013. Biochar as Yield Promoter for Agricultural Crops? Conjectures and Observations. http://www.oeaw.ac.at/forebiom/WS1lectures/SessionIII_Soja.pdf.

Spokas, K.A., Cantrell, K.B., Novak, J.M., Archer, D.A., Ippolito, J.A., Collins, H.P., Boateng, A.A., Lima, I.M., Lamb, M.C., McAloon, A.J., Lentz, R.D., Nichols, K.A., 2012. Biochar: A synthesis of its agronomic impact beyond carbon sequestration. J. Environ. Qual. 41, 973–989.

Steinbeiss, S., Gleixner, G., Antonietti, M., 2009. Effect of biochar amendment on soil carbon balance and soil microbial activity. Soil Biol. Biochem. 41 (6), 1301–1310.

Troy, S.M., Lawlor, P.G., O'Flynn, C.J., Healy, M.G., 2013. Impact of biochar addition to soil on greenhouse gas emissions following pig manure application. Soil Biol. Biochem. 60, 173–181.

Wang, C., Lu, H., Dong, D., Deng, H., Strong, P.J., Wang, H., Wu, W., 2013. Insight into the effects of biochar on manure composting: evidence supporting the relationship between N$_2$O emissions and denitrifying community. Environ. Sci. Technol. 47, 7341–7349.

Warnock, D.D., Lehmann, J., Kuyper, T.W., Rillig, M.C., 2007. Mycorrhizal responses to biochar in soil – concepts and mechanisms. Plant Soil 300, 9–20.

Woolf, D., Amonette, J.E., Street-Perrott, F.A., Lehmann, J., Joseph, S., 2010. Sustainable biochar to mitigate global climate change. Nat. Commun. 1 (56). http://dx.doi.org/10.1038/ncomms1053.

Zhang, A., Cui, L., Pan, G., Li, L., Hussain, Q., Zhang, X., Zheng, J., Crowley, D., 2010. Effect of Biochar Amendment on Yield and Methane and Nitrous Oxide Emissions from a Rice Paddy from Tai Lake Plain, China. Agric. Ecosyst. Environ. 139, 469–475.

Zhao, M., Xue, K., Wang, F., Liu, S., Bai, S., Sun, B., Zhou, J., Yang, Y., 2014. Microbial mediation of biogeochemical cycles revealed by simulation of global changes with soil transplant and cropping. ISME 8, 2045–2055.

CHAPTER

9

Potential Application of Biochar for Bioremediation of Contaminated Systems

H. Lyu, Y. Gong, R. Gurav, J. Tang

Nankai University, Tianjin, China

OUTLINE

INTRODUCTION

Biochar is a black carbon produced during the pyrolysis of biomass under oxygen-limited conditions and at relatively low temperatures (<700°C). Waste biomass, used widely in biochar production, includes: crop residues, forestry waste, animal manure, food processing waste, paper mill waste, municipal solid waste, and sewage sludge (Brick, 2010; Cantrell et al., 2012; Ahmad et al., 2014a). These cost-effective carbonaceous sorbents have a large surface area, several 1000-fold greater than uncharred material (Thies and Rillig, 2009), giving them a large affinity and capacity to adsorb inorganic and organic contaminants in different ecosystems (Brandli et al., 2008). When applied to the environment, biochar reduces the bioavailability of contaminants associated with accumulation and toxicity in plants and animals, with additional benefits of soil fertilization and mitigation of climate change (Sohi, 2012). The sorption capacity of biochar depends highly on parameters such as pyrolysis conditions, including residence time, feedstock types, temperature, and heat transfer rate. Similarly, the physicochemical properties such as surface area, surface charge, and chemical functionality are influential factors controlling the sorption of inorganic and organic contaminants (Zhu et al., 2005).

The estimated cost of different biochar productions is calculated in a range of $0.2–0.5/kg, whereas adsorbents such as ion exchange resins may cost up to $150/kg (Ahmed et al., 2015). Therefore biochar could be an effective adsorbent that can be produced potentially with a low carbon footprint, hence become highly beneficial for water treatment while minimizing other environmental damage such as greenhouse effects. Environmental remediation has been recognized as a promising area where biochar can be applied successfully (Cao et al., 2011; Ahmad et al., 2014a). Many laboratory and field experiments have been carried out to investigate the adsorption capacity of biochar for various inorganic and organic contaminants, including heavy metals, dyes, antibiotics, polycyclic aromatic hydrocarbons (PAHs), and organic pesticides (Tang et al., 2013; Tran et al., 2015).

ENVIRONMENTAL APPLICATION OF BIOCHAR IN HEAVY METAL-CONTAMINATED SYSTEMS

Biochar Applications in Heavy Metal Remediation

Heavy metals present serious health threats to different environments and organisms even at very low concentrations (Mohan et al., 2014). Some of the heavy metals are cumulative poisons, capable of assimilation storage and concentration by organisms exposed over long periods even at low concentrations (Gong et al., 2014). Also eventual metal build-up in tissues can have harmful physiological effects.

Biochar has been considered to be potentially effective at remediating heavy metals. Many reported studies have highlighted its potential effectiveness in removing heavy metals from aqueous solutions and soils (Table 9.1). Chen et al. (2015a,b,c,d) investigated the adsorption capacity of Cd(II) on biochar derived from municipal sewage sludge. The results showed that Cd(II) removal efficiency increased with an increase in biochar dosage with the maximum removal capacity of $42.80 \pm 2.38 \, mg \, g^{-1}$ obtained at the application rate of 0.2% (w/w). Initial pH was the most important factor affecting the adsorption process, with the removal capacity of more than $40 \, mg \, g^{-1}$ remaining when the solution initial pH was higher than 3. Tong et al. (2011) prepared biochars from peanut, canola, and soybean straw under oxygen-limited conditions (muffle furnace) at 300, 400, and 500°C for 3.75 h. All three biochars were used for the removal of Cu(II). The results showed that Cu(II) adsorption rates followed the order: peanut straw char ($89.0 \, mg \, g^{-1}$) > soybean straw char ($53.0 \, mg \, g^{-1}$) > canola straw char ($37.0 \, mg \, g^{-1}$). Also leguminous (peanut and soybean straw) chars had higher capacities than the nonleguminous canola straw char. Puga et al. (2015) produced sugar cane straw-derived biochar at 700°C, which was applied to a heavy metal-contaminated mine soil at 1.5%, 3.0%, and 5.0% (w/w). The application of biochar decreased the available concentrations of Cd(II), Pb(II), and Zn(II) by 56%, 50% and 54%, respectively. Furthermore, biochar reduced the uptake of Cd, Pb, and Zn by plants such as the jack bean, which translocates high proportions of metals (especially Cd(II)) to shoots. Karami et al. (2011) used green waste compost and biochar mixture amendments to evaluate their role in regulating the mobility of Cu(II) and Pb(II) and their resultant uptake into plants from soil. As a result of biochar and compost addition, both Cu(II) and Pb(II) concentrations in ryegrass shoot were reduced, especially for Pb(II).

Mechanisms of Heavy Metal Remediation by Biochar

According to current research, there are many adsorption mechanisms between biochar and heavy metals including electrostatic attraction, ion exchange, surface mineral adsorption, cation–π interactions and precipitation of surface functional groups (Inyang et al., 2012). Electrostatic attraction is attributed mainly to negatively charged surfaces (Yang et al., 2003). When applied to soil, the electrostatic attraction between heavy metals with positive charge and soil will be enhanced, which can increase the adsorption capacity of heavy metals (Uchimiya et al., 2011). The effect of four types of biochar derived from rice straw generated at different pyrolysis temperatures (300–600°C) on the removal efficiency of Pb(II) were investigated

TABLE 9.1 Adsorption capacities of different biochars for heavy metal removal

Type of biochars	Surface area ($m^2 g^{-1}$)	Targeted species	Adsorption study parameters		Adsorption capacity ($mg g^{-1}$)	References
			pH	Concentration range ($mg L^{-1}$)		
Oak wood char	2.73	Cr^{6+}	2.0	1–100	3.03	Mohan et al. (2011)
Oak wood char	2.73	Cr^{6+}	2.0	1–100	4.08	
Oak wood char	2.73	Cr^{6+}	2.0	1–100	4.93	
Oak bark char	1.88	Cr^{6+}	2.0	1–100	4.61	
Oak bark char	1.88	Cr^{6+}	2.0	1–100	7.43	
Oak bark char	1.88	Cr^{6+}	2.0	1–100	7.51	
Municipal sewage sludge	67.63	Cd^{2+}	3	~200	42.80	Chen et al. (2015a)
Municipal sludge	–	Cr^{6+}	–	50–200	25.27	Chen et al. (2015b)
Municipal sludge	–	Cr^{3+}	–	54.43–213.84	7	
Sludge	–	Pb^{2+}	–	1–31	16.8	Zhang et al. (2015)
Peanut straw biochar	–	Cu^{2+}	5.0	158–954	89.0	Tong et al. (2011)
Peanut straw biochar	–	Cu^{2+}	4.5	158–954	50.0	
Soybean straw char	–	Cu^{2+}	5.0	158–954	53.0	
Soybean straw char	–	Cu^{2+}	4.5	158–954	33.0	
Canola straw char	–	Cu^{2+}	5.0	158–954	37.0	
Canola straw char	–	Cu^{2+}	4.5	158–954	31.0	

Rice husk	155	As^{5+}	–	90	0.00259	Agrafioti et al. (2014)
Rice husk	155	Cr^{3+}	–	185	0.01513	
Municipal solid wastes	5	As^{5+}	–	90	0.00425	
Municipal solid wastes	5	Cr^{3+}	–	0.185	0.03012	
Sewage sludge	51	As^{5+}	–	0.09	0.0354	
Sewage sludge	51	Cr^{3+}	–	0.185	0.04237	
Soil	–	As^{5+}	7.02	0.09	0.01046	
Soil	–	Cr^{3+}	7.02	0.185	0.03953	
Pine wood biochar	–	Pb^{2+}	5.0	5–40	3.89	Liu and Zhang (2009)
Pine wood biochar	–	Pb^{2+}	5.0	5–40	4.03	
Pine wood biochar	–	Pb^{2+}	5.0	5–40	4.25	
Rice husk biochar	–	Pb^{2+}	5.0	5–40	1.84	
Rice husk biochar	–	Pb^{2+}	5.0	5–40	2.25	
Rice husk biochar	–	Pb^{2+}	5.0	5–40	4.40	
Digested dairy waste	555.2	Ni^{2+}	10.0	0.0587	0.0135	Inyang et al. (2012)
Digested dairy waste	555.2	Cu^{2+}	10.0	0.0635	0.0622	
Digested dairy waste	555.2	Cd^{2+}	10.0	0.112	0.0694	
Digested dairy waste	555.2	Pb^{2+}	10.0	0.106	0.104	
Digested whole sugar beet	128.5	Ni^{2+}	9.0	0.0587	0.0581	
Digested whole sugar beet	128.5	Cu^{2+}	9.0	0.0635	0.0616	
Digested whole sugar beet	128.5	Cd^{2+}	9.0	0.112	0.111	
Digested whole sugar beet	128.5	Pb^{2+}	9.0	0.106	0.103	

Continued

TABLE 9.1   Adsorption capacities of different biochars for heavy metal removal—cont'd

Type of biochars	Surface area ($m^2 g^{-1}$)	Targeted species	Adsorption study parameters		Adsorption capacity ($mg g^{-1}$)	References
			pH	Concentration range ($mg L^{-1}$)		
Hardwood biochar	118.98	Cu^{2+}	4.8	–	0.38	Inyang et al. (2012)
		Zn^{2+}	4.8	–	0.06	
	343.91	Cu^{2+}	4.8	–	19.50	
		Zn^{2+}	4.8	–	0.19	
	372.75	Cu^{2+}	4.8	–	7.05	
		Zn^{2+}	4.8	–	61.06	
	5.50	Cu^{2+}	4.8	–	1.08	
		Zn^{2+}	4.8	–	4.70	
Softwood biochar	94.74	Cu^{2+}	4.8	–	6.35	
		Zn^{2+}	4.8	–	5.94	
	383.66	Cu^{2+}	4.8	–	30.24	
		Zn^{2+}	4.8	–	4.44	
	362.33	Cu^{2+}	4.8	–	11.05	
		Zn^{2+}	4.8	–	256.42	
	6.72	Cu^{2+}	4.8	–	8.13	
		Zn^{2+}	4.8	–	1.56	
Switchgrass	–	Ca^{2+}	5.0	40	34	Regmi et al. (2012)
	–	Cu^{2+}	5.0	40	31	

by An et al. (2011). Biochars prepared at low temperature showed greater sorption capacity (RC300: 24.60 mg g^{-1}, RC400: 17.46 mg g^{-1}, RC 500: 4.90 mg g^{-1}, and RC600: 2.41 mg g^{-1}), which were attributed to the stronger electrostatic attraction of biochar with low pyrolysis temperatures.

Ion exchange depends mainly on the cation exchange capacity. Thus Yakkala et al. (2013) prepared buffalo weed biochars at 300, 500, and 700°C using 4 h of slow pyrolysis under N$_2$ for Cd(II) and Pb(II) removal and showed that more cation exchange capacity and surface availability for complexation leads to higher adsorption capacities of 11.63 and 333.3 mg g^{-1} for Cd(II) and Pb(II), respectively, by the 700°C biochar. Ion exchange and surface complexation were found to be the main mechanisms involved in the adsorption process.

Biomass material contains a variety of mineral components including Si, Ca, Mg, Fe, and Mn, which will accumulate on the biochar surface in the form of ash during the pyrolysis process (Gaskin et al., 2008). Compared to organic carbon, the mineral component of the resultant char has higher adsorption capacity and affinity for heavy metals. Chen et al. (2012a) compared the sorption capacity of biochar (RC350 and RC700) and acidified biochar (RC350-DA and RC700-DA) onto Pb(II) and showed that acidified biochar has lower Pb(II) adsorption capacity (RC350-DA: 29.6 ± 1.0 mg g^{-1} and RC700-DA: 38.2 ± 1.4 mg g^{-1}) than nonacidified biochar (RC350: 65.0 ± 6.9 mg g^{-1} and RC700: 76.3 ± 3.6 mg g^{-1}). Also there was no significant reduction of the organic functional group, whereas the amount of inorganic minerals (such as SiO$_2$) decreased, indicating that the inorganic mineral component played a key role in the removal of Pb(II). However, the adsorption of organic contaminants is attributed mainly to changes of organic carbon fractions during the pyrolysis process, rather than the mineral component (Chen et al., 2013).

Cation–π is a complex combination of electrostatic adsorption and π–π conjugation. Generally, the effect of cation–π is dependent mainly on the aromaticity of the biochar surface, which is an indicator of the number of π–π conjugations. Hence the more π-conjugated aromatic structure, the stronger the electron-donating ability (Ma and Dougherty, 1997). Li et al. (2012) prepared corn and straw biochars at 350 and 700°C, respectively, and investigated their effects on Cd(II) adsorption. A two-site Langmuir model was used to calculate the ion exchange adsorption and cation–π adsorption capacities. The results showed that the cation–π effect adsorption capacity (biochar 350: 33.8 mg g^{-1} and biochar 700: 52.3 mg g^{-1}) was significantly higher than that of ion exchange (biochar 350: 5.4 mg g^{-1} and biochar 700: 0.6 mg g^{-1}), indicating that cation–π interaction was the dominant mechanism for Cd(II) adsorption. The equation can be expressed as:

$$C\pi + 2H_2O \rightarrow C\pi - H_3O^+ + OH^-$$
$$C\pi - H_3O^+ + Cd^{2+} \rightarrow C\pi - Cd^{2+} + H_3O^+$$
$$\text{or } C\pi + Cd^{2+} \rightarrow C\pi - Cd^{2+}$$

Surface functional groups play a vital role in the environmental applications of biochar. There are abundant functional groups, ie, phenolic hydroxyl, carboxyl, on the biochar surface, which can easily form complexes with heavy metals (Chen et al., 2015c). For example, Yu et al. (2014) compared the removal efficiency of Cu(II) onto biochar and a biochar–manganese oxides composite and reported that the biochar–manganese oxides composite had a 509% greater Cu(II) removal efficiency, which was attributed to the increased amount of oxygen-containing groups on the red soil surface (ie, hydroxyl (—OH), magnesium–oxygen (Mg–O), silicon–oxygen (Si–O)). Sun et al. (2015) used hydrochars produced from sawdust wheat straw and corn stalk via hydrothermal carbonization and KOH modification to investigate their heavy metal removal efficiency. The surface oxygen-containing groups (ie, carboxyl groups) increased after KOH modification, resulting in about a 2–3 times increase of Cd(II) sorption capacity (30.40–$40.78\,mg\,g^{-1}$) compared to that of unmodified hydrochars (13.92–$14.52\,mg\,g^{-1}$).

Precipitation between biochar and heavy metals is mainly affected via two aspects. On the one hand, the markedly increased pH arising from biochar amendment may lead to decreased mobilization of heavy metal, thus it is easy to form heavy metal hydroxide precipitation (Shi and Zhou, 2014). On the other hand, various phosphate and carbonate precipitations would be formed under different conditions. For instance, a new precipitate was observed solely on Pb-loaded sludge-derived biochar as $5PbO\cdot P_2O_5\cdot SiO_2$ (lead phosphate silicate) at initial pH 5.0 (Lu et al., 2012). Xu et al. (2013) found the contribution rate of precipitation in the removal of Cu(II), Zn(II), and Cd(II) was above 75%, which is caused by the soluble phosphate and/or carbonate fractions in the animal dropping biochar, while heavy metal ions could combine to form phosphate/carbonate precipitates.

These mechanisms do not act separately when biochar is used to remove heavy metals; instead they can work together to achieve the remediation effect. Tong et al. (2011) reported that Cu(II) sorption by peanut, canola, and soybean straw biochars involved both electrostatic and nonelectrostatic adsorption. Thus Cu(II) sorption capacity rose as pH increased, where strong complexes formed between Cu–OH and char surface functions (—OH and —COOH). Higher phosphate contents of soybean and canola straw chars caused Cu–phosphate formation and precipitation compared to peanut straw char. Peanut straw char had a maximum Cu(II) capacity of $1.4\,mol\,kg^{-1}$ at pH 5.0 where the sorption occurred by both adsorption and surface precipitation. Furthermore, Lu et al. (2012) showed that a combination of organic hydroxyl and carboxyl functional groups accounted for 38.2–42.3% of the total sorbed Pb varying with pH, while coprecipitation or combining on mineral surfaces accounted for 57.7–61.8%. Also Dong et al. (2011) reported that Cr(VI) removal in the presence of biochar was caused by the joint action of electrostatic attraction, participation, and combination.

ENVIRONMENTAL APPLICATION OF BIOCHAR IN SYSTEMS IMPACTED BY ORGANIC POLLUTANTS

The release of organic pollutants from industrial, residential, agricultural, and commercial sources frequently contaminates soil ecosystems and water bodies, with dyes, antibiotics, PAHs, pesticides, and herbicides, causing the greatest concern (Ahmad et al., 2015). The biochar–organic contaminant interaction is likely to depend on the structural and chemical properties of the sorbates. Typically, biochars contain noncarbonized fractions, particularly O-containing carboxyl, hydroxyl, and phenolic surface functional groups, which may interact with and bind effectively to soil pollutants (Uchimiya et al., 2011). These multifunctional characteristics of biochar show its potential to be a very effective environmental sorbent for organic contaminants in soil and water.

Remediation of Synthetic Dyes

Synthetic dyes are released into the environment as components of effluents from the textile, leather, paper, and printing industries. Such dyes have adverse impacts in terms of total organic carbon, biological oxygen demand, and chemical oxygen demand and can cause severe ecological damage (Jadhav et al., 2012). They are difficult to treat by conventional methods since they are resistant to light and oxidizing agents, and resist aerobic remediation.

Biochar has been applied to aid dye removal with different mechanisms proposed by different authors. For example, some studies explored a cost-based system for the treatment of the released dyes using straw biochar (Hameed and El-Khaiary, 2008; Qiu et al., 2009; Xu et al., 2011). Hameed and EI-Khaiary (2008) prepared rice straw-derived biochar for C.I. Basic Green 4 (malachite green) removal and showed that its maximum sorption capacity ($148.74\,mg\,L^{-1}$) was reached at 30°C. Xu et al. (2011) compared the removal efficiency of methyl violet onto biochars derived from canola straw, peanut straw, soybean straw, and rice hull char and reported that the order of adsorption capacity was generally consistent with the amount of negative charge of the biochars (canola straw char > peanut straw char > soybean straw char > rice hull char), indicating that electrostatic attraction played an important role in the char-based adsorption of this dye. Similarly, Mui et al. (2010) used biochar prepared from waste bamboo scaffolding to treat wastewater-containing dyes and reported that the monolayer capacities of the bamboo char on acid blue 25, acid yellow

117, and methylene blue were 0.0406, 0.0416, and 0.998 mmol g^{-1}, respectively. Also Yang et al. (2013) reported that bamboo biochar prepared at 1000°C could remove dye acid black 172 from solutions effectively and the maximum sorption capacity is 401.88 mg g^{-1}.

Electrostatic attraction/repulsion between organic contaminants and biochar has been proposed as one of the possible dye adsorption mechanisms. Biochar surfaces are usually negatively charged facilitating the electrostatic attraction of positively charged cationic organic compounds. Such electrostatic attraction was described in adsorption studies of cationic dyes like methylene blue (Inyang et al., 2014), methyl violet (Xu et al., 2011), and rhodanine (Qiu et al., 2009) in water. However, according to Ahmad et al. (2014a), besides chemical functionality and electrostatic forces, the organic adsorbates can be affected considerably by pH and ionic strength. Surface charges on the biochar are generally controlled by pH. Hence at solution pH below the isoelectric point of char, the total external surface charge will be net positive, whereas at higher solution pH the surface will have a net negative charge (Mukherjee et al., 2011; Inyang and Dickenson, 2015). The sorption capacity of biochar derived from crop residue at 350°C for methyl violet increased sharply from pH 7.7 to 8.7 (Xu et al., 2011; Ahmad et al., 2014a). Similarly, the ionic strength of the solution also showed positive effects on the organic contaminant adsorption on biochar (Qiu et al., 2009; Xu et al., 2011). In particular, an increase in anionic brilliant blue dye adsorption on biochars was observed with an increase in ionic strength (Qiu et al., 2009). In general, the effect of ionic strength on adsorption onto biochar can be positive or negative depending on pH or the point of zero charge of the biochar (Bolan et al., 1999).

Pyrolysis of biomass at different temperatures greatly influenced the sorption of dyes as the surface area increased with higher pyrolysis temperatures. Overall, high heating rates resulted in higher surface areas, pore volumes and yields because of rapid depolymerization at char surfaces. Slow pyrolysis rice straw biochar prepared at 700°C under nitrogen and applied to a solution of 25 mg L^{-1} malachite green dye showed 95% removal at pH 5 and 30°C within 40 min (Hameed and El-Khaiary, 2008). In another study by Xu et al. (2011), canola straw, peanut straw, soybean straw, and rice hulls were pyrolyzed slowly for 4 h at 350°C to prepare respective biochars. These were then explored for their removal capacities for methyl violet from water. The sorption capacities were as follows: canola straw > peanut straw > soybean straw > rice hulls, which were parallel to their respective cation exchange capacities. Also methyl violet at concentrations higher than 1.0 mmol L^{-1} was adsorbed onto biochar because of its low water solubility.

Mui et al. (2010) prepared waste bamboo scaffolding biochars by slow pyrolysis under N$_2$ for 1–4 h at 400–900°C, with Brunauer–Emmett–Teller (BET) surface reaching 327 m^2 g^{-1} at 900°C. Acid blue 25, acid yellow 117, and methylene blue adsorption capacities were evaluated and found to be the highest for the methylene blue samples. Similarly, the bamboo biochar also adsorbed metal complex dye acid black 172 (Yang et al., 2013). *Hibiscus cannabinus* fiber char supplied by Kenaf Fiber Industries Sdn. Bhd., Malaysia, was pyrolyzed slowly to acid-treated biochar at 1000°C (Mahmoud et al., 2012). This adsorbed methylene blue in a honeycomb pore network as observed by scanning electron microscopy. Ates and Un (2013) prepared hornbeam sawdust biochars in a fixed-bed reactor at 500, 600, 700, and 800°C under an inert atmosphere to adsorb orange 30 dye. Although optimum adsorption occurred at pH 2.0 on all chars, the adsorption capacity was highest on chars made at 800°C.

Similarly, in a separate study by Qiu et al. (2009), straw biochar was treated with 0.1 M HCl to remove metals, followed by demineralization using hydrochloric or hydrofluoric acid. This treated biochar was studied in parallel with a commercial activated carbon (Darco G-60, Sigma Aldrich) for the removal of reactive brilliant blue and rhodamine B dyes from water. The biochar surface area ($1057 \, m^2 g^{-1}$) was higher than the activated carbon ($970 \, m^2 g^{-1}$), and adsorbed more rhodamine B than the commercial activated carbon as the dye molecules could more easily access the larger micropores of biochar than the fine pores of activated carbon.

Besides the chemical modification, previous studies (Zhang et al., 2012a; Inyang et al., 2014) have shown that biochar can serve as a low-cost support for nanoparticles such as graphene, carbon nanotubes, and metal oxides, which typically have extremely high specific surface area and/or tunable surface chemistry (Qu et al., 2013). As a result, grafting nanomaterials on biochar could enhance their sorption abilities by increasing the surface area. For example, the surface area of a carbon nanotube-modified bagasse biochar ($390 \, m^2 g^{-1}$) was increased by 40 times the original surface area of bagasse biochar ($9 \, m^2 g^{-1}$), and twice the amount of methylene blue was sorbed to the modified biochar compared to the unmodified biochar (Inyang et al., 2014).

Remediation of Antibiotics

The constant release and occurrence of pharmaceuticals, their metabolites, and transformation products in the environment is becoming a matter of a grave concern because of their extensive and increasing use in human and veterinary medicine (Długosz et al., 2015). Pharmaceutical residues, which are recognized as emerging contaminants, are detected frequently in treated wastewater, surface water, and groundwater worldwide.

Antibiotic removal has been reported widely using adsorbents including activated carbon, carbon nanotubes, bentonite, ion exchange resins, and biochar (Ahmed et al., 2015). The high material cost and potentially high regeneration costs associated with activated carbon and carbon nanotubes are circumvented by replacement with biochar prepared using the low-cost biomass. The sorption of antibiotics may vary greatly according to the properties of both the antibiotic itself and the biochar used (Yao et al., 2012a; Zhang et al., 2013a). Moreover, several authors have proven that biochar exhibited similar or even better adsorption capacity than commercially available activated carbon (Ahmed et al., 2015). In some cases the removal rate may be as high as 100% depending on antibiotic classes (Ahmed et al., 2015). Potentially different and diverse adsorption mechanisms by which antibiotics bind to carbon materials were proposed. Mostly, π–π-electron–donor–acceptor interaction (Teixido et al., 2011; Liu et al., 2012; Inyang et al., 2014), pore-filling, and hydrophobic interactions (Zheng et al., 2013) were the main mechanisms for the adsorption of antibiotics onto carbon-based adsorbents like biochar. The pH was an important factor, thus the optimum pH for the effective removal of specific antibiotics should be determined prior to use. The adsorption of sulfamethoxazole on giant reed-derived biochar reached equilibrium within 72h and neutral sulfamethoxazole was dominant at pH 1.0–6.0. Above pH 7.0, biochar surface became negatively charged while the sorption of negatively charged sulfamethoxazole species was increased with increasing pH (Zheng et al., 2013). Therefore the molecules being adsorbed and the extent of their adsorption are also highly dependent on the solution for those ionizable compounds.

In studying the adsorptive removal of sulfonamides (sulfamethoxazole and sulfapyridine), it was reported that pine wood and bamboo biochars, which were prepared under different thermochemical conditions, exhibited strong adsorption capacities (Yao et al., 2012a; Xie et al., 2014). The sulfamethoxazole was removed successfully at a maximum concentration of $50\,mg\,L^{-1}$ (Xie et al., 2014) while sulfapyridine was removed efficiently at $10\,mg\,L^{-1}$ (Yao et al., 2012a). In another study, the sulfamethoxazole removal ($125\,mg\,L^{-1}$) was executed using giant reed biochar prepared at 400°C (Zheng et al., 2013). Also Teixidó et al. (2011) pyrolyzed hardwood litter at 600°C and reported a high dissociation constant or value (K_d) of $106\,L\,kg^{-1}$ for the removal of sulfamethazine on the resultant biochar whereas Liu et al. (2012) observed a maximum adsorption capacity ($58.8\,mg\,g^{-1}$) for tetracycline on KOH-treated rice husk-derived biochar. Furthermore, removal efficiencies of up to 100% for the antibiotics florfenicol ($50\,mg\,L^{-1}$) and ceftiofur ($50\,mg\,L^{-1}$) were achieved on pine wood biochar within 24h from sewage and other contaminated effluents (Mitchell et al., 2015).

Remediation of Polycyclic Aromatic Hydrocarbons

PAHs are constituents of petroleum hydrocarbons that have become ubiquitous in the environment because of the persistent exploitation of crude oil and its derivatives. Such pollutants may undergo photolysis, chemical oxidation, volatilization, leaching, bioaccumulation, and/or adsorption in soil (Semple et al., 2003; Ogbonnaya et al., 2014).

The decrease in total and bioavailable PAH concentrations by biochar addition, because of its sorptive properties, has been reported previously (Beesley et al., 2010). For example, the adsorption of PAHs to wood chars has been shown to be assisted by π-electron interactions and pore-filling mechanisms (Chen and Chen, 2009), multilayer adsorption, surface coverage, condensation in capillary pores, and adsorption into the polymeric matrix (Werner and Karapanagioti. 2005). Other successful examples include the reduction of phenanthrene exposure to biota in soil following biochar addition (Ogbonnaya et al., 2014), and a decrease in soil pore water concentration of PAHs by 50% (Beesley et al., 2010). Benefits can also be observed to plants within contaminated systems. Thus Shi et al. (2011) added rice straw-derived biochar to phenanthrene-contaminated soil and noticed a significant reduction of phenanthrene uptake by maize seedlings. Chen et al. (2012b) also showed dissipation of PAHs in a contaminated soil amended with immobilized bacteria using biochar as a carrier (see also Chapter 4). Furthermore, the immobilized bacteria could degrade directly the biochar-sorbed PAHs. Hence the researchers proposed this as an effective bioaugmentation approach for enhancing bioremediation of PAH-contaminated soil. In a separate study, the presence of biochar reduced the concentration of heavier PAH accumulated within earthworm tissues by over 40% (Gomez-Eyles et al., 2011).

Although increased microbial degradation seems unlikely as the biochar sorbents reduce molecule bioavailability, the reductions in total PAH concentrations were greater for the heavier and more recalcitrant moieties (Rhodes et al., 2008). It is possible therefore that the heavier PAHs with a higher affinity to carbon increasingly sorbed to the biochar during the experiment forming nonextractable bound residues (Northcott and Jones, 2001).

Poultry litter and wheat straw biochar produced at 400°C for 120–420 min and hydrothermal poultry litter and swine solids chars produced at 250°C for 20 h have also been demonstrated to remove phenanthrene and other organic pollutants from water because of phenathrene adsorption via extensive π–π interactions (Sun et al., 2011a). An alternative adsorption mechanism of removal was reported by Obst et al. (2011) who found that phenanthrene was accumulated in the interconnected pore system along with primary cracks in the biochar particles. The mineral-rich biochars such as wastewater solids or animal-derived biochars with high Fe, Mg, Si, K, or Ca contents may participate in specific cation–π interactions with biochar PAHs (Oh et al., 2013), but hydrophobic or dispersive forces may be required to support such bondings.

Biochar adsorption and desorption of organic pollutants in soil is reportedly influenced greatly by pyrolysis temperature relative to the biomass used. James et al. (2005) determined the phenanthrene sorption capacity of wood biochars from the species *Pinus* and *Betula* and demonstrated that phenanthrene sorption was increased with the higher surface area of biochars produced at higher temperatures. Also the removal of phenanthrene was achieved with pine wood biochar produced at 350 and 700°C and showed PAH entrapment in the soil micro- or mesopores (Zhang et al., 2010). PAH removal in soil using hardwood biochar by sorption and biodegradation was studied further by Beesley et al. (2010), whereas Khan et al. (2013) used sewage sludge biochar (500°C) for the removal of PAHs from soil through a partitioning mechanism. In contrast, phenanthrene was removed from water using soybean stalk biochar pyrolyzed at 300–700°C via a partitioning mechanism, which was dependent mainly on the overall contribution of "compatibility," namely, the hydrophobic effect between phenanthrene and biochar, and "accessibility," namely, the rubbery domains of biochar (Kong et al., 2011). Similarly, studies by Chen et al. (2008, 2011), pyrolyzed pine needles at 100–700°C, and orange peel at 250, 400, and 700°C, to produce biochars for the removal of naphthalene from water via transitional adsorption and partition mechanisms.

Zhang et al. (2011) compared the pyrene sorption capacity of saw dust-derived biochar prepared at 400 and 700°C (denoted as 400BC and 700BC) and modified biochars (via hydrogen peroxide, hydroxylamine hydrochloride, and absolute ethyl alcohol). For 700BC, both biochars modified by hydrogen peroxide and hydroxylamine hydrochloride reduced pyrene sorption, with pyrene sorption capacity decreasing by 69.1–73.7% and 18.7–33.9%, respectively, whereas hydrolysis did not exert a significant influence. For 400BC, biochars modified by hydrogen peroxide and absolute ethyl alcohol reduced pyrene capacity by 2.28–25.9% and 29.2–33.9%, respectively, while biochar modified by hydroxylamine hydrochloride increased pyrene capacity by ~30%. In most cases, the change in sorption capacity could be explained by the changes in C content, surface area, and micropore volume of the biochars; however, the role of conformation (the accessibility to sorption sites) could not be ignored.

Remediation of Organic Pesticides

Organic pesticides are added to soil or other environmental compartments deliberately to control pests and disease in agriculture. The increased sorption and decreased dissipation of pesticides in the presence of biochar may lower the risk of environmental contamination and human exposure from the perspective of ecosystem and human health. Furthermore, the decreased bioavailability and plant uptake may increase crop yield and reduce organic pesticide residues in agricultural crops. However, since a pesticide that is intended to control specific pests or weeds needs to be bioavailable to be effective, the decreased efficacy of organic pesticides because of application of biochar is undesirable (Mesa and Spokas, 2011; Kookana et al., 2011).

The reduced efficacy of organic pesticides in the presence of biochar has been observed by some researchers (Kookana et al., 2011; Nag et al., 2011; Yu et al., 2009). Also Graber et al. (2011) reported a reduced efficacy of the fumigant 1,3-dichloropropene in a biochar-amended soil. The result showed that in the soil amended with 1% (w/w) biochar, the dose of 1,3-dichloropropene needed to be doubled to obtain full activity on nematode survival. Despite the decreased efficacy, adequate nematode control was still achieved at 0.5% and 1% biochar application at a 1,3-dichloropropene dose on the low end of the recommended rates range. However, if the adsorption strength of the biochar goes beyond a specific threshold, then adequate pest or weed control will not be obtained. In this case, biochar with higher surface area would be considered to be undesirable since the sorption capability of these biochars has been shown to be greater. Hence Graber et al. (2011) suggested a conservative value (up to 10 s of $m^2 g^{-1}$) of biochar surface area to achieve the pest control purpose. In this case, the negative effect of biochar on pest and weed control must not be neglected when it is used to remediate pesticide residues. Similarly, Nag et al. (2011) found a strong effect of biochar application to soil on the efficacy of atrazine to control ryegrass weed and noted that the dose for biochar-amended (1% w/w) soil may need to be increased by 3–4 times to achieve the desired weed control. They also noted that the extent of biochar's effect on herbicide efficacy was dependent on the chemistry of the herbicide molecule and its mode of action. However, it has been found that aging of biochars might reduce the sorption capacity with time, which may be important for herbicide efficacy control in biochar-amended soils (Martin et al., 2012). It is of vital importance therefore to balance the conflict between the potentially promising effect of biochar on pesticide remediation and its negative effect on pesticide efficacy.

ENVIRONMENTAL APPLICATION OF BIOCHAR TO GROUNDWATER SYSTEMS

Groundwater contamination by inorganic and organic pollutants has become a major environmental concern because of its enormous scale, intractable complexity, and severe effects on groundwater-based drinking water supply systems (Xie et al., 2015). Rapid industrialization, increased urbanization, modern agricultural practices, and inappropriate waste disposal approaches have led to increasing concentrations of a wide range of contaminants in the subsurface. Groundwater contaminants such as fertilizers, heavy metals, radionuclides, and organic compounds result in unacceptable environmental risks and affect both ecosystem and human health worldwide.

Biochar, carbonaceous residues of incomplete burning of carbon-rich biomass, has the potential to immobilize both inorganic and organic pollutants in groundwater because of its high specific surface area and cation exchange capacity, porous structure, active surface functional groups, and aromatic surfaces (Cao et al., 2009).

Remediation of Heavy Metals in Groundwater

Heavy metals in groundwater have been of great environmental concern because of their nonbiodegradable nature, toxicity, persistence, bioaccumulation in the food chain, and adverse effects on organisms and humans. It has been reported that biochar is an environmentally benign, effective, and green sorbent for a wide range of metal ions, including Mg(II), Ca(II) (Abdel-Fattah et al., 2015), As(III), As(V) (Samsuri et al., 2013), Cd(II) (Liang et al., 2014), Pb(II) (Mohan et al., 2007), Cu(II) (Tong et al., 2011), Zn(II) (Kołodyńska et al., 2012), Mn(II) (Zhang et al., 2014), Cr(III) (Agrafioti et al., 2014), and Cr(VI) (Dong et al., 2011) through electrostatic attraction, ion exchange, physical adsorption, surface complexation, and/or precipitation (Tan et al., 2015).

Biochar was shown to effectively reduce concentrations of Cu(II) in aqueous solutions (Tong et al., 2011; Zhang et al., 2014; Meng et al., 2014). For example, Tong et al. (2011) reported that biochars, produced from straws of canola, soybean, and peanut at 400°C by oxygen-limited pyrolysis, were rich in —COOH and phenolic hydroxyl groups and can specifically adsorb Cu(II) through surface complexes. The sorption isotherm data fitted the Langmuir model, which is a typical monolayer adsorption isotherm model, and the maximum adsorption capacities were in the ranges of 0.58–1.40 and 0.48–0.79 mol kg^{-1} at pH 5.0 and 4.5, respectively. Also Liang et al. (2014) demonstrated the simultaneous remediation of heavy metal-contaminated groundwater (50 mg L^{-1} of Pb(II), 50 mg L^{-1} of Zn(II), and 1.5 mg L^{-1} of Cd(II)) through a series of laboratory-scale column experiments packed with biochar-amended soils. It was concluded that after 160 L of groundwater leaching through the column, the soil supplemented with biochar removed Pb(II), Cd(II), and Zn(II) by 97.4%, 54.5%, and 53.4%, respectively.

According to Dong et al. (2011), biochar produced from sugar beet tailing removed Cr(VI) efficiently from acidic aqueous solutions with a maximum Langmuir sorption capacity of 123 mg g^{-1}. The optimum reaction pH was 2.0 and the attendant mechanisms included electrostatic attraction of Cr(VI) to positively charged biochar surface, reduction of Cr(VI) to Cr(III), and surface complexation between Cr(III) and biochar's functional groups. Furthermore, Hu et al. (2015) prepared an iron-impregnated biochar through direct hydrolysis of iron salt and showed a strong sorption ability to aqueous As(V). Batch sorption experiments suggested a Langmuir maximum sorption capacity of $2.16\,mg\,g^{-1}$ while fixed-bed columns recorded excellent As removal ability of 85%. Sorption of As onto the iron-impregnated biochar was mainly through chemisorption.

Biochar can also serve as an excellent sorbent to remove radionuclides such as uranium and cesium from groundwater, and can be used as a potentially competitive low-cost and effective permeable reactive barrier medium. Permeable reactive barrier, a structure consisting of an active filler, has received acceptance as a treatment technique for in situ remediation of groundwater (Kumar et al., 2011; Kimura et al., 2014; Zhang et al., 2013b). It was reported that at an initial biochar solid concentration of $4\text{–}5\,g\,L^{-1}$, 90% of $30\,mg\,L^{-1}$ U(VI) was removed from the groundwater within $8\,h$ through adsorption rather than precipitation. The sorption isotherm followed a Langmuir model and the maximum sorption capacity was determined to be $2.12\,mg\,g^{-1}$ (Kumar et al., 2011). The column experiment supported the use of biochar as a permeable reactive barrier medium, which showed that approximately $0.25\,g$ of biochar was packed into a 1.0×10-cm glass column as an interlayer of two quartz zones (0.6 and $5.4\,cm$). The U(VI) breakthrough from the permeable reactive barrier column indicated a biochar adsorption capacity of $0.53\,mg\,U\,g^{-1}$ of biochar. In general, when compared to the quartz column under identical conditions, U(VI) adsorbed to biochar was 473 times more on a unit mass basis.

Remediation of Organic Contaminants in Groundwater

Biochar shows high affinity for a wide range of pollutants including dyes, organic pesticides, herbicides, antibiotics, and organic contaminants. Some examples of specific ones reported in the literature are methylene blue dye (Mohan et al., 2014), trichloroethylene (Ahmad et al., 2013), atrazine (Cao et al., 2009; Delwiche et al., 2014), norflurazon (Sun et al., 2011b), fluridone (Yao et al., 2012b), sulfamethoxazole (Zheng et al., 2013), bisphenol A (Jung et al., 2013), ofloxacin and norfloxacin (Wu et al., 2013), phenanthrene and naphthalene (Chen et al., 2012c), N-nitrosodimethylamine (Chen et al., 2015d), and 17α-ethinyl estradiol (Sun et al., 2011c). In general, the adsorption mechanisms involved include partitioning, adsorption, and electrostatic interactions between organic contaminants and biochars (Ahmad et al., 2014b).

Many laboratory and field experiments have been carried out to investigate the adsorption capacity of biochar for organic contaminants from water/groundwater. Sun et al. (2013) compared the removal of cationic methylene blue dye by biochars produced from anaerobic digestion residue, palm bark, and eucalyptus. Within $2\,h$, $8\,g\,L^{-1}$ of the biochar produced from anaerobic digestion residue removed 96.4% of $5\,g\,L^{-1}$ methylene blue dye at 30°C, and the removal efficiencies for palm bark biochar and eucalyptus biochar were 89.8% and 78.3%, respectively. Ahmad et al. (2013) utilized pine needle-derived biochars at different pyrolysis temperatures (300, 500, and 700°C) to sorb trichloroethylene from water. The results showed that biochar produced at 700°C exhibited the highest trichloroethylene removal efficiency because of the high surface area, microporosity, and carbonized extent, and the prevailing sorption mechanism was pore filling. Delwiche et al. (2014) reported that pine chip-produced biochar (pyrolyzed between 300 and 550°C) could reduce cumulative atrazine leaching by 52% in homogenized (packed) soil columns. Field experimental results indicated that with the addition of $10 t ha^{-1}$ biochar, mean groundwater atrazine peak concentrations were 53% lower than that without the biochar. Sun et al. (2011a) reported that biochars produced at 400°C exhibited high sorption capacities for two fluorinated herbicides norflurazon and fluridone, thus preventing unwanted herbicide leaching into the groundwater. Zheng et al. (2013) stated that the adsorption of neutral antibiotic sulfamethoxazole molecules on biochar was dominant at pH 1.0–6.0, and that above pH 7.0, biochar and sulfamethoxazole were both negatively charged with H-bonds the primary sorption mechanisms.

Remediation of Other Pollutants in Groundwater

Apart from heavy metals and organic contaminants, biochar holds the potential to be an excellent sorbent for ammonium (Wang et al., 2015), nitrate (Zhang et al., 2012b), phosphate (Yao et al., 2011), fluoride (Oh et al., 2012; Mohan et al., 2012), and engineered nanoparticles (Inyang et al., 2013).

Groundwater contamination caused by fertilizer leaching is an important concern in many agricultural landscapes. Application of biochar has been reported to remove ammonium, nitrate, and phosphate from water and improve the fertilizer-retaining capacity of soil, thereby reducing nitrogen and phosphorus fertilizers leaching into groundwater (Wang et al., 2015; Zhang et al., 2012b; Zeng et al., 2013). Thus Kameyama et al. (2012) found that nitrate was adsorbed to biochar made from sugar cane bagasse, resulting in an increase in the residence time of nitrate in the root zone of crops with a greater opportunity for absorption of the nutrient. Similarly, Yao et al. (2012c) used column leaching experiments to assess the ability of Brazilian pepperwood-derived biochars produced at 600°C to hold nitrate, ammonium, and phosphate in a sandy soil. They reported that the biochar effectively reduced the total amount of nitrate, ammonium, and phosphate in the leachates by 34.0%, 34.7%, and 20.6%, respectively, compared to the soil-alone controls. Yao et al. (2011) investigated the phosphate removal abilities of biochars derived from anaerobically digested sugar beet tailings. The results suggested that nanosized MgO particles on the surface of biochar were primary sorption sites for aqueous phosphate, and that biochar was a promising alternative adsorbent to reclaim phosphate from water or reduce phosphate leaching from fertilized soils to groundwater.

Low-cost pine wood and pine bark chars, by-products from fast pyrolysis at 400 and 450°C, were used successfully by Mohan et al. (2012) for groundwater defluoridation. Pine chars successfully treated fluoride-contaminated groundwater at pH 2.0 within 48 h, and the Langmuir monolayer adsorption capacity of pink bark char was higher than pine wood char (9.77 vs. 7.66 mg g^{-1}) at 25°C. It should be noted that the fluoride adsorption capacity of these low surface area biochars (specific surface area (S_{BET}): 1–3 m^2 g^{-1}) were comparable to, or higher than, that of activated carbon (S_{BET}: 1000 m^2 g^{-1}), which was because adsorption occurred within the solid char walls. The chars with high oxygen content (8–11%) swelled in water, opening new internal pore volume. In addition, fluoride diffused into the chars' subsurface solid volume promoting further adsorption. Moreover, ion exchange and metal fluoride precipitation (from ash components) were responsible for adsorption.

Notwithstanding these positive outcomes, the widespread applications of engineered nanoparticles have raised concerns over their increased loading to the environment and potential toxic effects to soil and groundwater systems in particular. Inyang et al. (2013) found that biochar produced from hickory chips could be used as a low-cost filter material to remove three engineered nanoparticles, namely, silver nanoparticles, carbon nanotubes, and titanium dioxide from water. Among hickory chip biochar, activated carbon, iron-modified biochar, and iron-modified activated carbon, the iron-modified biochar was the best in filtering all the engineered nanoparticles. The interaction between Fe(OH)$_3$ and nanoparticles can increase intraparticle bridging and reduce the electron double layer repulsions to facilitate particle filtration in filter media. Furthermore, biochar can be used for other contaminant removal including pathogenic bacteria, such as *Escherichia coli* (Mohanty et al., 2014; Mohanty and Boehm, 2014). Specifically, Mohanty et al. (2014) investigated the efficacy of biochar-augmented model sand biofilters for *E. coli* removal under a variety of stormwater bacterial concentrations and infiltration rates. Thus the researchers replaced a native block of soil with a mixture of sand (60–70%) and softwood-derived biochar (30–40%) to increase the hydraulic conductivity of the block, thereby allowing rapid infiltration of stormwater into either the ground or an underdrain system connected to surface waters. They reported a ~96% *E. coli* removal in the biochar-augmented sand biofilters, which was not affected greatly by increases in stormwater infiltration rates and influent bacterial concentrations, particularly within the ranges expected in the field. Removal of fine biochar particles (<125 μm) from the biochar-sand biofilter decreased the removal capacity from

95% to 62%, indicating that the biochar size is important. Also the addition of compost to the biochar–sand biofilters not only lowered *E. coli* removal capacity but also increased the mobilization of deposited bacteria during intermittent infiltration. This result was attributed to exhaustion of attachment sites on biochar by the dissolved organic carbon leached from compost. In general, the authors concluded that biochar has potential to remove bacteria from stormwater under a wide range of field conditions, but for it to be effective, the size should be small and it should be applied without compost. Although the results aid in the optimization of biofilter design, further studies are needed to examine biochar potential in the field over an entire rainy season.

RECOMMENDATIONS FOR FURTHER RESEARCH

Their high adsorption capacities, environmental viability, and cost effectiveness have demonstrated that biochars obtained from waste biomass have high potential for removing heavy metals and organic contaminants such as synthetic dyes, antibiotics, PAHs, and organic pesticides from different environments. Adsorption mechanisms and capacities depend on the nature of the pollutants and preparation conditions for the biochars. Thus selectivity of suitable biochar is crucial and needs more attention.

Most investigations to date have focused on individual pollutants or single groups of environmental contaminants. Thus further work should be carried out to evaluate the adsorption capacity of biochar with multi-contaminants (ie, heavy metals–organic-polluted soils and water). Moreover, application of biochars has been limited to the laboratory scale. Large-scale studies on the application of biochars should be conducted and evaluated in terms of technical and economic aspects.

Acknowledgments

The research was supported by the National Natural Science Foundation of China (31270544, 41473070), a 863 Major Program (2013AA06A205).

References

Abdel-Fattah, T.M., Mahmoud, M.E., Ahmed, S.B., Huff, M.D., Lee, J.W., Kumar, S., 2015. Biochar from woody biomass for removing metal contaminants and carbon sequestration. J. Ind. Eng. Chem. 22 (0), 103–109.

Agrafioti, E., Kalderis, D., Diamadopoulos, E., 2014. Arsenic and chromium removal from water using biochars derived from rice husk, organic solid wastes and sewage sludge. J. Environ. Manage. 133 (0), 309–314.

Ahmad, M., Lee, S.S., Rajapaksha, A.U., Vithanage, M., Zhang, M., Cho, J.S., Lee, S.-E., Ok, Y.S., 2013. Trichloroethylene adsorption by pine needle biochars produced at various pyrolysis temperatures. Bioresour. Technol. 143 (0), 615–622.

Ahmad, M., Lee, S.S., Lim, J.E., Lee, S.E., Cho, J.S., Moon, D.H., Hashimoto, Y., Ok, Y.S., 2014a. Speciation and phytoavailability of lead and antimony in a small arms range soil amended with mussel shell, cow bone and biochar: EXAFS spectroscopy and chemical extractions. Chemosphere 95 (0), 433–441.

Ahmad, M., Rajapaksha, A.U., Lim, J.E., Zhang, M., Bolan, N., Mohan, D., Vithanage, M., Lee, S.S., Ok, Y.S., 2014b. Biochar as a sorbent for contaminant management in soil and water: a review. Chemosphere 99 (0), 19–33.

Ahmed, M.B., Zhou, J.L., Ngo, H.H., Guo, W., 2015. Adsorptive removal of antibiotics from water and wastewater: progress and challenges. Sci. Total Environ. 532, 112–126.

An, Z., Hou, Y., Cai, C., Xue, X., 2011. Lead (II) adsorption characteristics on different biochars derived from rice straw. Environ. Ehem. 30 (11), 1851–1857.

Ates, F., Un, U.T., 2013. Production of char from hornbeam sawdust and its performance evaluation in the dye removal. J. Anal. Appl. Pyrol. 103 (0), 159–166.

Beesley, L., Jimenez, E.M., Eyles, J.L.G., 2010. Effects of BC and greenwaste compost amendments on mobility, bioavailability and toxicity of inorganic and organic contaminants in a multi-element polluted soil. Environ. Pollut. 158 (0), 2282–2287.

Bolan, N.S., Naidu, R., Syers, J.K., Tillman, R.W., 1999. Surface charge and solute interactions in soils. Adv. Agron. 67 (0), 87–140.

Brandli, R.C., Hartnik, T., Henriksen, T., Cornelissen, G., 2008. Sorption of native polyaromatic hydrocarbons (PAH) to black carbon and amended activated carbon in soil. Chemosphere 73 (0), 1805–1810.

Brick, S., 2010. Biochar: Assessing the Promise and Risks to Guide US Policy. Natural Resource Defense Council USA.

Cantrell, K.B., Hunt, P.G., Uchimiya, M., Novak, J.M., Ro, K.S., 2012. Impact of pyrolysis temperature and manure source on physicochemical characteristics of biochar. Bioresour. Technol. 107 (0), 419–428.

Cao, X., Ma, L., Gao, B., Harris, W., 2009. Dairy-manure derived biochar effectively sorbs lead and atrazine. Environ. Sci. Technol. 43 (9), 3285–3291.

Cao, X., Ma, L., Liang, Y., Gao, B., Harris, W., 2011. Simultaneous immobilization of lead and atrazine in contaminated soils using dairy-manure biochar. Environ. Sci. Technol. 45 (0), 4884–4889.

Chen, B., Chen, Z., 2009. Sorption of naphthalene and 1-naphthol by biochars of orange peels with different pyrolytic temperatures. Chemosphere 76 (0), 127–133.

Chen, B., Chen, Z., Lv, S., 2011. A novel magnetic biochar efficiently sorbs organic pollutants and phosphate. Bioresour. Technol. 102 (0), 716–723.

Chen, Z., Fang, Y., Xu, Y., Chen, B., 2012a. Adsorption of Pb^{2+} by rice straw derived-biochar and its influential factors. Acta Sci. Circumstantiae 32 (4), 769–776 (China).

Chen, B., Yuan, M., Qian, L., 2012b. Enhanced bioremediation of PAH-contaminated soil by immobilized bacteria with plant residue and biochar as carriers. J. Soils. Sediments 12 (0), 1350–1359.

Chen, Z., Chen, B., Chiou, C.T., 2012c. Fast and slow rates of naphthalene sorption to biochars produced at different temperatures. Environ. Sci. Technol. 46 (20), 11104–11111.

Chen, B., Zhou, D., Zhu, L., 2008. Transitional adsorption and partition on nonpolar and polar aromatic contaminants by biochars of pine needles with different pyrolytic temperatures. Environ. Sci. Technol. 42 (0), 5137–5143.

Chen, T., Zhou, Z., Han, R., Meng, R., Wang, H., Lu, W., 2015a. Adsorption of cadmium by biochar derived from municipal sewage sludge: impact factors and adsorption mechanism. Chemosphere 134, 286–293.

Chen, T., Zhou, Z., Xu, S., Wang, H., Lu, W., 2015b. Adsorption behavior comparison of trivalent and hexavalent chromium on biochar derived from municipal sludge. Bioresour. Technol. 190, 388–394.

Chen, Z., Xiao, X., Chen, B., Zhu, L., 2015c. Quantification of chemical states, dissociation constants and contents of oxygen-containing groups on the surface of biochars produced at different temperatures. Environ. Sci. Technol. 49 (1), 309–317.

Chen, C., Zhou, W., Lin, D., 2015d. Sorption characteristics of N-nitrosodimethylamine onto biochar from aqueous solution. Bioresour. Technol. 179 (0), 359–366.

Chen, Z., Chen, B., Zhou, D., 2013. Composition and sorption properties of rice-straw derived biochars. Acta Sci. Circumstantiae 33 (1), 9–19 (China).

Delwiche, K.B., Lehmann, J., Walter, M.T., 2014. Atrazine leaching from biochar-amended soils. Chemosphere 95 (0), 346–352.

Długosz, M., Żmudzki, P., Kwiecień, A., Szczubiałka, K., Krzek, J., Nowakowska, M., 2015. Photocatalytic degradation of sulfamethoxazole in aqueous solution using a floating TiO$_2$-expanded perlite photocatalyst. J. Hazard. Mater. 298 (0), 146–153.

Dong, X., Ma, L.Q., Li, Y., 2011. Characteristics and mechanisms of hexavalent chromium removal by biochar from sugar beet tailing. J. Hazard. Mater. 190 (1–3), 909–915.

Gaskin, J., Steiner, C., Harris, K., Das, K., Bibens, B., 2008. Effect of low-temperature pyrolysis conditions on biochar for agricultural use. Trans. Asabe 51 (0), 2061–2069.

Gomez-Eyles, J.L., Sizmur, T., Collins, C.D., Hodson, M.E., 2011. Effects of biochar and the earthworm *Eisenia fetida* on the bioavailability of polycyclic aromatic hydrocarbons and potentially toxic elements. Environ. Pollut. 159 (0), 616–622.

Gong, Y., Liu, Y., Xiong, Z., Zhao, D., 2014. Immobilization of mercury by carboxymethyl cellulose stabilized iron sulfide nanoparticles: reaction mechanisms and effects of stabilizer and water chemistry. Environ. Sci. Technol. 48 (7), 3986–3994.

Graber, E.R., Tsechansky, L., Khanukov, J., 2011. Sorption, volatilization, and efficacy of the fumigant 1,3-dichloropropene in a biochar-amended soil. Soil Chem. 75 (0), 1365–1373.

Hameed, B.H., El-Khaiary, M.I., 2008. Kinetics and equilibrium studies of malachite green adsorption on rice straw-derived char. J. Hazard. Mater. 153 (0), 701–708.

Hu, X., Ding, Z., Zimmerman, A.R., Wang, S., Gao, B., 2015. Batch and column sorption of arsenic onto iron-impregnated biochar synthesized through hydrolysis. Water Res. 68 (0), 206–216.

Inyang, M., Dickenson, E., 2015. The potential role of biochar in the removal of organic and microbial contaminants from potable and reuse water: a review. Chemosphere 134 (0), 232–240.

Inyang, M., Gao, B., Wu, L., Yao, Y., Zhang, M., Liu, L., 2013. Filtration of engineered nanoparticles in carbon-based fixed bed columns. Chem. Eng. J. 220 (0), 221–227.

Inyang, M., Gao, B., Yao, Y., Xue, Y., Zimmerman, A.R., Pullammanappallil, P., Cao, X., 2012. Removal of heavy metals from aqueous solution by biochars derived from anaerobically digested biomass. Bioresour. Technol. 110 (0), 50–56.

Inyang, M., Gao, B., Zimmerman, A., Zhang, M., Chen, H., 2014. Synthesis, characterization, and dye sorption ability of carbon nanotube-biochar nanocomposites. Chem. Eng. J. 236 (0), 39–46. In: Lehmann, J., Joseph, S., (Eds.), Biochar Environ. Manag. Sci. Technol. Earthscan, London, 85–105.

Jadhav, S.B., Surwase, S.N., Kalyani, D.C., Gurav, R.G., Jadhav, J.P., 2012. Biodecolorization of azo dye remazol orange by *Pseudomonasaeruginosa* BCH and toxicity (oxidative stress) reduction in *Allium cepa* root cells. Appl. Biochem. Biotechnol. 168 (0), 1319–1334.

James, G., Sabatini, D.A., Chiou, C.T., Rutherford, D., Scott, A.C., Karapanagioti, H.K., 2005. Evaluating phenanthrene sorption on various wood chars. Water Res. 39 (4), 549–558.

Jung, C., Park, J., Lim, K.H., Park, S., Heo, J., Her, N., Oh, J., Yun, S., Yoon, Y., 2013. Adsorption of selected endocrine disrupting compounds and pharmaceuticals on activated biochars. J. Hazard. Mater. 263 (Part 2(0)), 702–710.

Kameyama, K., Miyamoto, T., Shiono, T., Shinogi, Y., 2012. Influence of sugarcane bagasse-derived biochar application on nitrate leaching in Calcaric dark red soil. J. Environ. Qual. 41 (4), 1131–1137.

Karami, N., Clemente, R., Jimenez, E., Lepp, N., Beesley, L., 2011. Efficiency of green waste compost and biochar soil amendments for reducing lead and copper mobility and uptake to ryegrass. J. Hazard. Mater. 191 (0), 41–48.

Khan, S., Chao, C., Waqas, M., Arp, H.P.H., Zhu, Y.G., 2013. Sewage sludge biochar influence upon rice (*Oryza sativa* L.) yield, metal bioaccumulation and greenhouse gas emissions from acidic paddy soil. Environ. Sci. Technol. 47 (0), 8624–8632.

Kimura, K., Hachinohe, M., Klasson, K.T., Hamamatsu, S., Hagiwara, S., Todoriki, S., Kawamoto, S., 2014. Removal of radioactive cesium (Cs-134 plus Cs-137) from low-level contaminated water by charcoal and broiler litter biochar. Food Sci. Technol. Res. 20 (6), 1183–1189.

Kołodyńska, D., Wnętrzak, R., Leahy, J.J., Hayes, M.H.B., Kwapiński, W., Hubicki, Z., 2012. Kinetic and adsorptive characterization of biochar in metal ions removal. Chem. Eng. J. 197 (0), 295–305.

Kong, H., He, J., Gao, Y., Wu, H., Zhu, X., 2011. Cosorption of phenanthrene and mercury(II) from aqueous solution by soybean stalk-based biochar. J. Agric. Food Chem. 59 (0), 12116–12123.

Kookana, R.S., Sarmah, A.K., Van Zwieten, L., Krull, K., Singh, B., 2011. Chapter three, biochar application to soil: agronomic and environmental benefits and unintended consequences. Adv. Agron. 104–143.

Kumar, S., Loganathan, V.A., Gupta, R.B., Barnett, M.O., 2011. An assessment of U(VI) removal from groundwater using biochar produced from hydrothermal carbonization. J. Environ. Manage. 92 (10), 2504–2512.

Li, L., Lu, Y., Liu, Y., Sun, H., Liang, Z., 2012. Adsorption mechanisms of cadmium(II) on biochars derived from corn straw. J. Agro-Environ. Sci. 31 (11), 2277–2283.

Liang, Y., Cao, X., Zhao, L., Arellano, E., 2014. Biochar- and phosphate-induced immobilization of heavy metals in contaminated soil and water: implication on simultaneous remediation of contaminated soil and groundwater. Environ. Sci. Pollut. R. 21 (6), 4665–4674.

Liu, P., Liu, W.J., Jiang, H., Chen, J.J., Li, W.W., Yu, H.Q., 2012. Modification of bio-char derived from fast pyrolysis of biomass and its application in removal of tetracycline from aqueous solution. Bioresour. Technol. 121 (0), 235–240.

Liu, Z., Zhang, F.S., 2009. Removal of lead from water using biochars prepared from hydrothermal liquefaction of biomass. J. Hazard. Mater. 167 (1–3), 933–939.

Lu, H., Zhang, W., Yang, Y., Huang, X., Wang, S., Qiu, R., 2012. Relative distribution of Pb^{2+} sorption mechanisms by sludge-derived biocha. Water Res. 46 (0), 854–862.

Ma, J., Dougherty, D., 1997. The cation-π interaction. Chem. Rev. 97 (5), 1303–1324.

Mahmoud, D.K., Salleh, M.A.M., Karim, W.A.W.A., Idris, A., Abidin, Z.Z., 2012. Batch adsorption of basic dye using acid treated Kenaf fibre char: equilibrium, kinetic and thermodynamic studies. Chem. Eng. J. 181–182 (1), 449–457.

Martin, S.M., Kookana, R.S., Van Zwieten, L., Krull, E., 2012. Marked changes in herbicide sorption-desorption upon ageing of biochars in soil. J. Hazard. Mater. 231–232, 70–78.

Meng, J., Feng, X.L., Dai, Z.M., Liu, X.M., Wu, J.J., Xu, J.M., 2014. Adsorption characteristics of Cu(II) from aqueous solution onto biochar derived from swine manure. Environ. Sci. Pollut. R. 21 (11), 7035–7046.

Mesa, A.C., Spokas, K.A., 2011. Impacts of biochar (black carbon) additions on the sorption and efficacy of herbicides. Herbic. Environ. 15 (0), 315–340.

Mitchell, S.M., Subbiah, M., Ullman, J.L., Frear, C., Call, D.R., 2015. Evaluation of 27 different biochars for potential sequestration of antibiotic residues in food animal production environments. J. Environ. Chem. Eng. 3 (0), 162–169.

Mohan, D., Pittman Jr., C.U., Bricka, M., Smith, F., Yancey, B., Mohammad, J., Steele, P.H., Alexandre-Franco, M.F., Gómez-Serrano, V., Gong, H., 2007. Sorption of arsenic, cadmium, and lead by chars produced from fast pyrolysis of wood and bark during bio-oil production. J. Colloid Interface Sci. 310 (1), 57–73.

Mohan, D., Rajput, S., Singh, V., Steele, P., Pittman, J., 2011. Modeling and evaluation of chromium remediation from water using low cost bio-char, agreen adsorbent. J. Hazard. Mater. 188 (1–3), 319–333.

Mohan, D., Sarswat, A., Ok, Y.S., Pittman Jr., C.U., 2014. Organic and inorganic contaminants removal from water with biochar, a renewable, low cost and sustainable adsorbent-A critical review. Bioresour. Technol. 160 (0), 191–202.

Mohan, D., Sharma, R., Singh, V.K., Steele, P., Pittman, C.U., 2012. Fluoride removal from water using bio-char, a green waste, low-cost adsorbent: equilibrium uptake and sorption dynamics modeling. Ind. Eng. Chem. Res. 51 (2), 900–914.

Mohanty, S.K., Boehm, A.B., 2014. *Escherichia coli* removal in biochar-augmented biofilter: effect of infiltration rate, initial bacterial concentration, biochar particle size, and presence of compost. Environ. Sci. Technol. 48 (19), 11535–11542.

Mohanty, S.K., Cantrell, K.B., Nelson, K.L., Boehm, A.B., 2014. Efficacy of biochar to remove *Escherichia coli* from stormwater under steady and intermittent flow. Water Res. 61 (0), 288–296.

Mui, E.L.K., Cheung, W.H., Valix, M., McKay, G., 2010. Dye adsorption onto char from bamboo. J. Hazard. Mater. 177 (0), 1001–1005.

Mukherjee, A., Zimmerman, A., Harris, W., 2011. Surface chemistry variations among a series of laboratory-produced biochars. Geoderma 163 (0), 247–255.

Nag, S.K., Kookana, R., Smith, L., Krull, E., Macdonald, L.M., Gill, G., 2011. Poor efficacy of herbicides in biochar-amended soils affected by their chemistry and mode of action. Chemosphere 84 (0), 1572–1577.

Northcott, G.L., Jones, K.C., 2001. Partitioning, extractability, and formation of nonextractable PAH residues in soil. 1. Compound differences in aging and sequestration. Environ. Sci. Technol. 35 (0), 1103–1110.

Obst, M., Grathwohl, P., Kappler, A., Eibl, O., Peranio, N., Gocht, T., 2011. Quantitative high-resolution mapping of phenanthrene sorption to black carbon particles. Environ. Sci. Technol. 45 (0), 7314–7322.

Ogbonnaya, O.U., Adebisi, O.O., Semple, K.T., 2014. The impact of biochar on the bioaccessibility of 14C-phenanthrene in aged soil. Environ. Sci. Process. Impacts 16 (0), 2635–2643.

Oh, S.Y., Son, J.G., Chiu, P.C., 2013. Biochar-mediated reductive transformation of nitro herbicides and explosives. Environ. Toxicol. Chem. 32 (0), 501–508.

Oh, T.K., Choi, B., Shinogi, Y., Chikushi, J., 2012. Effect of pH conditions on actual and apparent fluoride adsorption by biochar in aqueous phase. Water Air Soil Pollut. 223 (7), 3729–3738.

Puga, A.P., Abreu, C.A., Melo, L.C., Beesley, L., 2015. Biochar application to a contaminated soil reduces the availability and plant uptake of zinc, lead and cadmium. J. Environ. Manage. 159, 86–93.

Qiu, Y., Zheng, Z., Zhou, Z., Sheng, G.D., 2009. Effectiveness and mechanisms of dye adsorption on a straw-based biochar. Bioresour. Technol. 100 (0), 5348–5351.

Qu, X., Alvarez, P., Li, Q., 2013. Applications of nanotechnology in water and wastewater treatment. Water Res. 47 (0), 3931–3946.

Regmi, P., Garcia Moscoso, J.L., Kumar, S., Cao, X., Mao, J., Schafran, G., 2012. Removal of copper and cadmium from aqueous solution using switchgrass biochar produced via hydrothermal carbonization process. J. Environ. Manage. 109 (0), 61–69.

Rhodes, A.H., Carlin, A., Semple, K.T., 2008. Impact of black carbon in the extraction and mineralization of phenanthrene in soil. Environ. Sci. Technol. 42 (0), 740–745.

Samsuri, A.W., Sadegh-Zadeh, F., Seh-Bardan, B.J., 2013. Adsorption of As(III) and As(V) by Fe coated biochars and biochars produced from empty fruit bunch and rice husk. J. Environ. Chem. Eng. 1 (4), 981–988.

Semple, K.T., Morriss, A.W.J., Paton, G.I., 2003. Bioavailability of hydrophobic organic contaminants in soils: fundamental concepts and techniques for analysis. Eur. J. Soil Sci. 54 (0), 809–818.

Shi, H., Zhou, Q., 2014. Research progresses in the effect of biochar on soil-environmental behaviors of pollutants. Chin. J. Ecol. 33 (2), 486–494.

Shi, M., Hu, L.C., Huang, Z.Q., Dai, J.Y., 2011. The influence of bio-char inputting on the adsorption of phenanthrene by soils and by maize seedlings. J. Agro. Environ. Sci. 30 (0), 912–916.

Sohi, S.P., 2012. Carbon storage with benefits. Science 338, 1034–1035.

Sun, K., Keiluweit, M., Kleber, M., Pan, Z., Xing, B., 2011a. Sorption of fluorinated herbicides to plant biomass-derived biochars as a function of molecular structure. Bioresour. Technol. 102 (0), 9897–9903.

Sun, K., Keiluweit, M., Kleber, M., Pan, Z., Xing, B., 2011b. Sorption of fluorinated herbicides to plant biomass-derived biochars as a function of molecular structure. Bioresour. Technol. 102 (21), 9897–9903.

Sun, K., Ro, K., Guo, M., Novak, J., Mashayekhi, H., Xing, B., 2011c. Sorption of bisphenol A, 17α-ethinyl estradiol and phenanthrene on thermally and hydrothermally produced biochars. Bioresour. Technol. 102 (10), 5757–5763.

Sun, K., Tang, J., Gong, Y., Zhang, H., 2015. Characterization of potassium hydroxide (KOH) modified hydrochars from different feedstocks for enhanced removal of heavy metals from water. Environ. Sci. Pollut. Res. Int. 22 (21), 16640–16651.

Sun, L., Wan, S., Luo, W., 2013. Biochars prepared from anaerobic digestion residue, palm bark, and eucalyptus for adsorption of cationic methylene blue dye: characterization, equilibrium, and kinetic studies. Bioresour. Technol. 140 (0), 406–413.

Tan, X., Liu, Y., Zeng, G., Wang, X., Hu, X., Gu, Y., Yang, Z., 2015. Application of biochar for the removal of pollutants from aqueous solutions. Chemosphere 125 (0), 70–85.

Tang, J., Zhu, W., Kookana, R., Katayama, A., 2013. Characteristics of biochar and its application in remediation of contaminated soil. J. Biosci. Bioeng. 116 (6), 653–659.

Teixidó, M., Pignatello, J.J., Beltrán, J.L., Granados, M., Peccia, J., 2011. Speciation of the ionisable antibiotic sulfamethazine on black carbon (biochar). Environ. Sci. Technol. 45 (0), 10020–10027.

Thies, J.E., Rillig, M., 2009. Characteristics of Biochar: Biological Properties.

Tong, X., Li, J., Yuan, J., Xu, R., 2011. Adsorption of Cu(II) by biochars generated from three crop straws. Chem. Eng. J. 172 (2–3), 828–834.

Tran, V.S., Ngo, H.H., Guo, W., Zhang, J., Liang, S., Ton-That, C., Zhang, X., 2015. Typical low cost biosorbents for adsorptive removal of specific organic pollutants from water. Bioresour. Technol. 182 (0), 353–363.

Uchimiya, M., Klasson, K., Wartelle, L., Lima, I., 2011. Influence of soil properties on heavy metal sequestration by biochar amendment.1.Copper sorption isotherms and the release of cations. Chemosphere 82 (0), 1431–1437.

Wang, Z.H., Guo, H.Y., Shen, F., Yang, G., Zhang, Y.Z., Zeng, Y.M., Wang, L.L., Xiao, H., Deng, S.H., 2015. Biochar produced from oak sawdust by lanthanum (La)-involved pyrolysis for adsorption of ammonium (NH^{4+}), nitrate (NO^{3-}), and phosphate $\left(PO_4{}^{3-} \right)$. Chemosphere 119 (0), 646–653.

Werner, D., Karapanagioti, H.K., 2005. Comment on modelling-maximum adsorption capacities of soot and soot-like materials for PAHs and PCBs. Environ. Sci. Technol. 39 (0), 381–382.

Wu, M., Pan, B., Zhang, D., Xiao, D., Li, H., Wang, C., Ning, P., 2013. The sorption of organic contaminants on biochars derived from sediments with high organic carbon content. Chemosphere 90 (2), 782–788.

Xie, M., Chen, W., Xu, Z., Zheng, S., Zhu, D., 2014. Adsorption of sulfonamides to demineralized pine wood biochars prepared under different thermochemical conditions. Environ. Pollut. 186 (0), 187–194.

Xie, T., Reddy, K.R., Wang, C., Yargicoglu, E., Spokas, K., 2015. Characteristics and applications of biochar for environmental remediation: a review. Crit. Rev. Env. Sci. Technol. 45 (9), 939–969.

Xu, R.K., Xiao, S.C., Yuan, J.H., Zhao, A.Z., 2011. Adsorption of methyl violet from aqueous solutions by the biochars derived from crop residues. Bioresour. Technol. 102 (22), 10293–10298.

Xu, X., Cao, X., Zhao, L., Wang, H., Xu, H., Gao, B., 2013. Removal of Cu, Zn, and Cd from aqueous solutions by the dairy manure-derived biocha. Environ. Sci. Pollut. R. 30, 358–368.

Yakkala, K., Yu, M., Roh, H., Yang, J., Chang, Y., 2013. Buffalo weed (*Ambrosia trifida* L. *var. trifida*) biochar for cadmium (II) and lead (II) adsorption in single and mixed system. Desalin. Water Treat. 51, 7732–7745.

Yang, Y., Lin, X., Wei, B., Zhao, Y., Wang, J., 2013. Evaluation of adsorption potential of bamboo biochar for metal-complex dye: equilibrium, kinetics and artificial neural network modeling. Int. J. Environ. Sci. Technol. 11 (4), 1093–1100.

Yao, Y., Gao, B., Chen, H., Jiang, L., Inyang, M., Zimmerman, A.R., Cao, X., Yang, L., Xue, Y., Li, H., 2012a. Adsorption of sulfamethoxazole on biochar and its impact on reclaimed water irrigation. J. Hazard. Mater. 209 (0), 408–413.

Yao, H., Lu, J., Wu, J., Lu, Z., Wilson, P.C., Shen, Y., 2012b. Adsorption of fluoroquinolone antibiotics by wastewater sludge biochar: role of the sludge source. Water Air Soil Pollut. 224 (1), 1–9.

Yao, Y., Gao, B., Zhang, M., Inyang, M., Zimmerman, A.R., 2012c. Effect of biochar amendment on sorption and leaching of nitrate, ammonium, and phosphate in a sandy soil. Chemosphere 89 (11), 1467–1471.

Yao, Y., Gao, B., Inyang, M., Zimmerman, A.R., Cao, X., Pullammanappallil, P., Yang, L., 2011. Removal of phosphate from aqueous solution by biochar derived from anaerobically digested sugar beet tailings. J. Hazard. Mater. 190 (1–3), 501–507.

Yu, X.Y., Ying, G.G., Kookana, R.S., 2009. Reduced plant uptake of pesticides with biochar additions to soil. Chemosphere 76 (0), 665–771.

Yu, Z., Xie, L., Liu, S., Yang, S., Lian, F., Song, Z., 2014. Effects of biochar-manganese oxides composite on adsorption characteristics of Cu in red soil. Ecol. Environ. Sci. 23 (5), 897–903.

Zeng, Z., Zhang, S., Li, T., Zhao, F., He, Z., Zhao, H., Yang, X., Wang, H., Zhao, J., Rafiq, M., 2013. Sorption of ammonium and phosphate from aqueous solution by biochar derived from phytoremediation plants. J. Zhejiang Univ. Sci. B 14 (12), 1152–1161.

Zhang, G., Liu, X., Sun, K., He, Q., Qian, T., Yan, Y., 2013a. Interactions of simazine, metsulfuron-methyl, and tetracycline with biochars and soil as a function of molecular structure. J. Soils Sediments 1600–1610.

Zhang, Z., Cao, X., Liang, P., Liu, Y., 2013b. Adsorption of uranium from aqueous solution using biochar produced by hydrothermal carbonization. J. Radioanal. Nucl. Chem. 295 (2), 1201–1208.

Zhang, H., Lin, K., Wang, H., Gan, J., 2010. Effect of *Pinus* radiate derived biochars on soil sorption and desorption of phenanthrene. Environ. Pollut. 158 (0), 2821–2825.

Zhang, M., Gao, B., Yao, Y., Xue, Y., Inyang, M., 2012a. Synthesis, characterization, and environmental implications of graphene-coated biochar. Sci. Total Environ. 435–436, 567–572.

Zhang, M., Gao, B., Yao, Y., Xue, Y., Inyang, M., 2012b. Synthesis of porous MgO-biochar nanocomposites for removal of phosphate and nitrate from aqueous solutions. Chem. Eng. J. 210, 26–32.

Zhang, W., Wang, L., Sun, H., 2011. Modifications of black carbons and their influence on pyrene sorption. Chemosphere 85 (0), 1306–1311.

Zhang, W., Zheng, J., Zheng, P., Qiu, R., 2015. Atrazine immobilization on sludge derived biochar and the interactive influence of coexisting Pb(II) or Cr(VI) ions. Chemosphere 134, 438–445.

Zhang, Y.C., Tang, X.D., Luo, W.S., 2014. Metal removal with two biochars made from municipal organic waste: adsorptive characterization and surface complexation modeling. Toxicol. Environ. Chem. 96 (10), 1463–1475.

Zheng, H., Wang, Z., Zhao, J., Herbert, S., Xing, B., 2013. Sorption of antibiotic sulfamethoxazole varies with biochars produced at different temperatures. Environ. Pollut. 181 (0), 60–67.

Zhu, D., Kwon, S., Pignatello, J., 2005. Adsorption of single-ring organic compounds to wood charcoals prepared under different thermochemical conditions. Environ. Sci. Technol. 39, 3990–3998.

Interactions of Biochar and Biological Degradation of Aromatic Hydrocarbons in Contaminated Soil

G. Soja

AIT Austrian Institute of Technology GmbH, Tulln, Austria

INTRODUCTION

This chapter takes a closer look at the interactions between biochar and two classes of organic xenobiotics with widespread occurrences: polycyclic aromatic hydrocarbons (PAHs) and soil pesticides, with a focus on the model substance atrazine.

Biochar Application
http://dx.doi.org/10.1016/B978-0-12-803433-0.00010-2

Contamination of soils and sediments may occur as a result of a wide range of industrial production and commercial services as well as by waste treatment and disposal. An overview of the European Energy Agency based on national estimates identified about 250,000 contaminated sites in need of remediation in Europe. This number is expected to rise by about 50% by 2025, if the current contamination trends continue. The main pollutant classes of contaminants are heavy metals (35%), mineral oils (24%), and aromatic and polycyclic aromatic hydrocarbons (21%) as summarized by the European Environment Agency (2014). For heavy metal remediation, in situ immobilization by additives and phytotechnologies or ex-situ measures are frequently used. Whereas for organic pollutants biological in situ degradation and mineralization are desirable alternatives, provided there are favorable soil conditions and contaminant bioavailability. Biochar has high sorption capacity for different types of organic compounds. Therefore the interaction of biochar, applied as a soil amendment or environmental engineering technology, with existing organic soil pollutants is a growing research area. Also since biochar may act as a pollutant sorbent and potential inhibitor of mineralization, its effects on soil microorganisms capable of degrading organic pollutants have to be considered in depth. During pyrolysis the porosity of biochar and therefore the specific surface area increase considerably. On the one hand, plant-based feedstock materials provide anatomical structures by the xylem and phloem vessels; on the other hand, the pyrolysis process supports the loss of volatile matter that may fill micropores and thus increases the available pore space by several orders of magnitude (Mukherjee et al., 2011). Since the early days of biochar research, it was hypothesized that these micropores may serve as habitats for microorganisms where they are protected from grazers and/or desiccation (Steinbeiss et al., 2009). During the last several years it has been shown how biochar creates new habitats and changes the environment for soil microorganisms, thereby leading to changes in abundance and community structure and pollutant degradation capacity of microorganisms (Gul et al., 2015) (see Chapter 1).

A large proportion of global agriculture is dependent on pesticide use. Only about 1% of global agricultural land and 5.4% of farmland in the European Union are managed organically (Willer and Lernoud, 2015). This means that most of global food production is dependent on pest or disease control measures with widespread use of chemicals to minimize crop yield losses caused by weeds, animals, or plant diseases. If pesticide use stopped suddenly without replacement measures, it is estimated that the consequence could be about a 50% loss in crop production (Rice et al., 2007). Typically, these chemicals are applied directly to soil if they are utilized for control of weeds or soil pests. The success of the treatment is dependent on optimal efficacy to balance increased application dose with higher yield losses. The persistence of the pesticides should also be considered where after having reached and affected their target organisms, rapid degradation and mineralization is required to reduce the risk of erosion losses or groundwater contamination. Therefore if biochar is an additional component of the soil system, its interactions with both the pesticide and the potential microbiological degraders are of interest. Sorption of the pesticide to biochar may be negative because of reduced availability for the catabolic organism but also positive because of decreased leaching losses.

Usually indigenous microbiological communities adapt quickly to new conditions and start to mineralize compounds that can be used as carbon and energy sources (Mitchell et al., 2015). These natural attenuation processes may be as efficient as technical measures but, in the case of aromatic hydrocarbon contamination, they need extended amounts of time. Even if degradation is started and supported by measures that enhance natural attenuation processes, frequently this requires many years or decades of pollutant monitoring and restrictions in the uses of the affected areas. During this time, recalcitrant pollutants may diffuse into soil pore water and migrate toward groundwater and/or contaminate surface water during erosion events (Revitt et al., 2015; Xiao et al., 2015). Although many pesticides are degraded much quicker than high molecular weight hydrocarbons (Hoehener and Ponsin, 2014), there is some risk of dislocation by erosion events before they reach the target organisms (Ulrich et al., 2013; Lefrancq et al., 2014). Soil additives with significant sorption capacity like biochar may therefore assume an important role in risk minimization by potential contaminants. However, additional knowledge of the interactions between biochar, the xenobiotics, and the soil ecosystem beyond the mere contaminant sorption processes is required. This will provide a better basis for assessing the risks and benefits of applying biochar to contaminated soils.

PAH DEGRADATION

The sources of PAHs cover a wide range of processes where incomplete combustion of organic material takes place. Therefore PAHs may originate both from natural sources like forest fires or volcanoes and anthropogenic sources such as industrial activities, power generation, household heating, and combustion engines (Gupta et al., 2015). PAHs are environmental contaminants of considerable concern because of their mutagenic, teratogenic, and carcinogenic potential (Souza et al., 2014; Bansal and Kim, 2015). During the biochar production process, control is of utmost importance to minimize the PAH content of the resulting biochars. De la Rosa et al. (2016) found significant differences in PAH content of biochars produced with different reactor technologies. Also feedstock material and pyrolysis temperatures may have large effects on the PAH concentrations (Kloss et al., 2012).

PAHs in the environment are subject to sorption processes to particulate organic matter in soils and sediments because of their hydrophobic and apolar (or only slightly polar) nature. However, this decreases their availability for biological uptake and microbial metabolism (Mahmoudi et al., 2013; Maurathan et al., 2015; Peng et al., 2008). Biological degradation is an important pathway for the ecological recovery of PAH-impacted sites and may lead, after several steps of hydroxylation, ring cleavage, and oxidation, to complete mineralization as CO_2. This requires optimal conditions concerning nutrient, oxygen, and water supply for the microorganisms. Bioremediation techniques are designed to provide appropriate nutrient availability and ratios by biostimulation, occasionally supplemented by bioaugmentation that consists of inoculation with selected microorganisms that have the required catabolic abilities (Gupta et al., 2015). Also PAH-degrading bacteria can be enriched in contaminated soils by the immobilized-microorganism technique (Chen et al., 2012; Zhang et al., 2013b). The authors used biochar with high sorptive capability as a microbial carrier and thus promoted dissipation of PAHs from the contaminated soil.

A wide range of microbial communities consisting of bacteria, filamentous fungi, and even algae is able to metabolize aromatic hydrocarbons (Cerniglia, 1992). Both ligninolytic and nonligninolytic fungi that metabolize PAHs have been found with some removing this class of contaminants more efficiently than bacteria (Peng et al., 2008). Among the lignin degraders, white rot fungi such as *Phanerochaete chrysosporium*, *Irpes lacteus*, and *Armillaria* sp. have been identified as efficient degraders also under nonlimiting nitrogen conditions (Novotny et al., 2009; Chen et al., 2010; Hadibarata and Kristanti, 2012). For achieving maximum PAH degradation efficacy in contaminated soils, Garcia-Delgado et al. (2015) recommended a combined strategy consisting of a sequential application of biochar followed by the fungus *Pleurotus ostreatus*.

Bacterial strains that are efficient aerobic PAH degraders occur within the Proteobacteria, Actinobacteria, and Firmicutes families (Fuentes et al., 2014). Lu et al. (2011) stressed the importance of appropriate redox conditions to achieve optimal biodegradation rates. Bacteria capable of degrading hydrocarbons are not only located in soils and sediments but also within plants. Thus Kukla et al. (2014) found several endophytic *Rhodococcus* sp. and *Microbacterium* sp. in *Lolium perenne* that had the enzymatic equipment and potential for hydrocarbon degradation.

For any microbe, the availability of the hydrocarbons is a basic requirement to metabolize them successfully. Frequently this is a bottleneck during bioremediation efforts as the majority of PAHs partition into soil organic matter. For example, Mahmoudi et al. (2013) estimated that in most soils only 0–10% of PAHs are accessible to microorganisms thereby limiting successful biological site remediation. As a result the use of surfactants has been tested and significant improvements in PAH solubility and degradation success have been observed subsequently (Rodriguez and Bishop, 2008; Hadibarata and Tachibana, 2010). As an alternative to commercial surfactants like Tween 80 or Triton X-100, there have also been studies with vegetable oil that was supposed to be degraded faster when applied to the vadose zone of contaminated sites. Mellendorf et al. (2010) and Gartler et al. (2014) described experiments that deployed vegetable oil as an in situ soil washing technology for a PAH-contaminated soil on the one hand and a means to improve the biodegradation of residual PAHs on the other hand. When the use of soil additives must be avoided, soil microbes could be screened for their own production of biosurfactants that support the formation of micelles and improve bioavailability and hydrocarbon degradation (Souza et al., 2014). However, an increase in bioavailability may also be accompanied by higher soil pore water concentrations and an accompanying risk of PAH losses to seepage water or erosion. Therefore comprehensive considerations on how to avoid adverse environmental impacts by otherwise beneficial bioavailability increases are important (Ortega-Calvo et al., 2013). Consequently, the critical importance of bioavailability for pollution assessment has also been recognized in regulatory issues. The old concept of assessing risks by measuring total pollutant concentrations is being replaced gradually by the implementation of bioavailability results into risk assessment and environmental regulations (Ortega-Calvo et al., 2015).

For soil microorganisms the exudates of plant roots are an important nutrient source that leads to increased microbial abundance and activity in the rhizosphere. Thus hydrocarbon-degrading microbes are also supported, which then respond with increased degradation and PAH removal. Wei and Pan (2010) reported the development of optimized crop combinations of rape (*Brassica campestris* L.) and white clover (*Trifolium repens* L.) or alfalfa (*Medicago sativa* L.) as an efficient phytoremediation technology. A mixed cropping of these species was more successful in the removal of PAHs than single cropping. Also, in experiments with ryegrass Liu et al. (2013a,b) observed a distinct acceleration of PAH degradation for high molecular weight compounds like benzo(a)pyrene. The authors attributed this enhancement to the observed increase of degradation-relevant enzymes in the rhizosphere that contributed more to PAH removal than direct uptake by the plants.

The analyses of the metabolic pathways involved in the degradation of PAHs have delivered many findings essential for understanding the individual steps toward mineralization to CO_2. For example, Gupta et al.

(2015) distinguished three main pathways for PAH degradation by bacteria and fungi. During bacterial degradation, an initial hydroxylation step by dioxygenase enzymes is required before dehydrogenase enzymes catalyze ring fission of the resulting catechol at the *meta*- or *ortho*-position. The lignolytic fungal degradation starts with ring oxidation by lignin and manganese peroxidase enzymes. After the formation of PAH quinones, ring fission is possible. In the case of nonlignolytic fungal and bacterial degradation, cytochrome P_{450} monooxygenase enzymes oxidize the ring structure, leading to arene oxides. Subsequently, epoxide hydrolase catalyzed reactions or nonenzymatic rearrangements lead to dihydrodiols or phenolic compounds.

Remarkably, PAHs are degradable not only by aerobic pathways but also under anaerobic conditions when aromatic substrates donate electrons for primary metabolism. Widdel and Rabbus (2001) have shown that anaerobic methanogens may produce acetate and methane as end products after ring cleavage while Gupta et al. (2015) proposed the same metabolites via initial hydroxylation, followed by reduction and the β-oxidation pathway. Cometabolism may also play an important role for PAH degradation where the molecules are metabolized without serving as growth substrates but the bacteria grow in parallel on a compatible carbon energy source. Cometabolic degradation of PAHs is more efficient in nutrient-rich media and mixed PAH environments, with a rapid removal of metabolites (Gupta et al., 2015; Zhong et al., 2007).

Sorption of PAHs to Biochar

The high sorption capacity of biochar for organic pollutants is an attractive property to be used for in situ remediation of contaminated sites (Cornelissen et al., 2005). However, the physical and chemical sorbent characteristics are not the only determinants of the usability of biochar for soil or sediment remediation. Other organic soil constituents have to be considered as they may interact both with biochar and the contaminant of concern. Hale et al. (2015) reported that other nonpyrogenic carbon compounds in soil usually have a much lower sorption strength than biochar by a median factor of 17. This means that a gradual mass transfer of the contaminants from other organic soil fractions to biochar can be realistically expected. However, the sorption strength of biochar might become gradually weaker if natural organic matter or free lipids block the pores. Also Zhang et al. (2015b) observed a decrease of the adsorption capacity of biochar for phenanthrene when the soil organic carbon content increased. They explained this behavior with a blockage of the pores by dissolved organic matter. Although this can reduce the biochar sorption capacity by a factor of 5–10 (Hale et al., 2015), the effect may be reduced over time when the pollutants diffuse through the blocking deposits and reach the sorption sites (Hale et al., 2012a). Native pyrogenic carbon compounds like charcoal from forest fires or soot also have high sorption capacity. Hence they, too, will sorb organic contaminants more strongly than non-pyrogenic organic matter. However, the main purpose of decreasing the soil pore water concentrations of the contaminant of concern will not be affected by this competition. Instead the efficiency from biochar addition will decrease with probable effects on the cost effectiveness of the measure—in a soil with lower contents of native pyrogenic carbon the same amount of biochar will contribute more to the remediation success of the respective site.

Three mechanisms have been recognized to explain the sorption interactions between PAHs and biochar. First is the interaction of the π-electrons from the PAH aromatic rings and those of biochar surfaces (Gomez-Eyles et al., 2013). Second, biochar nanopores are important sorption sites for PAHs (Anyika et al., 2015; Cornelissen et al., 2005). Thus Han et al. (2015) observed that in different natural and engineered organic sorbents, including biochar, aromatic structures and nanopores contributed most to the sorption of PAHs. Third, and apart from surface adsorption, phase partitioning may play a role with a typically more linear dependence on aqueous concentration (Allen-King et al., 2002).

Considering that the basic structure of the lignin component of plant biomass used as feedstock for pyrolysis is essentially the same in different plant materials (Petridis et al., 2011; Gierlinger, 2014), one would expect similar aromatic chemical structures at the surface of different biochars. However, Sun et al. (2012) reported that differences in highest treatment temperature, heating rate, treatment time, and particle size may result in biochars with diverse chemical and physical structures, leading to different reactivity and sorption behavior in spite of the same feedstock. Although the majority of mesopores start to develop at about 450°C and micropores at about 550°C, feedstocks may differ in their specific modification intensity (Pituello et al., 2015). Porosity and specific surface areas can also be modified further by postpyrolysis treatment of the biochars. For example, Yakout et al. (2015) compared different chemical activations of rice straw biochar and found that KOH and H_2O_2 produced the highest surface areas and a more homogeneous micropore distribution than acid treatment. Furthermore, steam activation of biochar similar to activated carbon production may produce sorbents with high sorption capacities both for organic (Klasson et al., 2011; Rajapaksha et al., 2015) and inorganic (Niandou et al., 2013; Kimura et al., 2014) pollutants. It has been confirmed that the use of activated carbons is an effective strategy to sorb organic contaminants from soils and sediments (Hale et al., 2012b; Oen et al., 2012) and thereby reduce pore water concentrations of PAHs (Oleszczuk et al., 2012). Therefore these researchers suggested that the use of regionally grown input material for pyrolysis instead of coal or imported coconut shells would be a logical step to decrease the carbon footprint of such remediation technologies.

Bioavailability of Sorbed PAHs and Other Hydrophobic Organic Contaminants

The sorption of PAHs to biochar may have both positive and undesired consequences. The advantages of concentrating and immobilizing hydrophobic organic contaminants are a main reason for considering its application in soil and water remediation projects. Immobilized pollutants do not contribute to elevated soil pore water concentrations and thus reduce the risk of groundwater or surface water contamination. Additionally, phytotoxic effects and root uptake by crop plants may decrease (Sneath et al., 2013), leading to better food quality with lower contaminant concentrations (Brennan et al., 2014). However, natural attenuation processes might be retarded if the decrease in bioavailability also means a decrease in catabolic processes by microorganisms that are expected to degrade hydrocarbons (Juhasz et al., 2014; Rostami and Juhasz, 2013; Sihag et al., 2014).

Bioavailability studies require either living organisms to investigate the uptake of the contaminants or an artificial surrogate with similar sorption behavior. The use of polyoxymethylene (POM) samplers for passive sampling of organic contaminants has been confirmed as a useful monitoring tool also for bioavailable PAHs. These samplers essentially consist of sheets of organic polymers with a thickness of 0.015–0.1 mm. They are exposed in situ in contaminated soil or water for periods of days to months or ex situ under standardized laboratory conditions (US-EPA, 2012). Hale et al. (2012a) used the analytical results of POM samplers as a measure for bioavailable PAHs in comparison with total PAH concentration. Similarly, in a comparison study of three passive sampling methods for organic pollutants, Perron et al. (2013) revealed a superior PAH collection efficiency in the POM samplers whereas for polychlorinated biphenyl (PCB) bioavailability studies, polyethylene samplers proved more appropriate. Arp et al. (2015) recommended that the thickness of the used membrane and the extraction procedure should be adjusted to the respective contaminant to get comparable results for the specific POM–water partition coefficients. POM samplers that identify the freely dissolved PAH fraction were used by Oleszczuk et al. (2014) who showed that 5% biochar can reduce the bioavailable PAHs in sewage sludge by 17–58%. Compared to total PAH concentrations in sewage sludge, the freely dissolved fraction contained a higher proportion of smaller PAHs and a lower proportion of the five to six ring components. Also polydimethylsiloxane fibers were used by Jia and Gan (2014) as alternative samplers to measure freely dissolved polybrominated diphenyl ethers in sediments.

A 1% amendment rate of biochar reduced the freely dissolved contaminant concentration by 48–78%.

The use of water extracts of contaminated soils as a measure of bioavailability was favored by Xu et al. (2012) who studied the desorption of organic pollutants from soil amended with up to 5% (w/w) biochar and reported that the water concentrations were reduced by up to 65%. In experiments by Bastos et al. (2014) the water extracts of a soil with considerably high biochar addition rates (80 t ha^{-1}) were tested for toxicity to aquatic organisms. The authors observed that the bioluminescent bacterium *Vibrio fischeri* was sensitive to the soil–biochar water extracts with an EC50 of 76% of the extract concentration after an exposure of 5 min, and an EC50 of 91% of the extract concentration after 15 min. The mobility of *Daphnia magna* was only mildly affected, with up to 25% reduction when exposed to the undiluted biochar–soil extracts but no effects emerged on the growth of the alga *Pseudokirchneriella subcapitata* (Bastos et al., 2014). Ogbonnaya et al. (2014) used a water extract with cyclodextrin to determine the extractability of an aged phenanthrene-contaminated soil with up to 10% (w/w) biochar and found significant reductions. The water–cyclodextrin extraction was also used by Beesley et al. (2010) as a measure of PAH bioavailability and they observed reductions of >50% especially for high molecular weight compounds.

The use of biological organisms and the analyses of their uptake of a contaminant is an alternative way to assess bioavailability. The measurement of pollutant concentrations in plant root tissues shows the effectivity of uptake from plant available soil pore water without bias by reduced transport efficiency to above-ground plant parts. Therefore Denyes et al. (2012) observed a reduction of PCB concentrations in pumpkin roots by 77% when 2.8% (w/w) biochar was added to a contaminated soil. Also shoot reductions were lower but still statistically significant. Josko et al. (2013) analyzed the phytotoxicity mitigation of biochar in a contaminated sediment and reported a reduction of root growth inhibition in garden cress by 18–29%. The use of sewage sludge biochar reduced the PAH concentration in cucumber by 44–57% (Waqas et al., 2014). Notwithstanding these, Luo et al. (2014) recommended caution on the use of sewage sludge that was pyrolyzed at low to medium temperatures ($\leq$500°C) as the PAH concentrations exceeded the guideline values to prevent adverse effects on soil and groundwater. In their experiments maize-derived biochar was more appropriate for application as a soil amendment. Also Brennan et al. (2014) positively reported the use of maize biochar as it improved plant growth characteristics of maize and reduced PAH uptake into the shoots.

Similar to plant roots, soil-dwelling animals are in direct contact with the contaminants in an affected soil. Therefore uptake into the animals is also a measure for the bioavailability of the pollutants and the absorbability through the digestion tract. When studying sediments, benthic organisms like the chironomid larvae are reliable indicators of ecotoxic effects. This was exemplified by Shen et al. (2012) who observed a sharp decrease in the biota–sediment accumulation factor for PAHs when the carbonaceous additive was added up to 1.5% (w/w) in sediments. At higher application rates, the reduction rates of the accumulation factor became lower again, apparently because of increased particle ingestion rates of the larvae. A study of blackworms (*Lumbriculus variegatus*) in sediments that had been amended with 8% (w/w) magnetic-activated carbon or biochar for PAH remediation revealed a continuing ecotoxic effect even when 77% of the magnetic sorbents had been removed and aqueous PAH concentrations were decreased by 98% (Han et al., 2015).

Earthworms are effective bioindicators for ecotoxic effects in agricultural soils. Tammeorg et al. (2014) did not observe earthworm avoidance reactions in soil amended with 1.6% (w/w) biochar and found the highest earthworm densities and biomasses in a field soil where 30 t ha^{-1} had been added. When Gomez-Eyles et al. (2011) tested a PAH-contaminated soil, the bioavailable (cyclodextrin-extractable) PAH concentrations decreased both in soil and in earthworms as a result of 10% addition of biochar. The earthworms themselves increased the PAH bioavailability apparently because of digestion processes of the ingested soil. Also biochar amendments reduced the uptake of hexachlorobenzene by earthworms in studies by Song et al. (2012), even when the application rate was as low as 0.1% (see also Chapter 11).

The measurement of pollutant mineralization rates as an additional potential strategy to assess the bioavailability of pollutants because the breakdown of larger molecules to end products such as CO_2 is a main objective of any bioremediation strategy. In a comparison of different organic amendments on the desorption and mineralization of phenanthrene in different soils, Marchal et al. (2013a) observed reductions of both desorption and mineralization by biochar amendments. The authors suggested that desorption from the amended soil rather than impairments of bacterial activities were the limiting factor for mineralization. Based on experiments with ^{14}C-labeled phenanthrene, Marchal et al. (2013b) reported about 56% mineralization of the PAHs sorbed by different carbonized amendments. The desorbed contaminants did not limit directly their biodegradation but rather the rate of desorption. Similar results have been observed in a comparable experimental design by Ogbonnaya et al. (2014) although they used a different bacterial inoculum. The extent of mineralization was correlated significantly with the desorption rate as analyzed by the aqueous cyclodextrin extractions. Also the results by Quilliam et al. (2013) hinted at the importance of the increased sorption and reduced bioavailability that may limit catabolic processes and the mineralization of spiked PAHs. The results from these authors might also have been influenced by their use of biochars with inherently high PAH concentrations (up to $64\,mg\,kg^{-1}$) that could have affected microbial activity and therefore would not be deployed for soil remediation. When biochar was used as 15–30% component of a substrate mixture together with biogas digestates, Worzyk et al. (2014) found excellent degradation rates of aged mineral oil hydrocarbons and indicated that PAHs were predominantly sorbed to the biochar-containing substrate. Also reductions in plant uptake rates and soil leachate concentrations amounted to >50%. These results showed that the parallel supply of nutrients in the biochar substrate may improve the biodegrading activity of microorganisms especially for smaller and more labile hydrocarbons. More persistent molecules may be prevented from entering transfer pathways to food or groundwater, even if the substrate addition implies a high amount of dissolved organic carbon in the amended soils.

BIOCHAR AND ATRAZINE INTERACTIONS

Thousands of different organic chemicals may become soil or water pollutants if unintentionally or accidentally released into the environment. The case of pesticides is slightly different, however, where about 1100 different substances have been used globally and intentionally for the control of agricultural pests. Their typically long persistence times and exceedingly high application rates, combined with floods or erosion events, could make each pesticide a soil or water pollutant. Atrazine, a herbicide of the triazine class, is a prominent pesticide representative. Although it has been banned in the European Union since 2004 it continues to be used in other parts of the world and may still reach high concentrations in the groundwater of regions where it has long since been banned (Vonberg et al., 2014).

Studies on the interactions of biochar and atrazine have focused on the desired and undesired effects of atrazine sorption to biochar. In regions where it is still used, the reduction of herbicidal efficacy because of the high affinity of biochar for atrazine (and other herbicides) might be an incentive to use higher application rates or to switch to postemergence herbicides (Soni et al., 2015). The authors stated this as they observed a reduction of weed control by 75% following biochar application rates of $5\,t\,ha^{-1}$. Graber and Kookana (2015) warned that even higher application doses of soil herbicides might not offset the reduction in herbicide efficacy if soils were amended with biochar of high surface areas.

According to results from Hao et al. (2013), atrazine sorption is promoted by a high degree of aromatic condensation and hydrophobic structures. According to Zhang et al. (2013a), the π-electron interactions of biochar surfaces and atrazine also play an important role in its adsorption. If biochar was de-ashed, equivalent to a reduction in the inorganic moieties that could block organic sorption sites, the atrazine adsorption increased significantly. This could also explain the findings of Martin et al. (2012) who observed a decrease in herbicide sorption to aged biochars because organomineral complexes may form on biochar surfaces after prolonged contact with soil (Kumari et al., 2014), especially in soils with a high clay content.

Its occurrence in soil in regions where atrazine is banned turns it into a contaminant. Cao et al. (2009) found very little competition between the two contaminants for sorption to biochar in a soil that was impacted by both atrazine and Pb because of different sorption mechanisms. In contrast to activated carbon, which sorbed Pb poorly, biochar was effective in immobilizing both Pb and atrazine (Cao et al., 2011). However, if redox-sensitive heavy metals such as Cr(VI) co-occur with atrazine in soil, the herbicide adsorption may be suppressed (Zhang et al., 2015a). Apparently in this case atrazine is sorbed by weak H-bonds to the biochar surfaces and more prone to disturbances by the hydroxylic complexes of biochar with the heavy metal. Nevertheless, these findings were limited to sludge-derived biochar that may behave differently from more carbon-rich plant-based biochars. If soil contains high amounts of dissolved organic carbon, or is heterogeneously structured to include preferential flow paths, the sorption to biochar may decrease with a concomitant reduction in the leaching of atrazine (Delwiche et al., 2014). Also a low soil pH usually favors atrazine sorption to biochar (Deng et al., 2014; Zheng et al., 2010) and would therefore need to be monitored closely. Furthermore, pretreatment of the pyrolysis feedstock may support the production of biochar with increased specific surface area where high sorption capacity for atrazine is the main reason for biochar addition to soil. Thus Zhao et al. (2013) used ammonium dihydrogen phosphate to enhance the biochar sorption capacity for atrazine.

Jablonowski et al. (2013) reported that the mineralization of atrazine was not negatively affected in spite of its sorption to biochar. The authors concluded that the extent of mineralization was influenced more by the soil characteristics and an atrazine-adapted soil microbiome than by the addition of biochar. Although microorganisms were able to access atrazine for degradation, experiments by Cao et al. (2011) and Wang et al. (2014) reported that molecule uptake by earthworms was reduced. This suggested that the gut processes in earthworms are less effective in extracting atrazine from biochar than the accessibility for soil microorganisms.

CONCLUSIONS, OUTLOOK, AND FUTURE RESEARCH NEEDS

The growing global awareness of soil as a nonrenewable resource that deserves protection has highlighted the requirements for the detection, mitigation, and remediation of existing soil contaminations. Among the multitude of uses for biochar, its application in the remediation of contaminated sites is particularly based on the high sorption capacity of biochar as a key property. Because of the importance of soils for global food production, both the pollutant heritage from former industrial activities and the ongoing soil contamination by agricultural pesticide use may pose a risk to the integrity of groundwater and food quality. In general, combining the strengths that biochar and bioremediation technologies offer to mitigate and remediate contaminants in soil can contribute to cost-efficient and environmentally friendly solutions for problems that have arisen from soil pollution.

In the rectangle between soil characteristics, contaminant properties, soil microbiology, and biochar all mutual relations and interactions should be considered to optimize remediation success (Fig. 10.1). This requires detailed mechanistic and biological understanding of the underlying processes that take place in the soil both before and after the initiation of remediation efforts. As a result the relationships between three of these components were considered by Maurathan et al. (2015) with specific focus on soil microbial communities, naphthalene as a typical pollutant, and biochar.

Currently, our knowledge is based on many different case studies and phenomenological descriptions but we are still far from a comprehensive understanding of the interactions of factors affecting the remediation

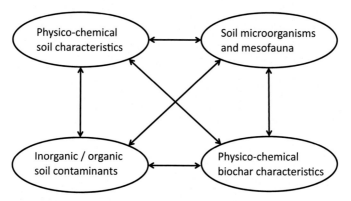

FIGURE 10.1 Soil as a habitat and microcosm for four main players whose interactions should be elucidated to understand the fate of pollutants in soil.

success. We might get closer to this long-term objective by focusing future research on the key knowledge gaps including the following:

- The physical and chemical characteristics of biochar are largely influenced by the type of input or feedstock material and the processing parameters during pyrolysis. However, we lack a detailed understanding of how biomass and waste materials from different sources used as feedstock for carbonization react to variations in pyrolysis temperature and duration with regard to the sorption and desorption characteristics of contaminants (see also Chapter 2).
- Among the multitude of different reactor types there are also low-cost production possibilities for biochar based on traditional kiln techniques. Better knowledge of the biochar characteristics originating from such alternative reactors in comparison with products from standardized reactor types would support the applicability of biochar as a soil remediation tool in less developed countries without the risk of inadvertent use of low-quality biochar.
- The activation of biochar might increase its efficacy greatly for some contaminants of concern. Better understanding of the soil and pollutant characteristics, where an activation of biochar is a relevant and justifiable alternative to activated carbon, will make it easier to decide if the activation efforts (and which types of activation) will be economically sensible and viable. For activated biochars, all questions concerning the interactions with soil microbiology are as relevant as for unactivated biochar.
- The particle size of biochar as soil amendment might be as important for the efficacy of its application as the technology to distribute it in and mix it with the soil to be remediated.
- A better understanding of the interactions of biochar with soil organic matter, clay minerals, organomineral complex formation, and preexisting biota will help to better understand problems such as: binding mechanisms and bond strengths; the persistence of sorbed contaminants; and the biological activity that determines the biological degradation efficiency (see also Chapter 9).
- The production of composite materials consisting of biochar with mineral or nanomaterial fractions might produce biochars with special binding characteristics for specific contaminants.
- Based on these knowledge bases, it will be easier to assess if remediation success will benefit from the use of biochar as a bioaugmentation tool by carrying immobilized microorganisms that underpin contaminant degradation.
- Detailed knowledge of soil characteristics, microbiological communities, and their reaction to biochar as a soil amendment will also support the assessment of whether biochar aging in soil will increase sorption capacity or reduce efficiency by biochar fouling.

- Special problems may arise if contaminated sites contain mixtures of inorganic and organic, and cationic and anionic pollutants. The potential of biochar combinations to address the specific binding requirements of different pollutant classes is currently a largely unexplored research area.
- Biochar as an environmental technology in soil remediation requires not only technical but also economic feasibility. Although site-specific conditions will always require a case-by-case assessment, current site assessments should be evaluated for comparisons of the cost efficiencies between traditional remediation methods and innovative biochar-based bioremediation strategies.
- The issue of potentially negative effects of biochar as a soil amendment on the efficacy of soil pesticides should receive special attention on how to reconcile these conflicting agricultural management measures. Economic assessments of the benefits of biochar application versus the costs of alternative preemergence herbicides might pave the way to rational and informed decisions that consider the site- and soil-specific situations.
- The topic of contaminant bioavailability should not be limited to scientific research questions when exploring biodegradation mechanisms but should also be considered critically within the legislative framework. Environmental policy governs authority decisions about remediation technologies and remediation objectives. More scientific answers to these questions will increase the confidence of authorities into the concept of bioavailability as a more reliable measure of ecotoxic effects than the reliance on total pollutant concentrations (see also Chapter 1).

References

Allen-King, R.M., Grathwohl, P., Ball, W.P., 2002. New modeling paradigms for the sorption of hydrophobic organic chemicals to heterogeneous carbonaceous matter in soils, sediments and rocks. Adv. Water Resour. 25, 985–1016.

Anyika, C., Majid, Z.A., Ibrahim, Z., Zakaria, M.P., Yahya, A., 2015. The impact of biochars on sorption and biodegradation of polycyclic aromatic hydrocarbons in soils-a review. Environ. Sci. Poll. Res. 22, 3314–3341.

Arp, H.P.H., Hale, S.E., Krusa, M.E., Cornelissen, G., Grabanski, C.B., Miller, D.J., Hawthorne, S.B., 2015. Review of polyoxymethylene passive sampling methods for quantifying freely dissolved porewater concentrations of hydrophobic organic contaminants. Environ. Toxicol. Chem. 34, 710–720.

Bansal, V., Kim, K.H., 2015. Review of PAH contamination in food products and their health hazards. Environ. Intern. 84, 26–38.

Bastos, A., Prodana, M., Abrantes, N., Keizer, J., Soares, A., Loureiro, S., 2014. Potential risk of biochar-amended soil to aquatic systems: an evaluation based on aquatic bioassays. Ecotoxicol. 23, 1784–1793.

Beesley, L., Moreno-Jimenez, E., Gomez-Eyles, J.L., 2010. Effects of biochar and greenwaste compost amendments on mobility, bioavailability and toxicity of inorganic and organic contaminants in a multi-element polluted soil. Environ. Pollut. 158, 2282–2287.

Brennan, A., Moreno Jimenez, E., Alburquerque, J.A., Knapp, C.W., Switzer, C., 2014. Effects of biochar and activated carbon amendment on maize growth and the uptake and measured availability of polycyclic aromatic hydrocarbons (PAHs) and potentially toxic elements (PTEs). Environ. Pollut. 193, 79–87.

Cao, X.D., Ma, L.N., Gao, B., Harris, W., 2009. Dairy-Manure derived biochar effectively sorbs lead and atrazine. Environ. Sci. Technol. 43, 3285–3291.

Cao, X.D., Ma, L.N., Liang, Y., Gao, B., Harris, W., 2011. Simultaneous immobilization of lead and atrazine in contaminated soils using dairy-manure biochar. Environ. Sci. Technol. 45, 4884–4889.

Cerniglia, C.E., 1992. Biodegradation of polycyclic aromatic hydrocarbons. Biodegradation 3, 351–368.

Chen, B.L., Wang, Y.S., Hu, D.F., 2010. Biosorption and biodegradation of polycyclic aromatic hydrocarbons in aqueous solutions by a consortium of white-rot fungi. J. Hazard. Mat. 179, 845–851.

Chen, B.L., Yuan, M.X., Qian, L.B., 2012. Enhanced bioremediation of PAH-contaminated soil by immobilized bacteria with plant residue and biochar as carriers. J. Soils Sediments 12, 1350–1359.

Cornelissen, G., Gustafsson, O., Bucheli, T.D., Jonker, M.T.O., Koelmans, A.A., Van Noort, P.C.M., 2005. Extensive sorption of organic compounds to black carbon, coal, and kerogen in sediments and soils: mechanisms and consequences for distribution, bioaccumulation, and biodegradation. Environ. Sci. Technol. 39, 6881–6895.

De la Rosa, J.M., Paneque, M., Hilber, I., Blum, F., Knicker, H., Bucheli, T., 2016. Assessment of polycyclic aromatic hydrocarbons in biochar and biochar-amended agricultural soil from Southern Spain. J. Soils Sediments 16, 557–565.

Delwiche, K.B., Lehmann, J., Walter, M.T., 2014. Atrazine leaching from biochar-amended soils. Chemosphere 95, 346–352.

Deng, H., Yu, H.M., Chen, M., Ge, C.J., 2014. Sorption of atrazine in tropical soil by biochar prepared from cassava waste. Bioresource 9, 6627–6643.

Denyes, M.J., Langlois, V.S., Rutter, A., Zeeb, B.A., 2012. The use of biochar to reduce soil PCB bioavailability to *Cucurbita pepo* and *Eisenia fetida*. Sci. Total Environ. 437, 76–82.

European Environment Agency, 2014. Progress in Management of Contaminated Sites. EEA, Copenhagen.

Fuentes, S., Mendez, V., Aguila, P., Seeger, M., 2014. Bioremediation of petroleum hydrocarbons: catabolic genes, microbial communities, and applications. Appl. Microbiol. Biotechnol. 98, 4781–4794.

Garcia-Delgado, C., Alfaro-Barta, I., Eymar, E., 2015. Combination of biochar amendment and mycoremediation for polycyclic aromatic hydrocarbons immobilization and biodegradation in creosote-contaminated soil. J. Hazard. Mater. 285, 259–266.

Gartler, J., Wimmer, B., Soja, G., Reichenauer, T.G., 2014. Effects of rapeseed oil on the rhizodegradation of polyaromatic hydrocarbons in contaminated soil. Int. J. Phytorem. 16, 671–683.

Gierlinger, N., 2014. Revealing changes in molecular composition of plant cell walls on the micron-level by Raman mapping and vertex component analysis (VCA). Front. Plant Sci. 5, 306.

Gomez-Eyles, J.L., Beesley, L., Moreno-Jimenez, E., Ghosh, U., Sizmur, T., 2013. The potential of biochar amendments to remediate contaminated soils. In: Ladygina, N., Rineau, F. (Eds.), Biochar and Soil Biota. CRC Press, Boca Raton, London, New York, pp. 100–133.

Gomez-Eyles, J.L., Sizmur, T., Collins, C.D., Hodson, M.E., 2011. Effects of biochar and the earthworm *Eisenia fetida* on the bioavailability of polycyclic aromatic hydrocarbons and potentially toxic elements. Environ. Pollut. 159, 616–622.

Graber, E.R., Kookana, S., 2015. Biochar and retention/efficacy of pesticides. In: Lehmann, J., Joseph, S. (Eds.), Biochar for Environmental Management. Earthscan/Routledge, London, New York, pp. 655–678.

Gul, S., Whalen, J.K., Thomas, B.W., Sachdeva, V., Deng, H.Y., 2015. Physico-chemical properties and microbial responses in biochar-amended soils: mechanisms and future directions. Agric. Ecosyst. Environ. 206, 46–59.

Gupta, S., Pathak, B., Fulekar, M.H., 2015. Molecular approaches for biodegradation of polycyclic aromatic hydrocarbon compounds: a review. Rev. Environ. Sci. Bio-Technol. 14, 241–269.

Hadibarata, T., Tachibana, S., 2010. Characterization of phenanthrene degradation by strain *Polyporus* sp. S133. J. Environ. Sci.-China. 22, 142–149.

Hadibarata, T., Kristanti, R.A., 2012. Fate and cometabolic degradation of benzo[a]pyrene by white-rot fungus *Armillaria* sp. F022. Biores. Technol. 107, 314–318.

Hale, S.E., Cornelissen, G., Werner, D., 2015. Sorption and remediation of organic compounds in soils and sediments by (activated) biochar. In: Lehmann, J., Joseph, S. (Eds.), Biochar for Environmental Management. Earthscan/Routledge, London, New York, pp. 625–654.

Hale, S.E., Lehmann, J., Rutherford, D., Zimmerman, A.R., Bachmann, R.T., Shitumbanuma, V., O'Toole, A., Sundqvist, K.L., Arp, H.P.H., Cornelissen, G., 2012a. Quantifying the total and bioavailable polycyclic aromatic hydrocarbons and dioxins in biochars. Environ. Sci. Technol. 46, 2830–2838.

Hale, S.E., Elmquist, M., Brandli, R., Hartnik, T., Jakob, L., Henriksen, T., Werner, D., Cornelissen, G., 2012b. Activated carbon amendment to sequester PAHs in contaminated soil: a lysimeter field trial. Chemosphere 87, 177–184.

Han, Z., Sani, B., Akkanen, J., Abel, S., Nybom, I., Karapanagioti, H.K., Werner, D., 2015. A critical evaluation of magnetic activated carbon's potential for the remediation of sediment impacted by polycyclic aromatic hydrocarbons. J. Hazard. Mater. 286, 41–47.

Hao, F.H., Zhao, X.C., Ouyang, W., Lin, C.Y., Chen, S.Y., Shan, Y.S., Lai, X.H., 2013. Molecular structure of corncob-derived biochars and the mechanism of atrazine sorption. Agron. J. 105, 773–782.

Hoehener, P., Ponsin, V., 2014. In situ vadose zone bioremediation. Curr. Opin. Biotechnol. 27, 1–7.

Jablonowski, N.D., Borchard, N., Zajkoska, P., Fernandez-Bayo, J.D., Martinazzo, R., Berns, A.E., Burauel, P., 2013. Biochar-mediated [C-14]atrazine mineralization in atrazine-adapted soils from Belgium and Brazil. J. Agric. Food Chem. 61, 512–516.

Jia, F., Gan, J., 2014. Comparing black carbon types in sequestering polybrominated diphenyl ethers (PBDEs) in sediments. Environ. Pollut. 184, 131–137.

Josko, I., Oleszczuk, P., Pranagal, J., Lehmann, J., Xing, B., Cornelissen, G., 2013. Effect of biochars, activated carbon and multiwalled carbon nanotubes on phytotoxicity of sediment contaminated by inorganic and organic pollutants. Ecol. Eng. 60, 50–59.

Juhasz, A.L., Aleer, S., Adetutu, E.M., 2014. Predicting PAH bioremediation efficacy using bioaccessibility assessment tools: validation of PAH biodegradation-bioaccessibility correlations. Intern. Biodeterior. Biodegrad. 95, 320–329.

Kimura, K., Hachinohe, M., Klasson, K., Hamamatsu, S., Hagiwara, S., Todoriki, S., Kawamoto, S., 2014. Removal of radioactive cesium (Cs-134 plus Cs-137) from low-level contaminated water by charcoal and broiler litter biochar. Food Sci. Technol. Res. 20, 1183–1189.

Klasson, K.T., Uchimiya, M., Lima, I.M., Boihem, L.L., 2011. Feasibility of removing furfurals from sugar solutions using activated biochars made from agricultural residues. Bioresources 6, 3242–3251.

Kloss, S., Zehetner, F., Dellantonio, A., Hamid, R., Ottner, F., Liedtke, V., Schwanninger, M., Gerzabek, M.H., Soja, G., 2012. Characterization of slow pyrolysis biochars: effects of feedstocks and pyrolysis temperature on biochar properties. J. Environ. Qual. 41, 990–1000.

Kukla, M., Plociniczak, T., Piotrowska-Seget, Z., 2014. Diversity of endophytic bacteria in *Lolium perenne* and their potential to degrade petroleum hydrocarbons and promote plant growth. Chemosphere 117, 40–46.

Kumari, K.G.I.D., Moldrup, P., Paradelo, M., de Jonge, L.W., 2014. Phenanthrene sorption on biochar-amended soils: application rate, aging, and physicochemical properties of soil. Water Air Soil Pollut. 225, 2105.

Lefrancq, M., Payraudeau, S., Verdu, A.J.G., Maillard, E., Millet, M., Imfeld, G., 2014. Fungicides transport in runoff from vineyard plot and catchment: contribution of non-target areas. Environ. Sci. Pollut. Res. 21, 4871–4882.

Liu, S.L., Cao, Z.H., Liu, H.E., 2013a. Effect of ryegrass (*Lolium multiflorum* L.) growth on degradation of phenanthrene and enzyme activity in soil. Plant Soil Environ. 59, 247–253.

Liu, S.L., Luo, Y.M., Cao, Z.H., Wu, L.H., Wong, M.H., 2013b. Effect of ryegrass (*Lolium multiflorum* L.) growth on degradation of benzo[a]pyrene and enzyme activity in soil. J. Food Agric. Environ. 11, 1006–1011.

Lu, X.Y., Zhang, T., Fang, H.H.P., 2011. Bacteria-mediated PAH degradation in soil and sediment. Appl. Microbiol. Biotechnol. 89, 1357–1371.

Luo, F., Song, J., Xia, W., Dong, M., Chen, M., Soudek, P., 2014. Characterization of contaminants and evaluation of the suitability for land application of maize and sludge biochars. Environ. Sci. Pollut. Res. 21, 8707–8717.

Mahmoudi, N., Slater, G.F., Juhasz, A.L., 2013. Assessing limitations for PAH biodegradation in long-term contaminated soils using bioaccessibility assays. Water Air Soil Pollut. 224, 1411.

Marchal, G., Smith, K.E., Rein, A., Winding, A., de Jonge, L.W., Trapp, S., Karlson, U.G., 2013a. Impact of activated carbon, biochar and compost on the desorption and mineralization of phenanthrene in soil. Environ. Pollut. 181, 200–210.

Marchal, G., Smith, K.E., Rein, A., Winding, A., Trapp, S., Karlson, U.G., 2013b. Comparing the desorption and biodegradation of low concentrations of phenanthrene sorbed to activated carbon, biochar and compost. Chemosphere 90, 1767–1778.

Martin, S.M., Kookana, R.S., Van Zwieten, L., Krull, E., 2012. Marked changes in herbicide sorption-desorption upon ageing of biochars in soil. J. Hazard. Mater. 231, 70–78.

Maurathan, N., Orr, C.H., Ralebitso-Senior, T.K., 2015. Biochar adsorption properties and the impact on naphthalene as a model environmental contaminant and microbial community dynamics – a triangular perspective. In: Borja, M.E.L. (Ed.), Soil Management: Technical Systems, Practices and Ecological Implications. Nova Publishers, New York, pp. 83–122.

Mellendorf, M., Soja, G., Gerzabek, M.H., Watzinger, A., 2010. Soil microbial community dynamics and phenanthrene degradation as affected by rape oil application. Appl. Soil Ecol. 46, 329–334.

Mitchell, P.J., Simpson, A.J., Soong, R., Simpson, M.J., 2015. Shifts in microbial community and water-extractable organic matter composition with biochar amendment in a temperate forest soil. Soil Biol. Biochem. 81, 244–254.

Mukherjee, A., Zimmermann, A.R., Harris, W., 2011. Surface chemistry variations among a series of laboratory-produced biochars. Geoderma 163, 247–255.

Niandou, M.A., Novak, J.M., Bansode, R.R., Yu, J., Rehrah, D., Ahmedna, M., 2013. Selection of pecan shell-based activated carbons for removal of organic and inorganic impurities from water. J. Environ. Qual. 42, 902–911.

Novotny, C., Cajthaml, T., Svobodova, K., Susla, M., Sasek, V., 2009. Irpex lacteus, a white-rot fungus with biotechnological potential - review. Folia Microbiol 54, 375–390.

Oen, A.M.P., Beckingham, B., Ghosh, U., Krusa, M.E., Luthy, R.G., Hartnik, T., Henriksen, T., Cornelissen, G., 2012. Sorption of organic compounds to fresh and field-aged activated carbons in soils and sediments. Environ. Sci. Technol. 46, 810–817.

Ogbonnaya, O., Adebisi, O., Semple, K., 2014. The impact of biochar on the bioaccessibility of C-14-phenanthrene in aged soil. Environ. Sci. Process. Impacts 16, 2635–2643.

Oleszczuk, P., Hale, S.E., Lehmann, J., Cornelissen, G., 2012. Activated carbon and biochar amendments decrease pore-water concentrations of polycyclic aromatic hydrocarbons (PAHs) in sewage sludge. Bioresour. Technol. 111, 84–91.

Oleszczuk, P., Zielinska, A., Cornelissen, G., 2014. Stabilization of sewage sludge by different biochars towards reducing freely dissolved polycyclic aromatic hydrocarbons (PAHs) content. Bioresour. Technol. 156, 139–145.

Ortega-Calvo, J., Tejeda-Agredano, M., Jimenez-Sanchez, C., Congiu, E., Sungthong, R., Niqui-Arroyo, J., Cantos, M., 2013. Is it possible to increase bioavailability but not environmental risk of PAHs in bioremediation? J. Hazard. Mater. 261, 733–745.

Ortega-Calvo, J.J., Harmsen, J., Parsons, J.R., Semple, K.T., Aitken, M.D., Ajao, C., Eadsforth, C., Galay-Burgos, M., Naidu, R., Oliver, R., Peijnenburg, W.J., Roembke, J., Streck, G., Versonnen, B., 2015. From bioavailability science to regulation of organic chemicals. Environ. Sci. Technol. 49, 10255–10264.

Peng, R.H., Xiong, A.S., Xue, Y., Fu, X.Y., Gao, F., Zhao, W., Tian, Y.S., Yao, Q.H., 2008. Microbial biodegradation of polyaromatic hydrocarbons. FEMS Microbiol. Rev. 32, 927–955.

Perron, M.M., Burgess, R.M., Suuberg, E.M., Cantwell, M.G., Pennell, K.G., 2013. Performance of passive samplers for monitoring estuarine water column concentrations: 1. Contaminants of concern. Environ. Toxicol. Chem. 32, 2182–2189.

Petridis, L., Pingali, S.V., Urban, V., Heller, W.T., O'Neil, H.M., Foston, M., Ragauskas, A., Smith, J.C., 2011. Self-similar multiscale structure of lignin revealed by neutron scattering and molecular dynamics simulation. Phys. Rev. E 83, 1911.

Pituello, C., Francioso, O., Simonetti, G., Pisi, A., Torreggiani, A., Berti, A., Morari, F., 2015. Characterization of chemical-physical, structural and morphological properties of biochars from biowastes produced at different temperatures. J. Soils. Sediments 15, 792–804.

Quilliam, R.S., Rangecroft, S., Emmett, B.A., Deluca, T.H., Jones, D.L., 2013. Is biochar a source or sink for polycyclic aromatic hydrocarbon (PAH) compounds in agricultural soils? Glob. Change Biol. Bioenergy 5, 96–103.

Rajapaksha, A.U., Vithanage, M., Ahmad, M., Seo, D.C., Cho, J.S., Lee, S.E., Lee, S.S., Ok, Y.S., 2015. Enhanced sulfamethazine removal by steam-activated invasive plant-derived biochar. J. Hazard. Mater. 290, 43–50.

Revitt, D.M., Balogh, T., Jones, H., 2015. Sorption behaviours and transport potentials for selected pharmaceuticals and triclosan in two sterilised soils. J. Soils Sediments 15, 594–606.

Rice, P.J., Arthur, E.L., Barefoot, A.C., 2007. Advances in pesticide environmental fate and exposure assessments. J. Agric. Food Chem. 55, 5367–5376.

Rodriguez, S., Bishop, P.L., 2008. Enhancing the biodegradation of polycyclic aromatic hydrocarbons: effects of nonionic surfactant addition on biofilm function and structure. J. Environ. Eng.-ASCE 134, 505–512.

Rostami, I., Juhasz, A.L., 2013. Bioaccessibility-based predictions for estimating PAH biodegradation efficacy - comparison of model predictions and measured endpoints. Intern. Biodeterior. Biodegrad. 85, 323–330.

Shen, M., Xia, X., Wang, F., Zhang, P., Zhao, X., 2012. Influences of multiwalled carbon nanotubes and plant residue chars on bioaccumulation of polycyclic aromatic hydrocarbons by *Chironomus plumosus* larvae in sediment. Environ. Toxicol. Chem. 31, 202–209.

Sihag, S., Pathak, H., Jaroli, D.P., 2014. Factors affecting the rate of biodegradation of polyaromatic hydrocarbons. Int. J. Pure Appl. Biosci. 2, 185–202.

Sneath, H.E., Hutchings, T.R., De Leij, F.A.A.M., 2013. Assessment of biochar and iron filing amendments for the remediation of a metal, arsenic and phenanthrene co-contaminated spoil. Environ. Poll. 178, 361–366.

Song, Y., Wang, F., Bian, Y., Kengara, F.O., Jia, M., Xie, Z., Jiang, X., 2012. Bioavailability assessment of hexachlorobenzene in soil as affected by wheat straw biochar. J. Hazard Mater. 217, 391–397.

Soni, N., Leon, R.G., Erickson, J.E., Ferrell, J.A., Silveira, M.L., 2015. Biochar decreases atrazine and pendimethalin preemergence herbicidal activity. Weed Technol. 29, 359–366.

Souza, E.C., Vessoni-Penna, T.C., Oliveira, R.P.D., 2014. Biosurfactant-enhanced hydrocarbon bioremediation: an overview. Intern. Biodeterior. Biodegrad. 89, 88–94.

Steinbeiss, S., Gleixner, G., Antonietti, M., 2009. Effect of biochar amendment on soil carbon balance and soil microbial activity. Soil Biol. Biochem. 41, 1301–1310.

Sun, H., Hockaday, W.C., Masiello, C.A., Zygourakis, K., 2012. Multiple controls on the chemical and physical structure of biochars. Ind. Eng. Chem. Res. 51, 3587–3597.

Tammeorg, P., Parviainen, T., Nuutinen, V., Simojoki, A., Vaara, E., Helenius, J., 2014. Effects of biochar on earthworms in arable soil: avoidance test and field trial in boreal loamy sand. Agric. Ecosyst. Environ. 191, 150–157.

Ulrich, U., Dietrich, A., Fohrer, N., 2013. Herbicide transport via surface runoff during intermittent artificial rainfall: a laboratory plot scale study. Catena 101, 38–49.

US-EPA, 2012. Guidelines for Using Passive Samplers to Monitor Organic Contaminants at Superfund Sediment Sites. OSWER Directive 9200.1-110 FS. Office of Research and Development, Washington, DC, USA.

Vonberg, D., Vanderborght, J., Cremer, N., Putz, T., Herbst, M., Vereecken, H., 2014. 20 years of long-term atrazine monitoring in a shallow aquifer in western Germany. Water Res. 50, 294–306.

Wang, F., Ji, R., Jiang, Z.W., Chen, W., 2014. Species-dependent effects of biochar amendment on bioaccumulation of atrazine in earthworms. Environ. Poll. 186, 241–247.

Waqas, M., Khan, S., Qing, H., Reid, B.J., Chao, C., 2014. The effects of sewage sludge and sewage sludge biochar on PAHs and potentially toxic element bioaccumulation in *Cucumis sativa* L. Chemosphere 105, 53–61.

Wei, S.Q., Pan, S.W., 2010. Phytoremediation for soils contaminated by phenanthrene and pyrene with multiple plant species. J. Soils Sediments 10, 886–894.

Widdel, F., Rabus, R., 2001. Anaerobic biodegradation of saturated and aromatic hydrocarbons. Curr. Opin. Biotechnol. 12, 259–276.

Willer, H., Lernoud, J., 2015. The World of Organic Agriculture. Statistics and Emerging Trends 2015. FiBL & IFOAM, Frick/CH, Bonn/D.

Worzyk, F., Schatten, R., Krüger, C., Terytze, K., Vogel, I., 2014. Auswirkungen von Biokohle-Substraten und Biokohle auf Bodenparameter und Pflanzenwachstum MKW- und PAK-kontaminierter Böden. AltlastenSpektrum 101–113 3/2014.

Xiao, F., Simcik, M.F., Halbach, T.R., Gulliver, J.S., 2015. Perfluorooctane sulfonate (PFOS) and perfluorooctanoate (PFOA) in soils and groundwater of a US metropolitan area: migration and implications for human exposure. Water Res. 72, 64–74.

Xu, T., Lou, L., Luo, L., Cao, R., Duan, D., Chen, Y., 2012. Effect of bamboo biochar on pentachlorophenol leachability and bioavailability in agricultural soil. Sci. Total Environ. 414, 727–731.

Yakout, S.M., Daifullah, A.E.H., El-Reefy, S.A., 2015. Pore structure characterization of chemically modified biochar derived from rice straw. Environ. Eng. Manag. J. 14, 473–480.

Zhang, P., Sun, H.W., Yu, L., Sun, T.H., 2013a. Adsorption and catalytic hydrolysis of carbaryl and atrazine on pig manure-derived biochars: impact of structural properties of biochars. J. Hazard. Mater. 244, 217–224.

Zhang, X., Wang, H., He, L., Lu, K., Sarmah, A., Li, J., Bolan, N.S., Pei, J., Huang, H., 2013b. Using biochar for remediation of soils contaminated with heavy metals and organic pollutants. Environ. Sci. Poll. Res. 20, 8472–8483.

Zhang, W.H., Zheng, J., Zheng, P.P., Qiu, R.L., 2015a. Atrazine immobilization on sludge derived biochar and the interactive influence of coexisting Pb(II) or Cr(VI) ions. Chemosphere 134, 438–445.

Zhang, X., McGrouther, K., He, L., Huang, H., Lu, K., Wang, H., 2015b. Biochar for organic contaminant management in soil. In: Ok, Y.S., Uchimiya, S.M., Cahng, S.X., Bolan, N. (Eds.), Biochar. Production, Characterization, and Applications. CRC Press, Boca Raton, London, New York, pp. 139–166.

Zhao, X.C., Ouyang, W., Hao, F.H., Lin, C.Y., Wang, F.L., Han, S., Geng, X.J., 2013. Properties comparison of biochars from corn straw with different pretreatment and sorption behaviour of atrazine. Bioresour. Technol. 147, 338–344.

Zheng, W., Guo, M.X., Chow, T., Bennett, D.N., Rajagopalan, N., 2010. Sorption properties of greenwaste biochar for two triazine pesticides. J. Hazard. Mater. 181, 121–126.

Zhong, Y., Luan, T.G., Wang, X.W., Lan, C.Y., Tam, N.F.Y., 2007. Influence of growth medium on cometabolic degradation of polycyclic aromatic hydrocarbons by *Sphingomonas* sp. strain PheB4. Appl. Microbiol. Biotechnol. 75, 175–186.

A Critical Analysis of Meso- and Macrofauna Effects Following Biochar Supplementation

X. Domene

Autonomous University of Barcelona, Bellaterra, Barcelona, Spain;
Centre for Ecological Research and Forestry Applications (CREAF),
Bellaterra, Barcelona, Spain

OUTLINE

Biochar Application
http://dx.doi.org/10.1016/B978-0-12-803433-0.00011-4

ROLE OF FAUNA IN SOIL ECOSYSTEM STRUCTURE AND FUNCTION

Soil fauna plays a central role in three key ecological functions of soil ecosystems, specifically: (1) organic matter mineralization and dynamics; (2) support and regulation of primary production; and (3) development and maintenance of soil structure (Verhoef, 2004). Also soil fauna has been shown to continuously redistribute organic matter from surface to subsoil (Wilkinson et al., 2009) and reseed the soil profile with microorganisms and their propagules (Coleman et al., 2012; Renker et al., 2005). Similarly, it has been demonstrated that earthworms can transport seeds along the soil profile, hence influencing seed bank dynamics (Forey et al., 2011).

Soil fauna is bottom-up regulated by microorganisms and plants, since energy transferred along food webs can either flow through the detritus or the plant-based energy channels (Bardgett, 2005). The detritus channel includes microbial feeders (fungivores and bacterivores) and detritus consumers (detritivores), while the plant-based channel involves root feeders (herbivores). Finally, carnivores feed on other animal species pertaining to any of these two channels. In turn, soil fauna exerts a top-down regulation of soil microorganisms by direct grazing or consumption of organic matter colonized by these microorganisms (Bardgett, 2005). Given the inability of animals to digest plant compounds such as cellulose and lignin, the ingestion of organic matter leads to their fragmentation, transportation, and release to soil, facilitating its decomposition and redistribution in the soil profile (Coleman and Wall, 2014). By means of this top-down regulation, faunal activity regulates nutrient release and primary production and, by extension, the whole ecosystem functioning (Verhoef, 2004). However, the expression and magnitude of such regulation effect have been suggested to be specific to every soil community and ecosystem (Cragg and Bardgett, 2001; Huhta et al., 1998).

In addition, soil structure is affected significantly by soil macrofauna, which constructs burrows (biopores) and promotes soil aggregation, in turn determining water retention and aeration. Soil aggregates result from the initial formation of organomineral complexes with clays, and consolidated by organic matter fragments and the adhesive properties of polysaccharides and proteins produced by microorganisms (Hayes, 1983;

Kögel-Knabner et al., 2008; Oades, 1993). Soil macrofauna, also designated as ecosystem engineers (Coleman and Wall, 2014), can promote soil aggregation by mixing surface organic matter with mineral particles from deep horizons, in the case of earthworms (Hayes, 1983), or soil particles transportation from deeper horizons to soil surface, in the case of termites and ants (Gouveia and Pessenda, 2000), but also through the top-down regulation of microbial populations. In the specific case of geophagous earthworms, the ingestion of organic matter and mineral particles followed by grinding and mixing with digestive secretions, later excreted as casts, promote soil aggregation and explain why soils containing earthworms are generally not compacted (Hayes, 1983). Many of the current agricultural management practices impact soil fauna communities, which can translate to effects on key soil ecological processes such as nutrient or water cycles that crop production relies on (Huhta et al., 1998; Neher, 1999). This is why soil fauna, generally more sensitive than microorganisms (Verhoef, 2004), is seen as a good bioindicator group for the assessment of soil quality impacts, despite the relatively high degree of functional redundancy in soil ecosystems (Naeem and Li, 1997; Van Straalen, 2004).

FAUNA EFFECTS ON BIOCHAR: BIOTURBATION AND PERSISTENCE

Biochar can affect soil fauna, but fauna itself can contribute to the fate of biochar in soil. Ingestion of charcoal has been widely reported in earthworms under field and laboratory conditions (Elmer et al., 2015; Ponge et al., 2006; Topoliantz and Ponge, 2003, 2005). Similarly, this behavior has been recorded for collembolans and enchytraeids in the laboratory, as easily observed in unpigmented species (Domene et al., 2015a), but also in slash-and-burn agricultural soils as demonstrated by the presence of enchytraeid and ant fecal pellets containing charcoal (Topoliantz et al., 2006). Although passive consumption cannot be excluded, these observations suggest an active feeding behavior driven by a complex mechanism, since biochars are not expected to have nutritive properties. Some authors have speculated that ingestion could be related to the microorganisms or microbial metabolites present on biochar's surface (Lehmann et al., 2011; Pietikäinen et al., 2000), their detoxifying properties and liming capacity (Topoliantz and Ponge, 2003), or the use of some recalcitrant carbon fractions by gut symbionts (Marks et al., 2014). During passage through the gut, biochar is inoculated by microorganisms (Augustenborg et al., 2012) and enzymes (Paz-Ferreiro et al., 2014). Thus physical reprocessing and inoculation of biochar particles after faunal ingestion unavoidably affects its particle size, distribution in the soil profile, and plausibly its persistence.

Bioturbation

Soil macrofauna, and to a lesser extent mesofauna, can transport organic matter vertically in the soil profile by mixing it with mineral particles mostly by constructing surface mounds and subsurface burrows and through the redistribution of ingested soil, in a process known as bioturbation (Wilkinson et al., 2009). Vertical transportation of charcoal by geophagus earthworms has been observed in several studies and deduced by a darker color in earthworm burrows surface (Major et al., 2010), biochar particles located at relatively high depths (Carcaillet, 2001; Eckmeier et al., 2007; Gouveia and Pessenda, 2000), and by its presence in earthworm casts and enchytraeid and ant fecal pellets (Topoliantz and Ponge, 2003; Topoliantz et al., 2006). For example, Hart and Luckai (2014) proposed that the freshness of charcoal particles in boreal forest soils compared to other ecosystems was explained by the low fauna activity and bioturbation combined with recombustion in the next fire. Furthermore, bioturbation has been proposed to explain the deep charcoal enrichment observed in *Terra Preta* soils in the Amazon (Glaser et al., 2000), together with intense human activity shown by abundant pottery in deep horizons (Glaser and Woods, 2004), and downward physical transport (Major et al., 2010). However, bioturbation is a slow process, as reported by Eckmeier et al. (2007), who indicated that only a few charcoal particles were incorporated into the soil matrix after a year of slash-and-burn in a forest temperate soil.

Persistence

It has been demonstrated that a fraction of fresh biochar is lost in soil as a result of biotic and abiotic mineralization processes (see review by Ameloot et al., 2013). Regarding the biotic mineralization, microorganisms have a principal role, especially shortly after biochar application to soil, related initially to the consumption of easily mineralizable carbohydrates and peptidic compounds (Mitchell et al., 2015). Soil fauna could facilitate mineralization by fragmenting biochar particles, along with the reported intrinsic biochar fragmentation related to graphitic sheet expansion (Spokas et al., 2014). Biochar grinding after ingestion unavoidably changes biochar particle size and total surface, potentially accelerating the mineralization of the less persistent fractions because of enhanced accessibility to microorganisms and hence potentially decreasing biochar persistence in soil (Ameloot et al., 2013). In turn, biochar soil physicochemical properties, such as total surface, porosity, cation exchange capacity, or water retention, might also be changed significantly by this grinding. On the other hand, soil fauna might increase biochar persistence by promoting: (1) organomineral complexes formation; (2) encapsulation within aggregates (Oades, 1993; Six et al., 2002); and (3) deep burial by bioturbation processes, which might increase the persistence of biochar because of the lower microbial activity at lower soil depths (Rumpel and Kögel-Knaber, 2011).

Charcoal grinding has been reported widely in earthworms (Ponge et al., 2006; Topoliantz et al., 2006), where casts consisting of small char fragments (10–100 mm) immersed in a mineral matrix have been observed in tropical soils under slash-and-burn practices. In their muscular gizzard, earthworms grind organic matter (and biochar) and mineral particles finely, and mix them with esophagus-secreted mucus (Hayes, 1983) in a process facilitated by the ingestion of sand particles (Marhan and Scheu, 2005). Other authors have suggested that biochar particles themselves might also help to grind organic matter (Lehmann et al., 2011). Once ground and mixed with mineral particles, biochar is later excreted as casts containing highly stable organomineral complexes (Hayes, 1983). For example, Marhan and Scheu (2005) showed that organic matter in casts was less mineralizable in soils with higher clay content. Casts act as soil aggregates, clearly visible as black and gray aggregates partly or totally made of powdered charcoal (Topoliantz, 2002), positively influencing soil tropical fertility in particular, in terms of physical and chemical properties (Glaser et al., 2002; Ponge et al., 2006).

Together with this organic matter stabilization capacity, soil fauna can, however, contribute to their mobilization (Kögel-Knaber et al., 2008) by breaking up aggregates, releasing physically protected organic matter that then becomes available for digestion in fauna gut or microorganisms. This phenomenon was reported by Topoliantz et al. (2006) who observed ant pellets containing charcoal probably derived from earthworm casts consumption. Despite this remobilization capacity, the released materials can easily reassociate and new organomineral complexes are formed, leading again to organic matter protection.

Furthermore, it has been demonstrated that biochar can be inoculated by microorganisms during passage through faunal guts (Augustenborg et al., 2012), something that might have consequences for its persistence in soil. Similarly, earthworms can stimulate soil enzymatic activities via their mucus and cast production, which are rich in nutrients, and by the release of their own enzymes (Paz-Ferreiro et al., 2014), in turn possibly affecting biochar persistence.

In summary, soil fauna might influence biochar persistence with two opposing mechanisms. On the one hand, fauna might reduce biochar persistence by grinding and remobilizing physically protected biochar within aggregates, by inoculating microorganisms, or by enhancing soil enzymatic activity. On the other hand, fauna might increase biochar persistence by promoting protection within aggregates. The role of physical protection for soil organic matter turnover has been proposed to be more influential than biochemical recalcitrance (Dungait et al., 2012). This is why this mechanism should also be taken into account for the estimation of biochar residence time in soil. While the role of earthworms is relatively well known, the contribution of other faunal groups on biochar physical protection in aggregates remains less clear (Lorenz and Lal, 2014) though their key role on soil aggregation has been reviewed (Maaß et al., 2015).

BIOCHAR EFFECTS ON SOIL FAUNA

Compared to other aspects of biochar research, relatively few data are available on soil fauna (Ameloot et al., 2013; Lehmann et al., 2011), with most studies focused on earthworms (Weyers and Spokas, 2011). The initial inferences from general soil biology and forest soils under slash-and-burn practices or wildfires are currently complemented with some empirical studies. Because of the bottom-up regulation of soil fauna, shifts in the abundance or composition of fungi and bacteria communities in response to biochar could translate to changes at higher trophic levels within the soil food web. Conversely, any direct effect of biochar on soil fauna could affect other biological groups and plant production through the top-down regulation carried out by soil animals.

Proposed Direct and Indirect Mechanisms

The variety of usable feedstocks and pyrolysis procedures leads to a wide range of biochars, including materials with undesirable properties, which may reduce soil quality. As a consequence, some authors have emphasized the need for demonstrating the benefits of biochar to soil health together with the absence of detrimental effects to the environment (Verheijen et al., 2010). The positive and negative effects on soil fauna might be separated into: (1) direct effects of biochar by its characteristic substances such as usable organic matter fractions, salts, pollutants, or volatile organic compounds (VOCs); and (2) indirect effects on the soil habitat conditions, in terms of pH, water retention (Abel et al., 2013), porosity, aeration and reactive surface (Ameloot et al., 2013; McCormack et al., 2013), or albedo (Genesio and Miglietta, 2012), as well as shifts in other soil biological groups interacting with fauna and caused by such habitat condition variation (McCormack et al., 2013).

Concerning the pollutant content of biochars, several studies have reported significant concentrations of heavy metals and polycyclic aromatic hydrocarbons (PAHs) (Enders et al., 2012; Hale et al., 2013; Luo et al., 2014; Oleszczuk et al., 2013), though generally not over the threshold concentrations as identified in some of the currently available guidelines (BBF, 2013; IBI, 2013; EBC, 2012). The release of heavy metals from biochars to water leachates has been demonstrated for heavy metals and PAHs (Bastos et al., 2014; Domene et al., 2015c). PAHs, which are known carcinogenic substances, were also suggested by Chakrabati et al. (2011) to explain the suppression of a gene encoding for a tumor suppressor protein in a nematode species exposed to *Terra Preta* soils. Also drying of biochars has been shown to decrease PAH content (Koltowski and Oleszczuk, 2015). On the other hand, the water-soluble fraction has been also hypothesized to contain aromatic organic substances such as phenolic compounds (Lievens et al., 2015; Smith et al., 2012) along with carbohydrates, peptides, amino acids, and short-chain carboxylic acids (Mitchell et al., 2015). It has been demonstrated that water washing might alleviate biochar's toxicity (Buss and Mašek, 2014; Lin et al., 2012). Although Smith et al. (2012) suggested the phenolic organic species, it remains unclear which soluble molecules are responsible for the toxicity observed in these biochars. Other studies have shown phytotoxicity of VOCs released from biochar (Buss and Mašek, 2014).

Some pioneer studies have adapted soil ecotoxicology standardized tests based on soil fauna that could be useful for the prospective ecological risk assessment of biochars. As well as measuring potential positive effects, the tests could also help to understand the mechanisms involved. Prospective risk assessments include short-term screening such as avoidance tests (Domene et al., 2015a; Hale, 2013) and long-term evaluations such as survival or reproduction tests (Domene et al., 2015c; Hale, 2013; Liesch et al., 2010; Marks et al., 2014). As a result, earthworm and enchytraeid avoidance tests, together with plant germination assays, have been proposed for the characterization of biochars before their application in pilot- or field-scale trials (Major, 2009). Therefore different OECD and ISO standardized ecotoxicological tests based on soil animals could be adapted easily to biochar. Notwithstanding this, ecotoxicity testing is not included in the current guidelines for biochar characterization and quality assessment (BBF, 2013; EBC, 2012), with the exception of plant germination inhibition assays (IBI, 2013). Similarly, soil biology and ecotoxicology field methods can be used for retrospective ecological risk assessment of biochar applications in real scenarios (Domene, 2015b; Domene et al., 2014).

Despite the limited literature, neutral, positive, and negative effects of biochars on soil fauna have been reported following biochar addition.

Observations in Charcoal-Enriched Soils

Because of the absence of studies, predictions of biochar effects on fauna were initially inferred from observations of char-enriched forest topsoils under slash-and-burn or after wildfires. Despite the known medium-term negative effects of wildfires on soil fauna (Gongalski et al., 2013), other studies have reported important soil fauna activity in the char-enriched layers in the short and medium term. As an example, Bunting and Lundberg (1987) and Phillips et al. (2000) studied boreal forests affected by wildfires and reported abundant faunal fecal pellets and fungal hyphae around biochar particles in char-enriched surface horizons, with further evidence of bioturbation (unlayered distribution of char in the soil profile). Also Ponge et al. (2006), Topoliantz and Ponge (2003), and Topoliantz et al. (2006) reported abundant earthworm casts, as well as enchytraeid and ant fecal pellets containing charcoal in tropical soils under slash-and-burn practices. Uvarov (2000) cited an unpublished study reporting higher mesofauna abundance and diversity in a temperate soil polluted by historical charcoal kiln smoke emissions. In addition, Matlack (2001) demonstrated a positive correlation between charcoal content and nematode numbers in 99 temperate forest soils.

Earthworms

Earthworms have a key role on soil structure and organic matter breakdown, burial and mixing with mineral soil. Earthworms are usually classified as (1) epigeic, when living on soil surface and consuming surface litter; (2) endogeic, when living in organic horizons ingesting organomineral materials by horizontal burrowing; and (3) anecic, when feeding on surface litter but inhabiting mineral soil horizons, accounting for deep soil aeration, water infiltration, and deep incorporation of organic matter through vertical burrowing activity (Coleman and Wall, 2014).

A variety of responses have been observed of earthworm preference/avoidance for soil–biochar mixtures. These have been explained by moisture or water availability, liming, or the release of toxic compounds from biochars such as ammonia, although trophic effects are also plausible according to the observed consumption of charcoal and biochar. Although they could not explain the mechanisms involved, Chan et al. (2008) reported earthworm's preference for a poultry litter char produced at 445°C but not the 550°C equivalent produced from the same feedstock. Some authors have attributed the positive effects on earthworms to the liming properties of biochars in acid soils (McCormack et al., 2013; Topoliantz and Ponge, 2005) within certain limits. For example, Liesch et al. (2010) reported high earthworm mortality following the application of a poultry litter biochar at rates over 4.7%, which was attributed to ammonia concentration and the significant increase in pH. In contrast, no toxicity was observed in a pine chip biochar applied at similar ratios. Van Zwieten et al. (2010) observed preference for soil–biochar mixtures in an acidic soil but not in an alkaline soil supporting liming as an explanation for earthworm behavior. Busch et al. (2011) reported preference for soil–biochar mixtures but avoidance to a peanut hull residue biochar and suggested that the response could be related to the liming effect of biochar that was lacking in the hydrochar. Li et al. (2011) demonstrated that *Eisenia fetida* avoided a slow pyrolysis wood biochar at ratios of 10% (w/w) and above. Topoliantz and Ponge (2005) reported preference of the tropical earthworm *Pontoscolex corethrurus* for charcoal–soil mixtures over soil, probably because of the increased pH after charcoal addition. Amaro (2013) reported avoidance of soil mixtures with a wood biochar applied at 4tha⁻¹ that was aged for 5 months but not to the initial mixtures. This behavior was hypothesized to result from the release of contaminants not previously bioavailable immediately after biochar application. Tammeorg et al. (2014) observed avoidance of *Aporrectodea caliginosa* to a spruce biochar after 14 days of exposure, which was attributed to a slight decrease in water availability.

Potential trophic effects of biochar on earthworms are plausible according to some studies that have shown biochar ingestion and preference for soil–biochar mixtures. It is, however, not clear yet if this is active behavior driven by the nutritive value of microbial communities present on the biochar surface, its liming capacity, or just passive behavior with inherent benefits. Thus in a 2-week laboratory experiment, Topoliantz and Ponge (2003) demonstrated *P. corethrurus* preference for soil rather than soil–charcoal mixtures. This was evidenced by reduced burrowing and casting when biochar was present. The authors suggested that burrows in the soil–charcoal mixtures were being constructed mostly by pushing aside charcoal particles rather than by ingesting them, which seemed to negate the hypothesis for a nutritive value of biochar ingestion. In contrast, Elmer et al. (2015) demonstrated active ingestion of biochar in *Lumbricus terrestris* despite the availability of other added food sources, and a preference for a biochar with the highest calcium content, which the authors hypothesized to be useful for the production of the mucilaginous gel used to coat their castings. However, no relationship was found with biochar's bacterial abundance to explain this consumption pattern.

If soil earthworms consume biochar for trophic purposes, an increased growth or reproduction might be expected, although this has not always been shown in the available literature. Topoliantz and Ponge (2005) reported that, in contrast to other organic amendments, the addition of charcoal to a tropical soil did not affect *P. corethrurus* reproduction and therefore did not demonstrate a positive trophic effect in this species. Li et al. (2011) did not report any positive effects on *Eisenia andrei* reproduction, while the 10% (w/w) addition of wood biochar increased growth. Liesch et al. (2010) reported no body weight improvements in *E. fetida* exposed to 2.3–9.4% (w/w) additions of wood and poultry litter biochars, but observed high survival in the first compared to the latter, with high toxicity. Therefore, besides trophic effects, habitat preferences or modifications of gut pH might be important to explain earthworms' response to biochar (Weyers and Spokas, 2011). Thus the exact mechanisms that explain biochar consumption or avoidance by earthworms remain unclear and justify further studies.

Despite the usefulness of laboratory trials, such short-term trends might not be representative of long-term effects in the field. As an example, Tammeorg et al. (2014) reported a higher earthworm density and biomass in a temperate arable soil 4 months after the application of a spruce wood biochar for which earthworm avoidance had been observed in laboratory trials. In another field study, Weyers and Spokas (2011) did not find statistically significant effects in total earthworm abundance 2 years after the addition of different biochars.

Enchytraeids

Soil enchytraeids, or potworms, are a relatively unknown group despite their predominant role in decomposition and mineralization regulation in some soil ecosystems (Cole et al., 2000; Coleman et al., 2004; Van Vliet et al., 2004). Therefore any detrimental effects of biochar addition on this group might be of concern. Enchytraeids ingest both mineral and organic particles (Coleman and Wall, 2014) and some species in particular have been shown to ingest biochar, clearly visible through their translucent bodies (Fig. 11.1).

FIGURE 11.1 *Enchytraeus crypticus* individuals showing biochar ingestion, easily observed as a dark gut content, after 24 h of exposure to a 1% corn stover biochar mixture with quartz sand (Domene et al., 20015a).

Since they thrive in acidic soils (McCormack et al., 2013), negative effects on members of this group have been observed after pH increases caused by ashes and pyrolysis products in soils after wildfires (Liiri et al., 2007; Lundkvist, 1998; Nieminen, 2008; Nieminen and Haimi, 2010). However, relatively important enchytraeid activity, shown as fecal pellets, has been reported in charcoal-enriched soil layers after wildfires (Topoliantz et al., 2006).

There is little evidence from the few studies available of any perceived negative effects in the short term following biochar addition. Using an agricultural alkaline soil under laboratory conditions, Marks et al. (2014) reported no effects on the survival and reproduction of *Enchytraeus crypticus* from the addition of several wood and sewage sludge biochars. However, negative effects were reported for a pine gasification biochar, associated largely to the considerable pH increase caused by this material. Similarly, Domene et al. (2015a), under laboratory conditions, found no negative effects in the survival and reproduction of the same species after the addition of a corn stover biochar to an agricultural temperate soil. The narrow range of biochar concentrations was unable to raise pH over the value in unamended soil (around 7). Instead a positive effect on reproduction was observed at 0.5–2% (w/w) additions.

Nematodes

Nematodes, or roundworms, are one of the most abundant groups in any soil ecosystem, with contrasting key roles as decomposition regulators and plant pathogens (Coleman and Wall, 2014). Nematodes have representatives in most trophic levels of soil ecosystems (bacterivores, fungivores, plant feeders, predators, and omnivores) and as a result can feed on a wide range of foods. This is why their communities are often described in terms of feeding groups, a trait easily assessed by observing the mouthparts of the individuals present (Yeates and Coleman, 1982). This, together with their prevalent distribution, explains why this group has been used widely for the assessment of impacts on soil quality (Bongers and Bogers, 1998; Neher, 2001).

The impact of biochar on this group was initially inferred from charcoal-enriched soils, where an unpublished study, as cited by Uvarov (2000), reported higher abundances of soil nematodes in forest soils polluted by charcoal kiln smoke emissions containing significant concentrations of aromatic hydrocarbons and PAHs (Fischer and Bieńkowski, 1999), as well as particulate charcoal, ashes, and resin wastes. Although smoke emission products have been shown to stimulate microorganisms and decomposition in the surrounding soils (Fischer and Bieńkowski, 1999; Stenier et al., 2008), as they might contain readily utilizable substrates (Hagner et al., 2008), the connection with mesofauna abundance in this study was unclear. Also this result contrasted with the study by Chakrabati et al. (2011), who demonstrated the suppression of an anticancer gene (*cep-1*) in *Caenorhabditis elegans* exposed to *Terra Preta* soil, which might be extrapolated to suggest a reduction in the protection capacity against carcinogenic PAHs. On the other hand, a positive correlation between charcoal content and increased nematode richness in 99 forest sites was reported by Matlack (2001), where plant feeder and omnivore nematode abundances decreased with soil pH. These results do not fully match with the plausible pH increase in acid soils after biochar addition, expected to favor bacterivore and predatory nematodes and decrease fungivore nematodes, since bacteria are stimulated over fungi at alkaline pH (McCormack et al., 2013).

Field studies on the effects of biochar application on nematodes are also scarce, though they do not generally agree with the expected trends previously indicated for acid soils, probably because they have been carried out in neutral to alkaline soils. For example, Zhang et al. (2013) did not observe a change in total nematode abundance after the addition of a wheat straw biochar to a soil microcosm, although maximum diversity was reported at the $2.4\,t\,ha^{-1}$ application rate. On the other hand, the same authors reported an increased abundance of fungivore nematodes at 12 and $48\,t\,ha^{-1}$ coupled to a decrease in plant feeders. The last trend was also reported 2 years after the addition of a pine gasification biochar to an alkaline soil in mesocosms cropped with barley (Domene et al., 2015b), where increased bacterivore abundance resulted but was coupled with a reduction in plant feeder and predator abundances at the highest application rate ($30\,t\,ha^{-1}$). The mechanism behind plant feeder nematode reduction is unknown, although this trend has also been observed after the application of raw manures (Nahar et al., 2006) where it was attributed to competition with bacterial and fungal feeders favored by the addition of organic matter, or to the toxic effects of decomposition products. Thus Rahman et al. (2014) reported a reduction in plant feeding nematodes in vineyard plots 2 years after the application of organic wastes (compost and rice hull) and a poultry litter biochar to a neutral soil, with the biochar treatment having the strongest effect. Plant feeder nematode suppression has an agronomical interest, either by their direct role as pests or as vectors of a variety of plant diseases (Fry, 2012).

Collembolans

Collembolans, together with mites, are the most abundant microarthropod group of any soil ecosystem (Coleman and Wall, 2014), where they carry out a key decomposition role through decomposer population regulation by facilitating the activity of breaking up organic matter particles (Bardgett, 2005), and by promoting soil aggregation (Maaß et al., 2015).

In char-enriched soil layers resulting from wildfires, Bunting and Lundberg (1987) reported intense biological activity and bioturbation, including abundant microarthropod fecal pellets, fungal activity, and fauna reprocessing of those pellets. Uvarov (2000) cited an unpublished study that reported higher abundance and diversity of collembolans in forest soils polluted by historical charcoal kiln smoke emissions. For acid soils, McCormack et al. (2013) predicted reduced collembolan abundance coupled to the expected fungal abundance decreases after biochar addition and liming.

Under laboratory conditions the consumption of biochar by collembolans is ostensible in the collembolan *Folsomia candida* (Fig. 11.2). The potential mechanisms underlying such behavior are unknown, although trophic effects could be potentially involved, as the species may feed on fungi-colonizing biochar (Lehmann et al., 2011). Salem et al. (2013a) reported the consumption of hydrochar particles in two collembolan species, *Coecobrya tenebricosa* and *Folsomia fimetaria*, provided as sole food source, although feeding and reproduction rates were lower than when yeast was provided as food. The relevance of such observations for pyrochar, with a clearly different composition, is unknown.

In Domene et al. (2015a), a generalized avoidance behavior of the collembolan *Folsomia candida* to a corn stover biochar (0.2–14% w/w) was observed, although after its modeling, such response was shown to be attenuated when a higher microbial biomass was present in soil–biochar mixtures compared to a soil-only control, agreeing with the plausible role of microorganisms in biochar consumption. Amaro (2013) reported no avoidance of *F. candida* to field samples where a

FIGURE 11.2   *Folsomia candida* individuals showing biochar in their gut after a 24-h exposure to 1% moist mixtures of a corn stover biochar with quartz sand (Domene et al., 20015a).

wood biochar had been applied (4 and 40 t ha^{-1}), both in the initial mixtures and those collected after 5 months of application. Similarly, Conti et al. (2014) did not record any avoidance of *F. candida* to different gasification biochars below 2–5% (w/w), but reported total avoidance at the 100% (w/w) application rate.

Regarding chronic effects, Hale et al. (2013) recorded no impact on reproduction of the collembolan *F. candida* by a corn stover biochar and two activated charcoals applied at 0.5, 2, and 5% (w/w). For a variety of slow pyrolysis biochars produced from different feedstocks and temperatures, Domene et al. (2015c) recorded no general negative effects on the reproduction of *F. candida*, with the exception of food waste biochars applied at similar and higher (0.2–14%) rates. The authors related the latter trend to this biochar's high soluble sodium content. Similarly, Marks et al. (2014) reported different impacts on *F. candida* reproduction at concentrations from 0.5% to 50% (w/w), in a variety of biochars in terms of feedstock and pyrolysis procedures. While general reproduction stimulation was observed for slow and fast pyrolysis wood biochars and a slow pyrolysis sewage sludge biochar, a strong inhibition was caused by a pine gasification char. The authors linked the significant negative effects of the latter to its high pH but could not explain the positive effects of the other biochars. Since the positive impacts could not be associated to microbial abundance increases, community shifts or the use of some biochar fractions by gut symbionts were hypothesized as explanations.

Regarding field effects, Conti et al. (2014) reported a significant reduction in microarthropod density, and increased diversity and evenness, after 2 years of 30 t ha^{-1} application of a gasification biochar to a poplar plantation in acidic soil. Specifically, density dropped to 659 collembolans m^{-2} in biochar plots from the 1035 collembolans m^{-2} present in controls, which agreed with the predictions of McCormack et al. (2013) despite the fact that collembolans exhibit a wide range of species-specific pH optima (Van Straalen and Verhoef, 1997). However, Prober et al. (2014) reported no variation in collembolan density after 2 years of the surface application of a green waste biochar to an acidic soil of a degraded grassy eucalyptus woodland. Similarly, and after 3 years of the application of a corn stover biochar to a corn field (3–30 t ha^{-1}), Domene et al. (2014) did not find variation in fauna activity, measured as feeding rates (partly carried out by collembolans), although modeling of this response together with other soil properties indicated a positive effect of biochar addition rate on fauna activity. On the contrary, Marks et al. (2016) reported feeding rates reduction of a pine gasification biochar 2 and 3 years after its application (12 and 30 t ha^{-1}) to arable land cropped with barley, in a material previously demonstrated to inhibit collembolans reproduction under laboratory conditions (Marks et al., 2014).

Mites

Mites are one of the largest and most ubiquitous faunal groups, including microbivores, detritivores, animal predators, and plant feeders (Walter and Proctor, 1999). Uvarov (2000) cited an unpublished study that reported higher abundance and diversity of oribatid mites in a forest soil historically polluted by charcoal kiln emissions. However, Conti et al. (2014) reported that only 8088 mites m^{-2} were found in biochar plots compared to the 21,606 mites m^{-2} found in the control plots after 2 years of a gasification biochar application at 30 t ha^{-1} Godfrey et al. (2014) reported avoidance of the herbivorous and fungivorous oribatid mite *Oppia nitens* to high concentrations (51.2% and 100% v/v) of three biochars, which contrasted a slight preference for a spruce biochar applied at 51.2%. On the other hand, Prober et al. (2014) did not find mite abundance variation in biochar plots (2 t ha^{-1} of a green waste biochar) after 2 years of application. The paucity of studies highlights the need for more research on biochar effects for this key group.

Macroarthropods

With body lengths of 10 mm to 15 cm, this group comprises large insects, spiders, myriapods, and others (Coleman and Wall, 2014). They have an important role in soil ecosystems functioning and structure development (earthworms, ants, and termites). Although ant fecal pellets have been observed in char-enriched soils after wildfires (Topoliantz et al., 2006), not much is known on biochar effects on this diverse group. A reduced arthropod diversity has been reported around charcoal kilns (Fontodji et al., 2009) and attributed to the decreased bulk density, soil organic matter levels, and higher pH in the surrounding soils. Amaro (2013) observed no avoidance of the isopod *Porcellionides pruinosus* to fresh and 5-month-aged field samples from plots where a wood biochar was applied at rates of 4 and 40 t ha^{-1}. In agricultural plots where different biochars were applied at a rate of 14 t ha^{-1}, Castracani et al. (2105) found no effects on macroarthropod communities sampled with pitfall traps during 3 months following the application. The only exception was the ant species *Tetramorium caespitum*, associated with disturbed habitats, which increased their abundance in biochar plots. The authors attributed the result to increased temperatures in the biochar plots.

INTERACTIONS BETWEEN BIOCHAR AND FAUNA

Despite their small body size, soil organisms carry out and are responsible for a variety of soil ecosystem services, ie, benefits that people derive from ecosystems (Brussaard, 2012). If biochar interacts with soil organisms then it is plausible for that application to end up affecting ecosystem services. Among others, soil fauna contribute to, and might boost, key ecosystem processes such as: (1) production services (soil structure development, nutrient cycling, and plant production), (2) regulation services (climate regulation, detoxification, or plant protection against pests), and (3) cultural services such as the consciousness of natural heritage (Lavelle et al., 2006).

Soil Structure Development

Soil macrofauna, such as earthworms, ants, or termites, contribute strongly to soil structure development. The role of mesofauna has also been highlighted (Maaß et al., 2015). The reported charcoal presence in earthworm casts and macro- and mesofauna fecal pellets (Ponge et al., 2006; Topoliantz, 2002; Topoliantz and Ponge, 2003) has unavoidable consequences for biochar interaction with soil mineral particles and the genesis of highly stable organomineral complexes (Hayes, 1983), and the associated benefits for soil fertility (Glaser et al., 2002). In one of the few available studies on this topic, Hardie et al. (2104) speculated the role of earthworms' burrowing activity to explain the observed soil bulk density reduction in plots where an acacia green waste biochar had been applied. Earthworms were visible in higher numbers in biochar plots and their activity related to the increased abundance of large macropores, not possibly explained by biochar's own porosity.

Nutrient Cycling and Plant Growth

By a top-down regulation, faunal activity influences nutrient release and plant production. As an example, earthworms have been positively linked to increased plant shoot biomass, and a variety of mechanisms have been proposed including increased mineralization, increased production of plant growth stimulatory substances, pathogen suppression, symbiont stimulation, and increased water retention and aeration (Scheu, 2003). Biochar could directly influence any of the previous mechanisms, and indirectly by their effects on earthworms and any other faunal groups. So additive, synergistic, and antagonistic effects might be expected because of the interaction between biochar and fauna, hence the scarcity of literature highlights the requirement for intense research of this topic.

Noguera et al. (2010) used soil microcosms and demonstrated short-term positive additive effects on rice growth from the concurrent presence of biochar and earthworms (*P. corethrurus*). Although this depended on the soil nutrient status, the authors proposed the enhancement of nutrient release and retention and the release of phytohormones by earthworms as explanatory mechanisms. Paz-Ferreiro et al. (2014) tested different biochars in two tropical soils and proposed interactions between the earthworm *P. corethrurus* with a sewage sludge biochar to explain the 47% increase in proso millet grain yield. On the contrary, Reibe et al. (2015) did not record any positive effects on spring wheat shoot and root growth when testing two biochars in the presence of the collembolan *Protaphorura fimata*. As a result of the decreased trend for growth in maize-derived biochar treatments when collembolans were present, the authors suggested that collembolans root grazing might have reduced the positive effects of biochar.

Climate Change Mitigation

The role of soil fauna in carbon sequestration associated with biochar application has already been discussed earlier in this chapter. Thus some faunal activities might increase biochar's persistence in soil while others might decrease it. On the other hand, soil microorganisms and fauna are significant contributors to soil greenhouse gas emissions, although some biochars have been demonstrated to mitigate CO_2 and N_2O emissions. Regarding the most powerful agricultural greenhouse gas, N_2O, Cayuela et al. (2014) used meta-analysis and estimated that biochar was able to reduce emissions to half on average. It is also known that earthworms account directly for around half of the soil N_2O emissions (Borken et al., 2000; Drake et al., 2006), although limited information is available on the interactive effect with fauna.

Augustenborg et al. (2012) assessed CO_2 and N_2O emissions in low and high organic matter soils, and the effect of the addition of biochars and earthworms. Without earthworms, biochars reduced CO_2 emission in the high organic matter soil, but not in the other soil, although the concurrent presence of earthworms and a peanut hull biochar increased CO_2 emissions. On the other hand, N_2O emissions increased in both soils in the presence of earthworms, but without earthworms such emissions decreased after biochar application. More precisely, the earthworm-induced emissions were reduced strongly with biochar application, especially in the low organic matter (70–95%) soil and, to a lesser extent, in the high organic matter (10–61%) equivalent. Conversely, Bamminger et al. (2014) reported 43% and 42% reductions in CO_2 and N_2O efflux, respectively, because of biochar application, while earthworms had no statistically significant effects on emissions. The different or contradictory trends observed in these studies emphasize the need for comprehensive future research on this area.

Detoxification

Soil pollutant biodegradation is essentially a microbial process, and less is known about the potential role of soil fauna in this process. Earthworms are known to increase heavy metal mobility and bioavailability by stimulating microbial activity and organic matter decomposition, acidifying soil pH, reducing the dissolved organic carbon (DOC) able to transport metals, and by direct uptake (Ma et al., 1995; Sizmur and Hodson, 2009). While this might contribute to soil detoxification, exportation of pollutants to other environmental compartments might occur. On the other hand, earthworm activity can decrease soil organic pollutants by stimulating microbial biodegradation (Schaefer and Filser, 2007), enhancing soil aeration, and by their own metabolism (Blouin et al., 2013) or by gut symbionts (Verma et al., 2006). The combination of earthworms and organic wastes and the likely consequences have also been studied, but contrasting results have been reported in terms of remediation efficiency (Ceccanti et al., 2006; Schaefer and Filser, 2007).

Biochar has been suggested as a valuable material for the remediation of polluted soils because of its large surface area, abundance of hydrophobic domains, high cation exchange and liming capacity, and ability to reduce pollutant bioavailability (Beesley et al., 2011; Kookana et al., 2010). Furthermore, biochars have the potential to enhance microbial activities and subsequently stimulate organic pollutant biodegradation (Quin et al., 2013). Pollutant retention in biochar can, in turn, hinder their biodegradation by limiting microorganism accessibility (Jones et al., 2011; Ogbonnaya and Semple, 2013), while fauna activity could also affect all these capacities. However, only a few studies have experimentally assessed how fauna might affect biochar detoxification efficiency. For example, Beesley and Dickinson (2011) reported that earthworms had little effect on DOC levels for a polluted soil, although biochar addition led to a decreased DOC and hence the associated heavy metal mobility, especially that of Cu and As. On the contrary, Sizmur et al. (2011) did not find any improvement in heavy metal mobility reduction caused by biochar with the addition of earthworms. Similarly, Gomez-Eyles et al. (2011) reported increased heavy metal mobility when earthworms were present, but did not find that earthworms enhanced the reported capacity of biochars to reduce heavy metal mobility. Even more scarce are studies about the combined effect of biochar and fauna on organic pollutant persistence. Gomez-Eyles et al. (2011) demonstrated a similar reduction in total and bioavailable PAH concentrations with a biochar-only application than when biochar and earthworms were applied in combination. Regarding the effects of fauna on detoxification of biochars themselves, it has been shown that the usual phytotoxicity observed in some hydrochars could be mitigated by the presence of earthworms (Salem et al., 2013b).

Plant Protection Against Pests

Biochar has been suggested to induce resistance to plant pathogens and promote plant growth through shifts in microbial communities in the rhizosphere by mechanisms not yet understood (Elad et al., 2011; Graber and

Elad, 2013; Thies et al., 2015). As reviewed previously in this chapter, some studies have also highlighted that biochar might contribute to the reduction of plant diseases transmitted by nematodes by suppressing plant-feeding nematodes (Domene et al., 2015b; Rahman et al., 2014; Zhang et al., 2013).

References

Abel, S., Peters, A., Trinks, S., Schonsky, H., Facklam, M., Wessolek, G., 2013. Impact of biochar and hydrochar addition on water retention and water repellency of sandy soil. Geoderma 202, 183–191.

Amaro, A., 2013. Optimised Tools for Toxicity Assessment of Biochar-Amended Soils (Master dissertation). Departamento de Biologia, Universidade de Aveiro, Portugal.

Ameloot, N., Graber, E.R., Verheijen, F.G.A., De Neve, S., 2013. Interactions between biochar stability and soil organisms: review and research needs. Eur. J. Soil Sci. 64, 379–390.

Augustenborg, C.A., Hepp, S., Kammann, C., Hagan, D., Schmidt, O., Müller, C., 2012. Biochar and earthworm effects on soil nitrous oxide and carbon dioxide emissions. J. Environ. Qual. 41, 1203–1209.

Bamminger, C., Zaiser, N., Zinsser, P., Lamers, M., Kammann, C., Marhan, S., 2014. Effects of biochar, earthworms, and litter addition on soil microbial activity and abundance in a temperate agricultural soil. Biol. Fertil. Soils 50, 1189–1200.

Bardgett, E., 2005. The Biology of Soil: A Community and Ecosystem Approach. Oxford University Press.

Bastos, A.C., Prodana, M., Abrantes, N., Keizer, J.J., Soares, A.M.V.M., Loureiro, S., 2014. Potential risk of biochar-amended soil to aquatic systems: an evaluation based on aquatic bioassays. Ecotoxicology 23, 1784–1793.

Beesley, L., Dickinson, N., 2011. Carbon and trace element fluxes in the pore water of an urban soil following greenwaste compost, woody and biochar amendments, inoculated with the earthworm *Lumbricus terrestris*. Soil Biol. Biochem. 43, 188–196.

Beesley, L., Moreno-Jiménez, E., Gomez-Eyles, J.L., Harris, E., Robinson, B., Sizmur, T., 2011. A review of biochars' potential role in the remediation, revegetation and restoration of contaminated soils. Environ. Pollut. 159, 3269–3282.

Blouin, M., Hodson, M.E., Delgado, E.A., Baker, G., Brussaard, L., Butt, K.R., Dai, J., Dendooven, L., Peres, G., Tondoh, J.E., Cluzeau, D., Brun, J.-J., 2013. A review of earthworm impact on soil function and ecosystem services. Eur. J. Soil Sci. 64, 161–182.

Bongers, T., Bongers, M., 1998. Functional diversity of nematodes. App. Soil Ecol. 10, 239–251.

Borken, W.S., Grundel, S., Beese, F., 2000. Potential contribution of *Lumbricus terrestris* L. to carbon dioxide, methane and nitrous oxide fluxes from a forest soil. Biol. Fertil. Soils 32, 142–148.

British Biochar Foundation (BBF), 2013. Biochar Quality Mandate. Version 1.0. http://www.geos.ed.ac.uk/homes/sshackle/BQM.pdf (accessed 24.04.14.).

Brussaard, L., 2012. Ecosystem services provided by the soil biota. In: Wall, D.H., Bardgett, R.D., Behan-Pelletier, V., Herrick, J.E., Jones, T.H., Six, J., Strong, D.R. (Eds.), Soil Ecology and Ecosystem Services. Oxford University Press, pp. 45–58.

Bunting, B.T., Lundberg, J., 1987. The humus profile – concept, class and reality. Geoderma 40, 17–36.

Busch, D., Kammann, C., Grünhage, L., Müller, C., 2011. Simple biotoxicity tests for evaluation of carbonaceous soil additives: establishment and reproducibility of four test procedures. J. Environ. Qual. 40, 1–10.

Buss, W., Mašek, O., 2014. Mobile organic compounds in biochar–A potential source of contamination–phytotoxic effects on cress seed (*Lepidium sativum*) germination. J. Environ. Manag. 137, 111–119.

Carcaillet, C., 2001. Soil particles reworking evidences by AMS ^{14}C dating of charcoal. C.R. Acad. Sci. Séries 2, Sci. de la Terre des Planètes 332, 21–28.

Castracani, C., Maienza, A., Grasso, D.A., Genesio, L., Malcevschi, A., Miglietta, F., 2015. Biochar–macrofauna interplay: Searching for new bioindicators. Science of the Total Environment 536, 449–456.

Cayuela, M.L., Van, Zwieten, L., Singh, B.P., Jeffery, S., Roig, A., Sánchez-Monedero, M.A., 2014. Biochar's role in mitigating soil nitrous oxide emissions: a review and meta-analysis. Agric. Ecosyst. Environ. 191, 5–16.

Ceccanti, B., Masciandaro, G., Garcia, C., Macci, C., Doni, S., 2006. Soil bioremediation: combination of earthworms and compost for the ecological remediation of a hydrocarbon polluted soil. Water Air Soil Pollut. 177, 383–397.

Chakrabarti, S., Kern, J., Menzel, R., Steinberg, C.E., 2011. Selected natural humic materials induce and char substrates repress a gene in *Caenorhabditis elegans* homolog to human anticancer P53. Ann. Environ. Sci. 5, 1–6.

Chan, K.Y., Van Zwieten, L., Meszaros, I., Downie, A., Joseph, S., 2008. Using poultry litter biochars as soil amendments. Aus. J. Soil Res. 46, 437–444.

Cole, L., Bardgett, R.D., Ineson, P., 2000. Enchytraeid worms (Oligochaeta) enhance mineralization of carbon in organic upland soils. Eur. J. Soil Sci. 51, 185–192.

Coleman, D.C., Crossley Jr., D.A., Hendrix, P.F., 2004. Fundamentals of Soil Ecology. Elsevier Academic Press.

Coleman, D.C., Vadakattu, G., Moore, J.C., 2012. Soil ecology and agroecosystem studies: a dynamic world. In: Cheeke, T., Coleman, D.C., Wall, D.H. (Eds.), Microbial Ecology in Sustainable Agroecosystems. CRC Press, Boca Raton, pp. 1–21.

Coleman, D.C., Wall, D.H., 2014. Soil fauna: occurrence, biodiversity, and roles in ecosystem function. In: Paul, E.A. (Ed.), Soil Microbiology, Ecology, and Biochemistry, fourth ed. Elsevier Publishers.

Conti, F.D., Gadi, C., Malcevschi, A., Panzacchi, P., Ventura, M., Visioli, G., Menta, C., December 2–5, 2014. How the edaphic microarthropod community reacts to the biochar application in soils. In: The First Global Soil Biodiversity Conference, Dijon, France.

Cragg, R.G., Bardgett, R.D., 2001. How changes in soil fauna diversity and composition within a trophic group influences decomposition processes. Soil Biol. Biochem. 33, 2073–2081.

Domene, X., Mattana, S., Hanley, K., Enders, A., Lehmann, J., 2014. Medium-term effects of corn biochar addition on soil biota activities and functions in a temperate soil cropped to corn. Soil Biol. Biochem. 72, 152–162.

Domene, X., Hanley, K., Enders, A., Lehmann, J., 2015a. Short-term mesofauna responses to soil additions of corn stover biochar and the role of microbial biomass. App. Soil Ecol. 89, 10–17.

Domene, X., Mattana, S., Sanchez-Moreno, S., 2015b. Soil nematode communities as indicators of medium-term ecological impacts of biochar application. In: 25th SETAC Europe Annual Meetting, 3–7 May, Barcelona, Spain.

Domene, X., Hanley, K., Enders, A., Lehmann, J., 2015c. Ecotoxicological characterization of biochars: role of feedstock and pyrolysis temperature. Sci. Total Environ. 512–513, 552–561.

Drake, H.L., Schramm, A., Horn, M.A., 2006. Earthworm gut microbial biomass: their importance to soil microorganisms, denitrification, and the terrestrial production of the greenhouse gas N_2O. In: König, H., Varma, A. (Eds.), Intestinal Microorganisms of Soil Invertebrates. Springer, Berlin, pp. 65–87.

Dungait, J.A., Hopkins, D.W., Gregory, A.S., Whitmore, A.P., 2012. Soil organic matter turnover is governed by accessibility not recalcitrance. Glob. Change Biol. 18, 1781–1796.

EBC, 2012. European Biochar Certificate - Guidelines for a Sustainable Production of Biochar. European Biochar Foundation (EBC), Arbaz, Switzerland. Version 4.8. http://www.europeanbiochar.org/en/download (accessed 24.04.14.).

Eckmeier, E., Gerlach, R., Skjemstad, J.O., Ehrmann, O., Schmidt, M.W.I., 2007. Minor changes in soil organic carbon and charcoal concentrations detected in a temperate deciduous forest a year after an experimental slash-and-burn. Biogeosciences 4, 377–383.

Elad, Y., Cytryn, E., Harel, Y.M., Lew, B., Graber, E.R., 2011. The biochar effect: plant resistance to biotic stresses. Phytopathol. Mediterr. 50, 335–349.

Elmer, W.H., Lattao, C.V., Pignatello, J.J., 2015. Active removal of biochar by earthworms (*Lumbricus terrestris*). Pedobiologia 58, 1–6.

Enders, A., Hanley, K., Whitman, T., Joseph, S., Lehmann, J., 2012. Characterization of biochars to evaluate recalcitrance and agronomic performance. Biores. Technol. 114, 644–653.

Fischer, Z., Bieńkowski, P., 1999. Some remarks about the effect of smoke from charcoal kilns on soil degradation. Environ. Monit. Assess. 58, 349–358.

Fontodji, J.K., Mawussi, G., Nuto, Y., Kokou, K., 2009. Effects of charcoal production on soil biodiversity and soil physical and chemical properties in Togo, West Africa. Int. J. Biol. Chem. Sci. 3, 870–879.

Forey, E., Barot, S., Decaëns, T., Langlois, E., Laossi, K.-R., Margerie, P., Scheu, S., Eisenhauer, N., 2011. Importance of earthworm-seed interaction for the composition and structure of plant communities: a review. Acta Oecol. 37, 594–603.

Fry, W.E., 2012. Principles of Plant Disease Management. Academic Press, London, UK.

Genesio, L., Miglietta, F., 2012. Surface albedo following biochar application in durum wheat. Environ. Res. Lett. 7, 014025.

Glaser, B., Balashov, E., Haumaier, L., Guggenberger, G., Zech, W., 2000. Black carbon in density fractions of anthropogenic soils of the Brazilian Amazon region. Org. Geochem. 31, 669–678.

Glaser, B., Woods, W.I., 2004. Amazonian Dark Earths: Explorations in Space and Time. Springer, Berlin.

Glaser, B., Lehmann, J., Zech, W., 2002. Ameliorating physical and chemical properties of highly weathered soils in the tropics with charcoal. A review. Biol. Fertil. Soils 35, 219–230.

Godfrey, R., Thomas, S., Gaynor, K., 2014. Biochar and Soil Mites: Behavioural Responses Vary with Dosage and Feedstock. Centre for Global Change Science (CGCS), University of Toronto. http://www.cgcs.utoronto.ca/Assets/CGCS+Digital+Assets/Robert+Godfrey.pdf (accessed 17.08.15.).

Gomez-Eyles, J.L., Sizmur, T., Collins, C.D., Hodson, M.E., 2011. Effects of biochar and the earthworm *Eisenia fetida* on the bioavailability of polycyclic aromatic hydrocarbons and potentially toxic elements. Environ. Pollut. 159, 616–622.

Gongalsky, K.B., Persson, T., 2013. Recovery of soil macrofauna after wildfires in boreal forests. Soil Biol. Biochem. 57, 182–191.

Gouveia, S.E.M., Pessenda, L.C.R., 2000. Datation par le 14C de charbons inclus dans le sol pour l'étude du rôle de la remontée biologique de matière et du colluvionnement dans la formation de latosols de l'état de São Paulo, Brésil. C.R. Acad. Sci. Séries 2, Sci. de la Terre de la Planètes 330, 133–138.

Graber, E.R., Elad, Y., 2013. Biochar impact on plant resistance to disease. In: Ladygina, N., Rineau, F. (Eds.), Biochar and Soil Biota. CRC Press.

Hagner, M., Pasanen, T., Lindqvist, B., 2008. Effects of birch tar oils on soil organisms and plants. Agric. Food Sci. 19, 13–23.

Hale, S.E., Jensen, J., Jakob, L., Oleszczuk, P., Hartnik, T., Henriksen, T., Okkenhaug, G., Martinsen, V., Cornelissen, G., 2013. Short-term effect of the soil amendments activated carbon, biochar, and ferric oxyhydroxide on bacteria and invertebrates. Environ. Sci. Technol. 47, 8674–8683.

Hardie, M., Clothier, B., Bound, S., Oliver, G., Close, D., 2014. Does biochar influence soil physical properties and soil water availability? Plant Soil 376, 347–361.

Hart, S.A., Luckai, N.J., 2014. Charcoal carbon pool in North American boreal forests. Ecosphere 5 Article 99.

Hayes, M.H.B., 1983. Darwin's "vegetable mould" and some modern concepts of humus structure and soil aggregation. In: Satchell, J.E. (Ed.), Earthworm Ecology from Darwin to Vermiculture. Chapman and Hall, London, pp. 19–33.

Huhta, V., Persson, T., Setälä, H., 1998. Functional implications of soil fauna diversity in boreal forests. Appl. Soil Ecol. 10, 277–288.

International Biochar Inititiative (IBI), 2013. Standardized Product Definition and Product Testing Guidelines for Biochar that Is Used in Soil. Version 1.1. http://www.biochar-international.org/characterizationstandard (accessed 24.04.14.).

Jones, D.L., Edwards-Jones, G., Murphy, D.V., 2011. Biochar mediated alterations in herbicide breakdown and leaching in soil. Soil Biol. Biochem. 43, 804–813.

Kögel-Knaber, I., Guggenberger, G., Kleber, M., Kandeler, M., Kalbitz, K., Scheu, S., Eusterhues, K., Leinweber, P., 2008. Organic-mineral associations in temperate soils: integrating biology, mineralogy, and organic matter chemistry. J. Plant Nutr. Soil Sci. 171, 61–62.

Kołtowski, M., Oleszczuk, P., 2015. Toxicity of biochars after polycyclic aromatic hydrocarbons removal by thermal treatment. Ecol. Eng. 75, 79–85.

Kookana, R.S., 2010. The role of biochar in modifying the environmental fate, bioavailability, and efficacy of pesticides in soils: a review. Soil Res. 48, 627–637.

Lavelle, P., Decaëns, T., Aubert, M., Barot, S., Blouin, M., Bureau, F., Margerie, P., Mora, P., Rossi, J.-P., 2006. Soil invertebrates and ecosystem services. Europ. J. Soil Biol. 42, S3–S15.

Lehmann, J., Rillig, M.C., Thies, J., Masiello, C.A., Hockaday, W.C., Crowley, D., 2011. Biochar effects on soil biota – a review. Soil Biol. Biochem. 43, 1812–1836.

Li, D., Hockaday, W.C., Masiello, C.A., Alvarez, P.J.J., 2011. Earthworm avoidance of biochar can be mitigated by wetting. Soil Biol. Biochem. 43, 1732–1737.

Liesch, A.M., Weyers, S.L., Gaskin, J.W., Das, K.C., 2010. Impact of two different biochars on earthworm growth and survival. Ann. Environ. Sci. 4, 1–9.

Lievens, C., Mourant, D., Gunawan, R., Hu, X., Wang, Y., 2015. Organic compounds leached from fast pyrolysis mallee leaf and bark biochars. Chemosphere 139, 659–664 Available online November 27, 2014.

Liiri, M., Ilmarinen, K., Setala, H., 2007. Variable impacts of enchytraeid worms and ectomycorrhizal fungi on plant growth in raw humus soil treated with wood ash. Appl. Soil Ecol. 35, 174–183.

Lin, Y., Munroe, P., Joseph, S., Henderson, R., Ziolkowski, A., 2012. Water extractable organic carbon in untreated and chemical treated biochars. Chemosphere 87, 151–157.

Lorenz, K., Lal, R., 2014. Biochar application to soil for climate change mitigation by soil organic carbon sequestration. J. Plant Nutr. Soil Sci. 177, 651–670.

Lundkvist, H., 1998. Wood ash effects on enchytraeid and earthworm abundance and enchytraeid cadmium content. Scand. J. For. Res. 13, 86–95.

Luo, F., Song, J., Xia, W., Dong, M., Chen, M., Soudek, P., 2014. Characterization of contaminants and evaluation of the suitability for land application of maize and sludge biochars. Environ. Sci. Pollut. Res. 21, 8707–8717.

Ma, W.C., Immerzeel, J., Bodt, J., 1995. Earthworm and food interactions on bioaccumulation and disappearance in soil of polycyclic aromatic hydrocarbons: studies on phenanthrene and fluoranthene. Ecotox. Environ. Saf. 32, 226–232.

Maaß, S., Caruso, T., Rillig, M.C., 2015. Functional role of microarthropods in soil aggregation. Pedobiologia 58, 59–63.

Major, J., Lehmann, J., Rondon, M., Goodale, C., 2010. Fate of soil-applied black carbon: downward migration, leaching and soil respiration. Glob. Change Biol. 16, 1366–1379.

Major, J., 2009. A Guide to Conducting Biochar Trials. International Biochar Initiative.

Marhan, S., Scheu, S., 2005. Effects of sand and litter availability on organic matter decomposition in soil and in casts of Lumbricus terrestris L. Geoderma 128, 155–166.

Marks, E.A.N., Mattana, S., Alcañiz, J., Domene, X., 2014. Biochars provoke diverse soil mesofauna reproductive responses in laboratory bioassays. Eur. J. Soil Biol. 60, 104–111.

Marks, E.A.N., Mattana, S., Alcañiz, J.M., Pérez-Herrero, E., Domene, X., 2016. Gasifier biochar effects on nutrient availability, organic matter mineralization, and soil fauna activity in a multi-year Mediterranean trial. Agric. Ecosyst. Environ. 215, 30–39.

Matlack, G.R., 2001. Factors determining the distribution of soil nematodes in a commercial forest landscape. Forest Ecol. Manag. 146, 129–143.

McCormack, S., Ostle, N., Bardgett, R.D., Hopkin, D.W., VanBergen, A.J., 2013. Biochar in bioenergy cropping systems: impacts on soil faunal communities and linked ecosystem processes. GCB Bioenergy 5, 81–95.

Mitchell, P.J., Simpson, A.J., Soong, R., Simpson, M.J., 2015. Shifts in microbial community and water-extractable organic matter composition with biochar amendment in a temperate forest soil. Soil Biol. Biochem. 81, 244–254.

Naeem, S., Li, S., 1997. Biodiversity enhances ecosystem reliability. Nature 390, 507–509.

Nahar, M.S., Grewal, P.S., Miller, S.A., Stinner, D., Stinner, B.R., Kleinhenz, M.D., Wszelaki, A., Doohan, D., 2006. Differential effects of raw and composted manure on nematode community, and its indicative value for soil microbial, physical and chemical properties. Appl. Soil Ecol. 34, 140–151.

Neher, D.A., 1999. Soil community composition and ecosystem processes: comparing agricultural ecosystems with natural ecosystems. Agroforest. Syst. 45, 159–185.

Neher, D.A., 2001. Nematode communities as ecological indicators of agroecosystem health. In: Gliessman, S.R. (Ed.), Agroecosystem Sustainability: Developing Practical Strategies. CRC Press, pp. 105–120.

Nieminen, J., Haimi, J., 2010. Body size and population dynamics of enchytraeids with different disturbance histories and nutrient dynamics. Basic Appl. Ecol. 11, 638–644.

Nieminen, J.K., 2008. Labile carbon alleviates wood ash effects on soil fauna. Soil Biol. Biochem. 40, 2908–2910.

Noguera, D., Rondón, M., Laossi, K.R., Hoyos, V., Lavelle, P., de Carvalho, M.H.C., Barot, S., 2010. Contrasted effect of biochar and earthworms on rice growth and resource allocation in different soils. Soil Biol. Biochem. 42, 1017–1027.

Oades, J.M., 1993. The role of biology in the formation, stabilization and degradation of soil structure. Geoderma 56, 377–400.

Ogbonnaya, U., Semple, K.T., 2013. Impact of biochar on organic contaminants in soil: a tool for mitigating risk? Agronomy 3, 349–375.

Oleszczuk, P., Jośko, I., Kuśmierz, M., 2013. Biochar properties regarding to contaminants content and ecotoxicological assessment. J. Hazard. Mat. 260, 375–382.

Paz –Ferreiro, J., Fu, S., Méndez, A., Gascó, G., 2014. Interactive effects of biochar and the earthworm Pontoscolex corethrurus on plant productivity and soil enzyme activities. J. Soils Sedim. 14, 483–494.

Phillips, D.H., Foss, J.E., Buckner, E.R., Evans, R.M., FitzPatrick, E.A., 2000. Response of surface horizons in an oak forest to prescribed burning. Soil Sci. Soc. Am. J. 64, 754–760.

Pietikäinen, J., Kiikkila, O., Fritze, H., 2000. Charcoal as habitat for microbes and its effects on the microbial community of the underlying humus. Oikos 89, 231–242.

Ponge, J.-F., Ballot, S., Rossi, J.-P., Lavelle, P., Betsch, J.-M., Gaucher, P., 2006. Ingestion of charcoal by the Amazonian earthworm Pontoscolex corethrurus: a potential for tropical soil fertility. Soil Biol. Biochem. 38, 2008–2009.

Prober, S.M., Stol, J., Piper, M., Gupta, V.V.S.R., Cunningham, S.A., 2014. Enhancing soil biophysical condition for climate for climate-resilient restoration in mesic woodlands. Ecol. Eng. 71, 246–255.

Qin, G., Gong, D., Fan, M.Y., 2013. Bioremediation of petroleum-contaminated soil by biostimulation amended with biochar. Int. Biodeterior. Biodegrad. 85, 150–155.

Rahman, L., Whitelaw-Weckert, M.A., Orchard, B., 2014. Impact of organic soil amendments, including poultry-litter biochar, on nematodes in a Riverina, New South Wales, vineyard. Soil Res. 52, 604–619.

Reibe, K., Götz, K.P., Roß, C.L., Döring, T.F., Ellmer, F., Ruess, L., 2015. Impact of quality and quantity of biochar and hydrochar on soil Collembola and growth of spring wheat. Soil Biol. Biochem. 83, 84–87.

Renker, C., Otto, P., Schneider, K., Zimdars, B., Maraun, M., Buscot, F., 2005. Oribatid mites as potential vectors for soil microfungi: study of mite-associated fungal species. Microb. Ecol. 50, 518–528.

Rumpel, C., Kögel-Knaber, I., 2011. Deep soil organic matter – a key but poorly understood component of terrestrial C cycle. Plant Soil 338, 143–158.

Salem, M., Kohler, J., Rillig, M.C., 2013a. Palatability of carbonized materials to Collembola. Appl. Soil Ecol. 64, 63–69.

Salem, M., Kohler, J., Wurst, S., Rillig, M.C., 2013b. Earthworms can modify effects of hydrochar on growth of *Plantago lanceolata* and performance of arbuscular mycorrhizal fungi. Pedobiologia 56, 219–224.

Schaefer, M., Filser, J., 2007. The influence of earthworms and organic additives on the biodegradation of oil contaminated soil. App. Soil Ecol. 36, 53–62.

Scheu, S., 2003. Effects of earthworms on plant growth: patterns and perspectives. Pedobiologia 47, 846–856.

Six, J., Conant, R.T., Paul, E.A., Paustian, K., 2002. Stabilization mechanisms of soil organic matter: implications for C-saturation of soils. Plant Soil 241, 155–176.

Sizmur, T., Hodson, M.E., 2009. Do earthworms impact metal mobility and availability in soil? A review. Environ. Pollut. 157, 1981–1989.

Sizmur, T., Wingate, J., Hutchings, T., Hodson, M.E., 2011. *Lumbricus terrestris* L. does not impact on the remediation efficiency of compost and biochar amendments. Pedobiologia 54, S211–S216.

Smith, C.R., Buzan, E.M., Lee, J.W., 2012. Potential impact of biochar water-extractable substances on environmental sustainability. ACS Sustain. Chem. Eng. 1, 118–126.

Spokas, K.A., Novak, J.M., Masiello, C.A., Johnson, M.G., Colosky, E.C., Ippolito, J.A., Trigo, C., 2014. Physical disintegration of biochar: an overlooked process. Environ. Sci. Technol. Lett. 1, 326–332.

Steiner, C., Das, K.C., Garcia, M., Förster, B., Zech, W., 2008. Charcoal and smoke extract stimulate the soil microbial community in a highly weathered xanthic Ferralsol. Pedobiologia 51, 359–366.

Tammeorg, P., Parviainen, T., Nuutinen, V., Simojoki, A., Vaara, E., Helenius, J., 2014. Effects of biochar on earthworms in arable soil: avoidance test and field trial in boreal loamy sand. Agric. Ecosyst. Environ. 191, 150–157.

Thies, J.E., Rillig, M.C., Graber, E.R., 2015. Biochar effects on the abundance, activity and diversity of the soil biota. In: Lehmann, J., Joseph, S. (Eds.), Biochar for Environmental Management: Science, Technology and Implementation, second ed. Earthscan Publications Ltd., London, UK.

Topoliantz, S., Ponge, J.F., Lavelle, P., 2006. Humus components and biogenic structures under tropical slash-and-burn agriculture. Eur. J. Soil Sci. 57, 269–278.

Topoliantz, S., Ponge, J.F., 2003. Burrowing activity of the geophagous earthworm *Pontoscolex corethurus* (Oligochaeta: Glossoscolecidae) in the presence of charcoal. Appl. Soil Ecol. 23, 267–271.

Topoliantz, S., Ponge, J.F., 2005. Charcoal consumption and casting activity by *Pontoscolex corethurus* (Glossoscolecidae). Appl. Soil Ecol. 28, 217–224.

Topoliantz, S., 2002. Réponse fonctionnelle de la pédofaune à la mise en culture itinérante et permanente des sols du sud-ouest de la Guyane française (Ph.D. thesis). Museum National d'Histoire Naturelle, Paris.

Uvarov, A.V., 2000. Effects of smoke emissions from a charcoal kiln on the functioning of forest soil systems: A microcosm study. Env. Monit. Assess. 60, 337–357.

Van Straalen, N.M., Verhoef, H.A., 1997. The development of a bioindicator system for soil acidity based on arthropod pH preferences. J. App. Ecol. 34, 217–232.

Van Vliet, P.C.J., Beare, M.H., Coleman, D.C., Hendrix, P.F., 2004. Effects of enchytraeids (Annelida: Oligochaeta) on soil carbon and nitrogen dynamics in laboratory incubations. App. Soil Ecol. 25, 147–160.

Van Zwieten, L., Kimber, S., Morris, S., Chan, K.Y., Downie, A., Rust, J., Joseph, S., Cowie, A., 2010. Effects of biochar from slow pyrolysis of papermill waste on agronomic performance and soil fertility. Plant Soil 327, 235–246.

Verheijen, F.G.A., Jeffery, S., Bastos, A.C., van der Velde, M., Diafas, I., 2010. Biochar Application to Soils— a Critical Scientific Review of Effects on Soil Properties, Processes and Functions. Office for the Official Publications of the European Communities (EUR 24099 EN).

Verhoef, H., 2004. Soil biota and activity. In: Doelman, P., Eijsackers, H. (Eds.), Vital Soil: Function Value and Properties. Developments in Soil Science, vol. 29. Elsevier Science, Amsterdam, Netherlands.

Verma, K., Agrawal, N., Farooq, M., Misra, R.B., Hans, R.K., 2006. Endosulfan degradation by a *Rhodococcus* strain isolated from earthworm gut. Ecotox. Env. Saf. 64, 377–381.

Walter, D.E., Proctor, H.C., 1999. Mites: Ecology, Evolution, and Behaviour. NSW Press, Sydney.

Weyers, S.l., Spokas, K.A., 2011. Impact of biochar on earthworm populations: a review. Appl. Environ. Soil Sci. 2011 Article ID 541592.

Wilkinson, M.T., Richards, P.J., Humphreys, G.S., 2009. Breaking ground: pedological, geological and ecological implications of soil bioturbation. Earth-Sci. Rev. 97, 257–272.

Yeates, G.W., Coleman, D.C., 1982. Nematodes in decomposition. In: Freckman, D.W. (Ed.), Nematodes in Soil Ecosystems. University of Texas, Austin, pp. 55–80.

Zhang, X.-K., Li, Q., Liang, W.-J., Zhang, M., Bao, X.-L., Xie, Z.-B., 2013. Soil nematode response to biochar addition in a Chinese wheat field. Pedosphere 23, 98–103.

12

Summation of the Microbial Ecology of Biochar Application

C.H. Orr, T. Komang Ralebitso-Senior

Teesside University, Middlesbrough, United Kingdom

BOOK RATIONALE

Although biochar is receiving unprecedented global interest as a powerful cost-effective climate change mitigation tool, with its application to soil potentially providing a long-term carbon sink, major terrestrial benefits also accrue. These include: soil structure modification (bulking agent addition, moisture/nutrient/microbiota retention, and proliferation); organic/inorganic compound adsorbent addition; fertilizer supplementation; carbon sequestration; and contaminant amelioration. These are being investigated but there is a considerable paucity of work on microbial diversity, structure, and function response to char type (feedstock/production conditions) and application.

To date, several biochar critical reviews (eg, Atkinson et al., 2010; Verheijen et al., 2010) and books have considered largely its: production

Biochar Application
http://dx.doi.org/10.1016/B978-0-12-803433-0.00012-6

293

(eg, Scholz et al., 2014); physicochemical properties (eg, Steiner, 2015); potential for climate change mitigation (eg, Bates, 2010; Bruges, 2010; Green and Wayman, 2013); application in different ecosystems (eg, Taylor, 2010); life cycle assessment (eg, Scholz et al., 2014); and, to a lesser extent (eg, Ladygina and Rineau, 2013; Lehmann and Joseph, 2009, 2015; Woods et al., 2009), impacts on occurring microbial communities at cellular and molecular levels. All environmental biotechnologies, including agricultural productivity and contaminated ecosystem amelioration, are underpinned by complex interacting microbial populations (multispecies gene pools), exemplifying the need for specific focus on these associations and their functional, structural, and compositional dynamics following biochar application. Emerging research reflects this with knowledge furtherance predominating peer-reviewed papers on understanding the responses of microbial communities in soil ecosystems following applications of different biochars—feedstock, pyrolysis conditions—sites and application regimes (Chapters 1 and 2). These studies have entailed the use of different physicochemical analyses that are now complemented by microecophysiology tools, which target DNA (in particular), RNA, lipids, proteins, and specific functional genes with denaturing gradient gel electrophoresis (DGGE), terminal restriction fragment length polymorphism (T-RFLP), (automated) ribosomal intergenic spacer analysis, phospholipid-derived fatty acids (PLFA), quantitative polymer chain reaction (qPCR), and next-generation sequencing.

Therefore this book has aimed at addressing a paucity in the literature by providing an advanced compilation with a specific perspective on these cutting-edge, topical, and relevant approaches. Thus Chapter 2 gave an overview of the impacts of different production parameters on the properties of the generated biochars, and initiated discussion of their effects and potential applications, particularly for the soil ecosystem. Undoubtedly, these impacts will differ relative to the specific application ecosystem or context such as agronomy and contaminant attenuation. Aspects of these were therefore explored in more detail in the subsequent chapters as relevant to the individual contexts.

Multiple biomolecules are targeted in microbial ecology to understand how different perturbations affect ecosystem function. Therefore the use of phospholipid fatty acid-based analyses in biochar-impacted soil, although still in its nascent stage, was explored critically in Chapter 3 with consideration also of carbon flow determinations that are reliant on natural isotope and radioisotope labeling. Overall, this approach has revealed that biochar carbon molecule biodegradability and impacts on microbial community composition and structure differ with specific microbial clades, ie, Gram-positive and Gram-negative bacteria, fungi, and actinomycetes. Method development and optimization, data management, and subsequent interpretation were critical aspects of the discourse. Thus multivariate analysis, for example, was recommended to ensure statistically robust and appropriate PLFA-based determinations of biochar ecological impacts on total and functional soil microbial communities.

In Chapter 4 we considered analysis of culturable communities, with current knowledge gaps and likely potentials then identified toward exploring mechanistic processes around biochar addition using simple mono-/cocultures and soilless systems (hydroponic) initially, and soil/soil-like models subsequently. Although not discussed in detail, the concept of the "charosphere" (Quilliam et al., 2013) was introduced with regards to the potential use of cultivable microbial strains/communities to then elucidate fully how the immediate biochar environment affects soil ecosystem function. These approaches could be considered for future studies in combination with those discussed in Chapters 7 and 8, which focused on the nitrogen-cycling communities. Thus the intricacies of how biochar affects the soil functional capacity can be elucidated within the context of important biogeochemical cycles, such as the N-cycle, and these could be achieved potentially by use of simple model communities that are culturable on commercial and bespoke habitat media.

Attenuation of organic and inorganic contaminants is identified as one of the three key applications of biochar via physicochemical interactions including sorption (eg, Li et al., 2016) and ion exchange, and biological transformation. Examples of the molecules and contexts where this potential is being explored, the knowledge gaps, and future research needs have therefore been discussed in Chapters 9 and 10. Some interesting related deliberations on biochar properties or attenuation potential have then been considered quite organically in different chapters. Examples include: the role of feedstock, production parameters, and consequently biochar physico- and biochemical characteristics (Chapter 2); the use of culture-based studies to elucidate the underpinning mechanisms (Chapter 4); and bioprospecting for specific contaminant degrading strains and communities in Amazonian Dark Earths (ADE, Chapter 5).

In proximity to the soil microbiome are the meso- and macrofaunal communities, which are also affected by biochar addition, with immediate and long-term implications for the soil ecosystem and its functional capacities (eg, Lone et al., 2015). Therefore this chapter explores the effects of biochar supplementation on these important soil populations and identifies the need to understand biochar retention parameters, distribution processes, and trophic-level dynamics.

As in all microbial ecology research, cutting-edge and high-throughput platforms are central to elucidating biochar-impacted (eco)systems. Thus Chapters 5 and 6 discuss the application of T-RFLP and next-generation sequencing in ecosystems with historical (ADE) and contemporary biochar augmentation. While pyrosequencing in particular, confirmed findings of diverse functional capacities, including aromatic hydrocarbon degradation (in ADE), other/novel platforms such as stable isotope probing (SIP) and metatranscriptomics, should be considered to expand knowledge of the temporal occurrence of specific microbial community functions. These must also be directly and fully cognizant of abiotic processes that dictate (micro)biological function expression in biochar-impacted soils.

METHODOLOGICAL STATE OF THE ART

The microecophysiology methods employed so far in biochar research have been summarized in Chapter 1 with those discussed specifically in this book presented in Table 12.1. Parallel to deliberations in specific research articles, relative to respective contexts, we presented an expansive consideration of additional tools that have potential to address key knowledge gaps in the microbial ecology of biochar application (Ennis et al., 2012). Some of the techniques were fluorescence in situ hybridization (FISH), stable isotope probing, genechips/microarrays, Raman spectroscopy, and brdU (bromodeoxyuridine) labeling. To date, no literature has yet emerged on the application of these tools to elucidate biochar effects specifically on soil microbial associations. Nonetheless, studies such as Ye et al. (2015) have generated interesting knowledge that can be transferred to the wider biochar context. Thus this research team used biochar produced (450°C) from bamboo stems, kaoline, and iron mineral as a complex medium to test a newly developed protocol. Gold nanoparticle probes were used with in situ hybridization—replacing FISH with genomic in situ hybridization—for subsequent electron microscopy visualization of surface-attached bacteria (*Escherichia coli* and *Neisseria sicca*). Although neither designed nor developed specifically for biochar research, the study reflects an elegant approach that has direct relevance for the application of FISH-based tools in biochar-based microbial ecology.

Essential discourse continues on microbial-based carbon cycling or carbon efflux from soils in general, and following biochar application in particular. For example, Gougoulias et al. (2014) presented a critical perspective of the contribution of microbial metabolic activities, on plant-derived carbon particularly in agronomic soils, to atmospheric carbon dioxide. The authors also identified multiple molecular ecology tools that could be applied to understand the underpinning processes and mechanisms of soil-to-atmosphere carbon efflux. These included: SIP based on DNA, RNA, protein, phospholipid fatty acids, and neutral lipid fatty acids; and FISH with Raman spectroscopy, microautoradiography, and secondary ion mass spectrometry, most of which we have proposed previously for biochar application (Ennis et al., 2012; Orr and Ralebitso-Senior, 2014).

TABLE 12.1 Summary of the in-chapter discussions regarding microecophysiology tools used in biochar-impacted ecosystems, existing knowledge, current paucities, and new research opportunities

Biochar source	Study ecosystems or contexts	Primary microbial ecology techniques used/considered	Developed knowledge	Knowledge gaps, potential new tools, and future research	Chapters
A wide range from the literature	Cranbourne sandy soil; Werribee clayey soil; Cranbourne soil; several from the literature	PLFA	1. Technique has potential to provide insights into microbial community composition relatively rapidly, quantitatively, and in a standardized manner 2. Trends revealed specific microbial community responses and uptake of biochar C	1. As in other contexts, full cognizance must be made of PLFA strengths and limitations specifically for biochar research 2. Appropriate data handling and statistical analysis must underpin PLFA/microecophysiology analysis of biochar-augmented soil ecosystems	3
Hardwood	Domestic garden soil	Culture-based analysis; DGGE	1. Commercial, selective, and habitat soil media have potential for culture-based analysis to differentiate biochar-impacted and nonimpacted soil 2. Combination of viable cell counts and DGGE facilitates more comprehensive analysis	1. Bespoke habitat media development may enhance knowledge development on biochar impacts in soil-/site-specific contexts 2. The relevance of culture-based analyses may be explored as part of the toolkit to elucidate underlying biochar-based mechanisms in different contexts, with and without hydroponics	4

Continued

TABLE 12.1 Summary of the in-chapter discussions regarding microecophysiology tools used in biochar-impacted ecosystems, existing knowledge, current paucities, and new research opportunities—cont'd

Biochar source	Study ecosystems or contexts	Primary microbial ecology techniques used/considered	Developed knowledge	Knowledge gaps, potential new tools, and future research	Chapters
	Amazonian Dark Earths	Next-generation sequencing	1. ADE support: heterogeneous and functionally diverse bacterial communities; populations with aromatic hydrocarbon degradative potential 2. Biochar has a key role in biogeochemical functioning	1. Metagenomics and enrichments with stable isotope-based bioassays to probe microbial enzyme activity 2. Metatranscriptomics to expand existing knowledge by providing identity and function information of occurring microorganisms in a "punctuated" time	5
A wide range from the literature	A wide range of soils from literature including ADE	Metagenomics, ie, pyrosequencing	1. Metagenomics have demonstrable potential to characterize biochar-impacted microbial community function 2. The role of abiotic processes *re* biochar production conditions, eg, fast versus slow pyrolysis, and physicochemical properties	1. RNA-based metagenomics can facilitate further understanding of biochar supplementation 2. Metagenomic tools have limitations that need to be considered for the biochar context 3. The tools must be used in conjunction with physicochemical determinations and robust statistical data handling approaches, eg, multivariate analysis 4. Focused research with these combinations can help elucidate knowledge gaps on rhizosphere interactions, cell signaling, and soil health	6

| A wide range from the literature | N-cycle; agro-ecosystems | A range from the literature, eg, enzyme activity, qPCR, 16S rRNA gene sequencing, T-RFLP | 1. Soil microbial community nitrogen-cycling dynamics, including in the rhizosphere
2. Response of some specific functional communities and genes in ammonification, nitrification, and denitrification
3. The role of biochar's physicochemical properties | 1. Fundamental understanding of mechanisms that determine biochar's interactions with nitrogen-cycling microbial communities at micro-, macro-, bulk, and rhizospherescale, as well as the "charosphere"
2. Biochar's feedstock, production conditions, and subsequently its physicochemical properties mandate further investigation regarding the processes/mechanisms as above
3. Application of other approaches for knowledge furtherance, eg, stable isotope probing and soil (micro)electrodes for N_2O emissions, microscopy and spectroscopy techniques for temporal biochar physicochemical property analysis, and culture-based studies to elucidate electron shuttling mechanisms
4. Recommendations for experimental designs/analysis to ensure cross-study and cross-treatment comparisons, eg, standardization for biochar property determinations; appropriate controls including original feedstocks relative to the used biochar; normalization strategies such as biochar dry matter content, and "% water filled pore space" particularly for oxygen concentration impact on nitrogen biotransformation |

Continued

TABLE 12.1 Summary of the in-chapter discussions regarding microecophysiology tools used in biochar-impacted ecosystems, existing knowledge, current paucities, and new research opportunities—cont'd

Biochar source	Study ecosystems or contexts	Primary microbial ecology techniques used/considered	Developed knowledge	Knowledge gaps, potential new tools, and future research	Chapters
Hardwood	Agronomy: clover rhizosphere; N-cycle	DGGE	1. Depending on the feedstock and production conditions, biochar addition can alter several soil bio-/physicochemical parameters including N and C dynamics, pH, P availability, and heavy metal availability impacting the rhizosphere bacterial community 2. Reported changes in bacterial communities are often limited to individual phyla 3. Most studies focus on bulk soil rather than the rhizosphere-specific activity	1. The impact that biochar-based changes to the rhizosphere community could have on plant growth and disease, and nutrient cycling 2. Impact of biochar addition on key functional groups 3. How biochar impacts the microbial community within the rhizosphere of different plant species 4. Knowledge furtherance can be achieved with novel techniques such as genechips/microarrays, FISH, and SIP	8
A range from the literature	Biochar-based remediation of organic and inorganic contaminants in soil and groundwater	–	1. Demonstrable potential of waste biomass-derived biochars to ameliorate different ecosystems impacted by a range of pollutants 2. Most knowledge developed from laboratory-scale studies	1. Feedstock and biochar selection can ensure enhanced remediation efficacy and must therefore be investigated closely 2. Biochar amelioration potential needs to be investigated for multicontaminant waste streams, which are typical of most industrial discharges 3. Investigations need to be scaled up to pilot and field/industrial applications 4. Potential of biochar as inoculum carrier alluded to and would necessitate further research for organic molecule remediation	9

| A wide range from the literature | Organic pollutant-impacted ecosystems | Enzyme activity; mesofaunal activity | 1. Pollutant removal capability is biochar (feedstock and production conditions) dependent
2. Molecule (PAH) partitioning into biochar affects bioavailability directly
3. Differences in bacterial and fungal communities to remediate organic pollutants in the presence of biochar | 1. The four-way relationship between soil characteristics, contaminant properties, soil microbiology, and biochar, which should be studied closely to optimize remediation success
2. Biochar properties (*re* feedstock and production conditions), architecture (*re* particle size and postproduction activation), and interactions with different soil types require further investigation regarding microbial functional response
3. Contaminant bioavailability and microbial strains and communities in toxicity biosensor analyses should be used to inform policy development | 10 |
| A range from the literature | Meso- and macrofauna in laboratory microcosms and forest soils | Microscopy; dissolved organic carbon; fauna feeding, metabolic and reproductive activity | 1. Emerging knowledge development is underpinned by species that are representative of different soul fauna groups
2. Biochar properties including: inherent pollutant concentration and bioavailability, feedstock and production parameters, affect soil fauna differently
3. Some of the meso- and macrofauna roles in biochar distribution impact on soil physicochemical properties, eg, aggregation
4. Different effects result relative to soil properties such as pH and biochar addition ratio | 1. Considerable paucities on: mechanisms underpinning preferential ingestion of biochar by soil faunal; potential trophic effects of biochar addition; and impacts of meso-/macrofauna on microbial community dynamics | 11 |

DGGE, Denaturing gradient gel electrophoresis; ADE, Amazonian Dark Earths; PAH, polycyclic aromatic hydrocarbon; T-RFLP, terminal restriction fragment length polymorphism; PLFA, phospholipid-derived fatty acids; FISH, fluorescence in situ hybridization; SIP.

BOOK/KNOWLEDGE PAUCITIES AND RECOMMENDATIONS FOR FUTURE WORK

Despite the original planned scope of this book, the important topic of fungi/biochar interactions has not been considered in a designated chapter. The omission is not intended to undermine the relevance of research in this field. The reader is referred therefore to several seminal discussions and critical reviews including those cited in different chapters (eg, Ennis et al., 2012; Lehmann et al., 2011; Warnock et al., 2007) with specific reference to the roles of fungi in:

- Soil. Mycorrhizal fungi in: crops for enhanced yield (Mau and Utami, 2014) and plant growth and salinity stress amelioration (Hammer et al., 2015); and
- Agriculture. Since fungi are often pathogenic to economically important crops, biochar ability to suppress fungal pathogens.

Studies that have examined biochar/fungal community interactions have been considered in this chapter and in different chapters such as 3, 6, and 10. Thus mechanisms by which biochar impacts fungal diversity and activity are being elucidated or investigated, and future fungi-targeted research needs have been identified.

Although explored briefly in Chapter 4, an exciting and relevant agronomy theme, where biochar is emerging as a potential tool to control crop disease and minimize wastage, will probably benefit further from the application of microbial ecology analysis to elucidate the mechanisms involved in controlling soil- and air-borne plant pathogens. This has been reflected in a few studies and has warranted sections in reviews or book chapters (eg, Elad et al., 2010; Graber et al., 2014; Zwart and Kim, 2012). The current literature reflects a great deal of potential concomitant with several questions/unknowns, which will inevitably drive extensive research of the topic.

Biochar impacts on microbial intercellular communication, including with plant systems and hence wider ecosystem function, have only recently been considered. Thus an elegant exemplary study by Masiello et al. (2013) was designed to initiate elucidation of the response to biochar binding capacity of an acyl-homoserine lactone (AHL), a common quorum-sensing molecule involved in gene expression regulation. Briefly, the researchers used green fluorescent protein (GFP) expressing *Escherichia coli* XL1-Blue and *E. coli* BLIM to test the effects of mesquite wood (*Prosopis glandulosa*)-derived biochars, pyrolyzed at seven different temperatures (300–700°C), on N-3-oxo-dodecanoyl-L-homoserine lactone. They recorded differences in GFP expression relative to the degree of AHL sorption by, and in proximity to, different temperature biochars and thus their effects on cellular communication. The workers emphasized existing knowledge gaps where cellular mechanisms need to be elucidated generally, and at microbially relevant biogeographical and time scales in particular. Also, although not intended for biochar application, Mason et al. (2005) used a mixed model microbial assemblage (community) and measured the impacts of specific biofilm architecture on AHL dynamics. Thus combinations of approaches like those of the Masiello and Mason teams should be considered to explore cell-to-cell communication of microbial biofilm processes, and structure and function on the rhizoplane and in the rhizosphere of biochar-impacted (agronomic) soils. The potential use of simple systems, ie, monocultures or model communities, for detailed investigations with controllable parameters and hence culture-based approaches to elucidate the mechanisms proposed by Masiello et al. (2013) and the role of biochar-based abiotic effects were therefore alluded to in Chapters 4 and 6.

Biochar has a recognized potential to address multiple contemporary concerns including those generated by, and/or ameliorated with, key environmental biotechnologies. Consequently, its use in novel contexts such as biofiltration and wastewater treatment has been proposed (eg, Orr and Ralebitso-Senior, 2014) but is yet to emerge in published literature. In contrast, biochar application potential in anaerobic digestion, including for enhanced methane production, was exemplified by Chen et al. (2014). Other emerging areas, and those yet to be explored, where microbial ecology analyses would have direct relevance include biochar potential application and role in: microbial fuel cells (Huggins et al., 2014); biofertilizers (De Oliveira Mendes et al., 2015; Fox et al.,. 2014); holistic, economically feasible, and ecologically sustainable sanitation systems (De Gisi et al., 2014); and, as also demonstrated for contaminated groundwater using permeable reactive barriers (see Chapter 9), malodorant gas biofiltration (Ralebitso-Senior et al., 2012).

An emerging technique, where pyrolyzed biomass is mixed with magnetic precursors to create magnetic char, is being investigated for subsequent use in applied and laboratory settings. Several researchers such as Chen et al. (2011), Zhang et al. (2013), Shan et al. (2016), and Trakal et al. (2016) have proposed that this method could be used in the remediation of common pollutants such as arsenic, heavy metals, and biopharmaceuticals. For example, arsenic is a highly toxic element constituent of some pesticides and antibiotics used in agriculture and medicine, and hence it migrates into aquifers via many routes. Also exhausted biochar absorbents may contain pollutants, which can account for secondary contamination. Thus magnetic biochar facilitates the recycling of the pollutant-laden adsorbent following its collection, magnetically, from contaminated ecosystems (Chen et al., 2011). Char magnetization would also enable recovery from liquid or solid media for subsequent detailed qualitative analysis of surface-adsorbed microbial strains/communities. The effects of magnetized biochar on functional microbial populations, including its potential biotoxicity, are yet to be reported and consequently mandate focused investigations with specific scope for microecophysiology analyses.

Overall, the book chapters have highlighted the current level of understanding and knowledge gaps of the microbial ecology of biochar application, with alternative analyses also identified for some contexts. These have been consolidated in Table 12.1, which has been adapted in part from Ennis et al. (2012). A more comprehensive analysis would consider, for each scenario, the chars that have been used, their feedstocks, application ratios, and complementary physicochemical techniques. Therefore this has been developed quite appropriately in the tables and figures in Chapters 1, 2, 3 and 6, for example, but with specific focus on microecophysiology-based studies. The reader is referred to the wealth of literature in peer-reviewed articles and relevant compilations that considers, strictly, the feedstocks, production parameters, and resultant biochar properties (eg, Zhang and Wang, 2016).

Recognition has also been made of the important role of policy in biochar research and application. As a result, we presented a succinct analysis and discourse on this in Chapter 1, which was based on the relatively limited information that is available in the public domain. Furthermore, discussions in Chapter 10 concluded with recommendations for legislative framework and policy development especially for contaminant bioavailability and biotoxicity caused by biochar addition. Generally, the existing policy statements and/or proposals for policy frameworks reflect little or no cognizance of soil biota. For example, the International Biochar Initiative (IBI) Biochar Standards refer to microorganisms within the context of biochar "postprocessing" with biological analysis limited to toxicity determinations based largely on (plant) germination inhibition. Therefore our discussion attempted to highlight the fundamental policy needs for the sustainable maintenance of ecosystem function following biochar application, which must be informed also by relevant microecophysiology findings on the impacts of the material on soil microbial communities as well as the meso- and macrofauna. Overall, we propose that future (micro)biology-based biochar research, application, and analyses, at the community, cellular, and molecular levels, should be cognizant of and be used to develop policy as follows: (1) reflect existing policy statements, however few; (2) start to identify those that are likely to be drawn, eg, in the immediate/near future; (3) indicate those that need to be made in the long term; and (4) explore their differences and relevance to different countries, economies, societies, and possibly geopolitical regions. Ultimately, these will need to be presented and considered even if only to highlight the challenge of reconciling the different needs of various stakeholders including, but not limited to, policy developers, governmental bodies/initiatives, and researchers.

Furthermore, the discourse was informed by statements in the literature on the critical need for considered policy statements and/or guidelines regarding quality control for biochars relative to substrates/feedstocks, but especially standardized physicochemical criteria for analytical, research, and application requirements. Therefore existing key criteria that are recommended by different bodies are compared to those measured in

TABLE 12.2 Overview of recommended and typical target parameters for physicochemical characterization of biochar with examples of microbial ecology-informed soil investigations

Required (R) or determined (D)	Moisture or water content (%)	C content	H:C_org (molar)	Total ash	pH	Electrical conductivity	Liming if pH <7 (% CaCO3)	Particle size distribution (% <0.5 to % >50 mm)	O:C (molar)	Macronutrients (NPK)	Bulk density	Surface area	Heavy metals, metalloids, and other elements	Volatile matter	Water-holding capacity	CEC	C/N ratio
International Biochar Initiative (R)	✓	Organic carbon (C_{org}; %)	✓	✓	✓	✓ ($dS\,m^{-1}$)	✓	✓	X	Total N	X	Optional	✓	Optional				
European Biochar Certificate (R)	✓	Total C	✓	✓	✓	✓ ($\mu S\,cm^{-1}$)	X	X	✓	Total N, P, K, Mg, Ca	✓	✓	✓	✓				
British Biochar Foundation (R)	✓	✓ (C_{org}) and total C	✓	✓	✓	✓ ($dS\,m^{-1}$)	✓	✓		Total P, N, K	✓	✓	✓	✓	✓	✓	✓
Gaskin et al. (2008)		✓		✓	✓					✓			✓				✓
Inyang et al. (2010)		✓		✓	✓					✓		✓	✓			✓	✓
Bird et al. (2011)		✓		✓	✓	✓ ($ms\,cm^{-1}$)				✓			✓			✓	✓
Hossain et al. (2011)	✓	✓		✓	✓	✓				✓			✓			✓	✓

Continued

TABLE 12.2 Overview of recommended and typical target parameters for physicochemical characterization of biochar with examples of microbial ecology-informed soil investigations—cont'd

Required (R) or determined (D)	Moisture or water content (%)	C content	H:C$_{org}$ (molar)	Total ash	pH	Electrical conductivity	Liming if pH <7 (% CaCO$_3$)	Particle size distribution (% <0.5 to % >50 mm)	O:C (molar)	Macronutrients (NPK)	Bulk density	Surface area	Heavy metals, metalloids, and other elements	Volatile matter	Water-holding capacity	CEC	C/N ratio
Kolton et al. (2011)		✓	✓	✓					✓	✓							✓
Lehmann et al. (2011)		✓	✓	✓	✓				✓	✓		✓		✓		✓	✓
Santos et al. (2012)		✓	✓						✓	✓							✓
Ameloot et al. (2013)		✓		✓	✓								✓		✓		✓
Chen et al. (2013)		✓		✓	✓					✓			✓				
Fox et al. (2014)		✓								✓					✓		
Hale et al. (2014)					✓	✓ (mScm^{-1})					✓				✓		
Purakayastha et al. (2015)		✓			✓	✓ (dSm^{-1})					✓	✓			✓	✓	✓
Rosas et al. (2015)		✓		✓	✓	✓ (dSm^{-1})				✓			✓			✓	✓
Ahmad et al. (2016)		✓		✓	✓	✓						✓				✓	✓

✓ Designates required (R) or determined (D) while X designates not required.

different publications specifically within the context of biochar-directed microbial ecology (Table 12.2). Although some commonalities are emerging, with some researchers specifying their alignment with existing guidelines (eg, Hale et al., 2014 for IBI; Rosas et al., 2015 for European Biochar Certificate and IBI), considerable differences in approaches still remain.

SUMMARY

Emerging studies elucidating ecosystem structural and functional dynamics in response to biochar application have focused on DNA-based analyses, with increasing examples of next-generation sequencing. Few have used FISH (a well-established technique to explore functional dynamics in soils and biofilms), RNA, genechips, proteomics, transcriptomics, and comprehensive probing of specific functional genes underpinning key biogeochemical cycles, specifically carbon and nitrogen. Most biochar-related studies, published before January 2016, focused on soils with sediments, wastewaters, and malodorant gases yet to be considered. Also, although magnetic biochar is recognized as having particular potential for postattenuation recovery (including for contaminated soil washing), subsequent recycling, and/or appropriate disposal, no studies have been published in the microbial ecology literature on this novel aspect.

Although the benefits of biochar have been publicized widely there are also findings of numerical decreases in some microbial community members with potential implications of reduced ecosystem function. Similarly, recognized probable and/or unintended negative impacts have been reflected upon and presented in either concluding sections of original research articles or as emphatic statements in critical reviews (eg, Kuppusamy et al., 2016; Lone et al., 2015). As exemplified by these two research teams in particular, the deliberations often conclude with the identification of future biochar research scope such as, for example: standardization of feedstocks and pyrolysis conditions; development of production guidelines and quality standards; investigation of the adverse effects of biochar on soil flora and fauna; and exploration of soil-specific enzyme activity following biochar addition. These therefore establish directly and indirectly the importance of comprehensive and relevant microbial ecology tools as components of informed decision-making strategies before, during, and after biochar application.

We have used Table 12.1 to present a focused collated synthesis of the in-chapter deliberations on specific contexts, ecosystems, and contaminants. These include paucities in literature, current knowledge gaps, and application opportunities that therefore require and mandate systematic future microbial ecology-based research on biochar supplementation. In general, these can be encapsulated and realized fully through the adoption of, for example, Laird's (2008) "charcoal vision" where, beyond mainstream biochar, the author recognized the requirement to resolve the fundamental mechanisms underpinning its reactivity in soil ecosystems and how it can be integrated into other systems, processes, and agendas. Thus the proposed biochar-based "Win–Win–Win Scenario" comprised energy production by pyrolysis, carbon sequestration by soil supplementation, and increased environmental quality of soil and groundwater.

Similarly, recommendations such as by De Gisi et al. (2014), to revisit and adopt the *Terra Preta* Sanitation Strategy, to combine human waste management, char addition, and anaerobic lactic acid fermentation, would further this win–win–win potential. Indeed, studies such as Dil et al. (2014) and Schmidt et al. (2015), who primed biochar with urea ammonium nitrate to test its impacts on maize and pumpkin biomass, N uptake and utilization efficiency, and microbial activity for different soils, may be adapted with human waste urea providing the ammonium nitrate. This probably exemplifies the "55 uses of biochar" compiled by Schmidt and Wilson (2014) who, in alignment with Laird's vision, suggested biochar-derived cascades to create simultaneously potential agroeconomic benefits and relevant research opportunities. Ultimately, the realization of the win–win–win scenarios, and potential for economically and ecologically closed-loop systems, will be dependent on appropriately robust policy guidelines on biochar production and application, and requisite cognizance of the intended and unintended long-term impacts of its application to soil ecosystems. It is only through directed, concerted, and strategic efforts that biochar augmentation of soils can then continue to realize expectations as evoked in its captive descriptors such as the "black future" and "black is the new gold."

References

Ahmad, M., Ok, Y.S., Kim, B.-Y., Ahn, J.-H., Lee, Y.H., Zhang, M., Moon, D.H., Al-Wabel, M.I., Lee, S.S., 2016. Impact of soybean stover- and pine needle-derived biochars on Pb and as mobility, microbial community, and carbon stability in a contaminated agricultural soil. J. Environ. Manage. 166, 131–139.

Ameloot, N., De Neve, S., Jegajeevagan, K., Yildiz, G., Buchan, D., Funkuin, Y.N., Prins, W., Bouckaert, L., Sleutel, S., 2013. Short-term CO_2 and N_2O emissions and microbial properties of biochar amended sandy loam soils. Soil Biol. Biochem. 57, 401–410.

Atkinson, C.J., Fitzgerald, J.D., Hipps, N.A., 2010. Potential mechanism for achieving agricultural benefits from biochar application to temperate soils: a review. Plant Soil 337, 1–18.

Bates, A., 2010. The Biochar Solution: Carbon Farming and Climate Change. New Society Publishers, British Columbia.

Bird, M., Wurster, C., de Paula Silva, P., Bass, A., de Nys, R., 2011. Algal biochar production and properties. Bioresour. Technol. 102, 1886–1891.

Bruges, J., 2010. The Biochar Debate: Charcoal's Potential to Reverse Climate Change and Build Soil Fertility. Chelsea Green Publishing, Vermont.

Chen, B., Chen, Z., Lv, S., 2011. A novel magnetic biochar efficiently sorbs organic pollutants and phosphate. Bioresour. Technol. 102, 716–723.

Chen, J., Liu, X., Zheng, J., Zhang, B., Lu, H., Chi, Z., Pan, G., Li, L., Zheng, J., Zhang, X., Wang, J., Yu, X., 2013. Biochar soil amendment increased bacterial but decreased fungal gene abundance with shifts in community structure in a slightly acid rice paddy from Southwest China. Appl. Soil Ecol. 71, 33–44.

Chen, S., Rotaru, A.-E., Shrestha, P.M., Malvankar, N.S., Liu, F., Fan, W., Nevin, K.P., Lovley, D.R., 2014. Promoting interspecies electron transfer with biochar. Sci. Rep. 4 Art. no. 5019.

De Gisi, S., Petta, L., Wendland, C., 2014. History and technology of Terra preta sanitation. Sustainability 6 (3), 1328–1345.

De Oliveira Mendes, G., Zafra, D.L., Vassilev, N.B., Silva, I.R., Ribeiro Jr., J.I., Costaa, M.D., 2015. Biochar enhances Aspergillus niger rock phosphate solubilization by increasing organic acid production and alleviating fluoride toxicity. Appl. Environ. Microbiol. 80 (10), 3081–3085.

Dil, M., Oelbermann, M., Xue, W., 2014. An evaluation of biochar pre-conditioned with urea ammonium nitrate on maize (Zea mays L.) production and soil biochemical characteristics. Can. J. Soil Sci. 94 (4), 552–562.

Elad, Y., David, D.R., Harel, Y.M., Borenshtein, M., Kalifa, H.B., Silber, A., Graber, E.R., 2010. Induction of systemic resistance in plants by biochar, a soil-applied carbon sequestration agent. Phytopathology 100, 913–921.

Ennis, C.J., Evans, G.A., Islam, M., Ralebitso-Senior, T.K., Senior, E., 2012. Biochar: carbon sequestration, land remediation and impacts on soil microbiology. Crit. Rev. Environ. Sci. Technol. 42 (22), 2311–2364.

European Biochar Certificate. http://www.european-biochar.org/en/ebc-ibi (accessed 15.01.16.).

Fox, A., Kwapinski, W., Griffiths, B.S., Schmalenberger, A., 2014. The role of sulphur- and phosphorus-mobilizing bacteria in biochar-induced growth promotion of Lolium perenne. FEMS Microbiol. Ecol. 90, 78–91.

Gaskin, J.W., Steiner, C., Harris, K., Das, K.C., Bibens, B., 2008. Effect of low-temperature pyrolysis conditions on biochar for agricultural use. Biol. Eng. Trans. 51, 2061–2069.

Gougoulias, C., Clark, J.M., Shaw, L.J., 2014. The role of soil microbes in the global carbon cycle: tracking the below-ground microbial processing of plant-derived carbon for manipulating carbon dynamics in agricultural systems. J. Sci. Food Agric. 94, 2362–2371.

Graber, E.R., Frenkel, O., Jaiswal, A.K., Elad, Y., 2014. How may biochar influence the severity of diseases caused by soilborne pathogens? Carbon Manag. 5 (2), 169–183.

Green, W.A., Wayman, L.G., 2013. Carbon Consideration: Biochar, Biomass, Biopower and Sequestration. Nova Science Publishers, New York.

Hale, L., Luth, M., Kenny, R., Crowley, D., 2014. Evaluation of pinewood biochar as a carrier of bacterial strain Enterobacter cloacae UW5 for soil inoculation. Appl. Soil Ecol. 84, 192–199.

Hammer, E.C., Forstreuter, M., Rillig, M.C., Kohler, J., 2015. Biochar increases arbuscular mycorrhizal plant growth enhancement and ameliorates salinity stress. Appl. Soil Ecol. 96, 114–121.

Hossain, M., Strezov, V., Chan, K., Ziolkowski, A., Nelson, P., 2011. Influence of pyrolysis temperature on production and nutrient properties of wastewater sludge biochar. J. Environ. Manage. 92, 223–228.

Huggins, T., Wang, H., Kearns, J., Jenkins, P., Ren, Z.J., 2014. Biochar as a sustainable electrode material for electricity production in microbial fuel cells. Bioresour. Technol. 157, 114–119.

Inyang, M., Gao, B., Pullammanappallil, P., Ding, W., Zimmerman, A., 2010. Biochar from anaerobically digested sugarcane bagasse. Bioresour. Technol. 101, 8868–8872.

Kolton, M., Harel, Y.M., Pasternak, Z., Graber, E.R., Elad, Y., Cytryn, E., 2011. Impact of biochar application to soil on the root-associated bacterial community structure of fully developed greenhouse pepper plants. Appl. Environ. Microbiol. 77 (14), 4924–4930.

Kuppusamy, S., Thavamani, P., Megharaj, M., Venkateswarlu, K., Naidu, R., 2016. Agronomic and remedial benefits and risks of applying biochar to soil: current knowledge and future research directions. Environ. Int. 87, 1–12.

Ladygina, N., Rineau, F., 2013. Biochar and Soil Biota. CRC Press, Boca Raton.

Laird, D.A., 2008. The charcoal vision: a win–win–win scenario for simultaneously producing bioenergy, permanently sequestering carbon, while improving soil and water quality. Agron. J. 100 (1), 178–181.

Lehmann, J., Joseph, S., 2009. Biochar for Environmental Management: Science and Technology. Earthscan, London.

Lehmann, J., Joseph, S., 2015. Biochar for Environmental Management: Science, Technology and Implementation, second ed. Routledge, Abingdon.

Lehmann, J., Rillig, M.C., Thies, J., Masiello, C.A., Hockaday, W.C., Crowley, D., 2011. Biochar effects on soil biota – a review. Soil Biol. Biochem. 43 (9), 1812–1836.

Li, H., Ye, X., Geng, Z., Zhou, H., Guo, X., Zhang, Y., Zhao, H., Wang, G., 2016. The influence of biochar type on long-term stabilization for Cd and Cu in contaminated paddy soils. J. Hazard. Mater. 304, 40–48.

Lone, A.H., Najar, G.R., Ganie, M.A., Sofi, J.A., Ali, T., 2015. Biochar for sustainable soil health: a review of prospects and concerns. Pedosphere 25 (5), 639–653.

Masiello, C.A., Chen, Y., Gao, X., Liu, S., Cheng, H.-Y., Bennett, M.R., Rudgers, J.A., Wagner, D.S., Zygourakis, K., Silberg, J.J., 2013. Biochar and microbial signalling: production conditions determine effects on microbial communication. Environ. Sci. Technol. 47, 11496–11503.

Mason, V.P., Markx, G.H., Thompson, I.P., Andrews, J.S., Manefield, M., 2005. Colonial architecture in mixed species assemblages affects AHL mediated gene expression. FEMS Microbiol. Lett. 244, 121–127.

Mau, A.E., Utami, S.R., 2014. Effects of biochar amendment and arbuscular mycorrhizal fungi inoculation on availability of soil phosphorus and growth of maize. J. Degrad. Min. Land. Manage. 1 (2), 69–74.

Orr, C.H., Ralebitso-Senior, T.K., January 18, 2014. Tracking N-cycling genes in biochar-supplemented ecosystems: a perspective. OA Microbiol. 2 (1), 1.

Purakayastha, T.J., Chauhan, S.K., Pathak, H., 2015. Characterisation, stability, and microbial effects of four biochars produced from crop residues. Geoderma 239–240, 293–303.

Quilliam, R.S., Glanville, H.C., Wade, S.C., Jones, D.L., 2013. Life in the 'charosphere' – does biochar in agricultural soil provide a significant habitat for microorganisms? Soil Biol. Biochem. 65, 287–293.

Ralebitso-Senior, T.K., Senior, E., Di Felice, R., Jarvis, K., 2012. Waste gas biofiltration: advances and limitations of current approaches in microbiology. Environ. Sci. Technol. 46 (16), 8542–8573.

Rosas, J., Gamez, N., Cara, J., Ubalde, J., Sort, X., Sanchez, M., 2015. Assessment of sustainable biochar production for carbon abatement from vineyard residues. J. Anal. Appl. Pyrol. 113, 239–247.

Santos, F., Torn, M.S., Bird, J.A., 2012. Biological degradation of pyrogenic organic matter in temperate forest soils. Soil Biol. Biochem. 51, 115–124.

Schmidt, H.-P., Pandit, B., Martinsen, V., Cornelissen, G., Conte, P., Kammann, C., 2015. Fourfold increase in pumpkin yield in response to low-dosage root zone application of urine-enhanced biochar to a fertile tropical soil. Agriculture 5 (3), 723.

Schmidt, H.-P., Wilson, K., 2014. The 55 uses of biochar. Biochar J. www.biochar-journal.org/en/ct/2.

Scholz, S.M., Sebres, T., Roberts, K., Whitman, T., Wilson, K., Lehmann, J., 2014. Biochar Systems for Smallholders in Developing Countries. World Bank Publications, Washington DC.

Shan, D., Deng, S., Zhao, T., Wang, B., Wang, Y., Huang, J., Yu, G., Winglee, J., Wiesner, M.R., 2016. Preparation of ultrafine magnetic biochar and activated carbon for pharmaceutical adsorption and subsequent degradation by ball milling. J. Hazard. Mater. 305, 156–163.

Steiner, C., 2015. Considerations in Biochar Characterization. Agricultural and Environmental Applications of Biochar: Advances and Barriers. Soil Science Society of America Inc., Madison.

Taylor, P., 2010. The Biochar Revolution: Transforming Agriculture & Environment. Global Publishing Group, Denver.

Trakal, L., Veselská, V., Šafařík, I., Vítková, M., Číhalová, S., Komárek, M., 2016. Lead and cadmium sorption mechanisms on magnetically modified biochars. Bioresouce Technol. 203, 318–324.

Verheijen, F., Jeffery, S., Bastos, A.C., van der Velde, M., Diafas, I., 2010. Biochar Application to Soils: A Critical Scientific Review of Effects on Soil Properties, Processes and Functions. JRC Scientific and Technical Reports. European Communities, Luxembourg.

Warnock, D.D., Lehmann, J., Kuyper, T.W., Rillig, M.C., 2007. Mycorrhizal responses to biochar in soil – concepts and mechanisms. Plant Soil 300, 9–20.

Woods, W.I., Teixeira, W.G., Lehmann, J., Steiner, C., WinklerPrins, A., Rebellato, L., 2009. Amazonian Dark Earths: Wim Sombroek's Vision. Springer.

Ye, J., Nielsen, S., Joseph, S., Thomas, T., 2015. High-resolution and specific detection of bacteria on complex surfaces using nanoparticle probes and electron microscopy. PLoS One 10 (5), e0126404.

Zhang, M., Gao, B., Varnoosfaderani, S., Hebard, A., Yao, Y., Inyang, M., 2013. Preparation and characterization of a novel magnetic biochar for arsenic removal. Bioresour. Technol. 130, 457–462.

Zhang, J., Wang, Q., 2016. Sustainable mechanisms of biochar derived from brewers' spent grain and sewage sludge for ammonia nitrogen capture. J. Clean. Prod. 112, 3927–3934.

Zwart, D.C., Kim, S.-H., 2012. Biochar amendment increases resistance to stem lesions caused by *Phytophthora* spp. in tree seedlings. HortScience 47 (12), 1736–1740.

Index